AF356379

Advances in Low-Temperature Nitriding and Carburizing of Stainless Steels and Metallic Materials: Formation and Properties

Advances in Low-Temperature Nitriding and Carburizing of Stainless Steels and Metallic Materials: Formation and Properties

Shinichiro Adachi

Francesca Borgioli

Thomas Lindner

Basel • Beijing • Wuhan • Barcelona • Belgrade • Novi Sad • Cluj • Manchester

Editors
Shinichiro Adachi
Metal Finishing & Analysis
Osaka Research Institute
of Industrial Science
and Technology
Osaka
Japan

Francesca Borgioli
Department of Industrial
Engineering (DIEF)
Università di Firenze
Florence
Italy

Thomas Lindner
Institute of Materials Science
and Engineering
Chemnitz University
of Technology
Chemnitz
Germany

Editorial Office
MDPI AG
Grosspeteranlage 5
4052 Basel, Switzerland

This is a reprint of articles from the Special Issue published online in the open access journal *Metals* (ISSN 2075-4701) (available at: www.mdpi.com/journal/metals/special_issues/Nitriding_Carburizing).

For citation purposes, cite each article independently as indicated on the article page online and as indicated below:

Lastname, A.A.; Lastname, B.B. Article Title. *Journal Name* **Year**, *Volume Number*, Page Range.

ISBN 978-3-7258-2520-2 (Hbk)
ISBN 978-3-7258-2519-6 (PDF)
doi.org/10.3390/books978-3-7258-2519-6

Contents

Francesca Borgioli, Shinichiro Adachi and Thomas Lindner
Advances in Low-Temperature Nitriding and Carburizing of Stainless Steels and Metallic
Materials: Formation and Properties
Reprinted from: *Metals* **2024**, *14*, 1179, doi:10.3390/met14101179 . . . . **1**

Francesca Borgioli
The "Expanded" Phases in the Low-Temperature Treated Stainless Steels: A Review
Reprinted from: *Metals* **2022**, *12*, 331, doi:10.3390/met12020331 . . . . **6**

Stephan Mändl and Darina Manova
Comparison of Nitriding Behavior for Austenitic Stainless Steel 316Ti and Super Austenitic
Stainless Steel 904L
Reprinted from: *Metals* **2024**, *14*, 659, doi:10.3390/met14060659 . . . . **45**

Shinichiro Adachi, Takuto Yamaguchi and Nobuhiro Ueda
Formation and Properties of Nitrocarburizing S-Phase on AISI 316L Stainless Steel-Based WC
Composite Layers by Low-Temperature Plasma Nitriding
Reprinted from: *Metals* **2021**, *11*, 1538, doi:10.3390/met11101538 . . . . **61**

Thomas Lindner, Martin Löbel, Maximilian Grimm and Jochen Fiebig
Cold Gas Spraying of Solution-Hardened 316L Grade Stainless Steel Powder
Reprinted from: *Metals* **2021**, *12*, 30, doi:10.3390/met12010030 . . . . **74**

**Bruna C. E. Schibicheski Kurelo, Gelson Biscaia De Souza, Silvio Luiz Rutz Da Silva, Carlos
Maurício Lepienski, Clodomiro Alves Júnior and Rafael Fillus Chuproski et al.**
Tribo-Mechanical Behavior of Films and Modified Layers Produced by Cathodic Cage and Glow
Discharge Plasma Nitriding Techniques
Reprinted from: *Metals* **2023**, *13*, 430, doi:10.3390/met13020430 . . . . **85**

Cyprian Illing, Zhe Ren, Anna Agaponova, Arthur Heuer and Frank Ernst
Rapid Alloy Surface Engineering through Closed-Vessel Reagent Pyrolysis
Reprinted from: *Metals* **2021**, *11*, 1764, doi:10.3390/met11111764 . . . . **104**

**Hyunseok Cheon, Kyu-Sik Kim, Sunkwang Kim, Sung-Bo Heo, Jae-Hun Lim and Jun-Ho
Kim et al.**
Effect of Deformation Structure of AISI 316L in Low-Temperature Vacuum Carburizing
Reprinted from: *Metals* **2021**, *11*, 1762, doi:10.3390/met11111762 . . . . **122**

**Bruna C. E. Schibicheski Kurelo, Carlos M. Lepienski, Willian R. de Oliveira, Gelson B. de
Souza, Francisco C. Serbena and Rodrigo P. Cardoso et al.**
Identification of Expanded Austenite in Nitrogen-Implanted Ferritic Steel through In Situ
Synchrotron X-ray Diffraction Analyses
Reprinted from: *Metals* **2023**, *13*, 1744, doi:10.3390/met13101744 . . . . **135**

**Yamid Nuñez de la Rosa, Oriana Palma Calabokis, Vladimir Ballesteros-Ballesteros, Cristian
Lozano Tafur and Paulo C. Borges**
Assessment of the Pitting, Crevice Corrosion, and Mechanical Properties of Low-Temperature
Plasma-Nitrided Inconel Alloy 718
Reprinted from: *Metals* **2023**, *13*, 1172, doi:10.3390/met13071172 . . . . **150**

Luiz Henrique Portela de Abreu, Muhammad Naeem, Renan Matos Monção, Thercio H. C. Costa, Juan C. Díaz-Guillén and Javed Iqbal et al.
The Effect of Cathodic Cage Plasma TiN Deposition on Surface Properties of Conventional Plasma Nitrided AISI-M2 Steel
Reprinted from: *Metals* **2022**, *12*, 961, doi:10.3390/met12060961 . . . . **169**

Jiwen Yan, Minghao Shao, Zelong Zhou, Zhehao Zhang, Xuening Yi and Mingjia Wang et al.
Electrochemical Corrosion Behavior of Ti-N-O Modified Layer on the TC4 Titanium Alloy Prepared by Hollow Cathodic Plasma Source Oxynitriding
Reprinted from: *Metals* **2023**, *13*, 1083, doi:10.3390/met13061083 . . . . **183**

Sławomir Maksymilian Kaczmarek, Jerzy Michalski, Tadeusz Fraczek, Agata Dudek, Hubert Fuks and Grzegorz Leniec
Mechanical and Magnetic Investigations of Balls Made of AISI 1010 and AISI 1085 Steels after Nitriding and Annealing
Reprinted from: *Metals* **2023**, *13*, 1060, doi:10.3390/met13061060 . . . . **198**

Shoufeng Zhang, Chao Zhou, Guiqian Sun, Xin Wang, Kuo Bao and Pinwen Zhu et al.
Electronic Structure and Hardness of Mn_3N_2 Synthesized under High Temperature and High Pressure
Reprinted from: *Metals* **2022**, *12*, 2164, doi:10.3390/met12122164 . . . . **210**

Editorial

Advances in Low-Temperature Nitriding and Carburizing of Stainless Steels and Metallic Materials: Formation and Properties

Francesca Borgioli [1,*] , Shinichiro Adachi [2] and Thomas Lindner [3]

1   Department of Industrial Engineering (DIEF), Università di Firenze, via di S. Marta 3, 50139 Firenze, Italy
2   Osaka Research Institute of Industrial Science and Technology, 2-7-1 Ayumino, Izumi 594-1157, Osaka, Japan; shinadachi@orist.jp
3   Materials and Surface Engineering, Institute of Materials Science and Engineering, Chemnitz University of Technology, 09107 Chemnitz, Germany; th.lindner@mb.tu-chemnitz.de
*   Correspondence: francesca.borgioli@unifi.it; Tel.: +39-055-275-8734

1. Introduction

Surface engineering techniques are currently used to overcome the limitations of metal alloys and improve their surface hardness, tribological properties, fatigue resistance and corrosion resistance in specific environments [1,2]. From this perspective, stainless steels have represented a challenge for many years, since nitriding and carburizing treatments, as carried out on low-alloy steels or tool steel, cannot be directly applied to steel of these grades. In fact, when nitrogen (N) and/or carbon (C) surface alloying is performed through thermochemical processes at temperatures comparable to those used for low-alloy steels and tool steels (500–580 °C for nitriding, 850–950 °C for carburizing), chromium (Cr)-based compounds can form in stainless steels. Thus, the Cr-depleted matrix becomes unable to maintain a protective passive film and corrosion may occur [3].

The drawbacks of "traditional" nitriding and carburizing processes can be overcome by "low-temperature" thermochemical treatments. By using sufficiently low treatment temperatures to allow the diffusion of interstitial atoms (N, C), while Cr (substitutional) diffusion is significantly slowed down, the formation of Cr compounds is hindered. Using these so-called para-equilibrium conditions for the surface treatment of stainless steels, supersaturated solid solutions in austenite, ferrite or martensite are obtained [4]. Forming these "expanded phases", as they are often called, can benefit and prolong the life of stainless steel components. These phases enable an increase in the surface hardness, improvements to the wear and fatigue resistance, and the maintenance of the good corrosion resistance of these alloys, or even enhancements of it, in particular in chloride-ion-containing solutions [4].

The formation of expanded austenite, or S-phase, in austenitic stainless steels was first reported in the mid-1980s, and since then, many studies have investigated this phase, its properties and the treatment conditions usable for its production without Cr and/or Fe compound precipitation in different austenitic stainless steels.

Low-temperature treatments have also proven to be suitable for producing expanded phases in other grades of stainless steels (ferritic, martensitic, duplex and precipitation hardening), and the formation of expanded ferrite and expanded martensite has been reported [4,5].

A further step has been the application of these treatments to non-ferrous alloys. The formation of expanded austenite has been reported in nickel-based [5], cobalt-based [5] and high-entropy alloys [6].

Another area of research has focused on the different process techniques used to produce expanded phases. The suitability of many techniques has been studied, from the more "traditional" gaseous treatments, to ion implantation processes, to plasma-based processes [5]. Recently, the active screen or cathodic cage plasma-based process has gained particular attention, since it avoids the edge effect typical of glow discharge DC plasma processes and may enable the combination of the diffusion treatment with surface film deposition [7,8]. An additional step is represented by low-temperature thermochemical treatment application to additively manufactured parts or to powders used for additive manufacturing techniques.

The purpose of this Special Issue is to gather original research and critical review articles on recent advances in all aspects of low-temperature thermochemical treatments using N and/or C applied to stainless steels and ferrous and non-ferrous alloys.

2. Overview of the Published Articles

A total of thirteen articles have been published, which cover different aspects of low-temperature thermochemical treatments, from microstructure analysis and evaluation of the properties of surface-modified alloys, to new proposals for performing thermochemical processes, to applications in different stainless steel grades and other ferrous and non-ferrous alloys, both in the massive and powder state.

The formation of expanded phases in different stainless steel grades was reviewed by Borgioli (Contribution 1). The formation conditions and the characteristics of expanded austenite, expanded ferrite and expanded martensite were discussed, and their properties were considered, focusing on the microhardness, load-bearing capacity, tribological properties and corrosion resistance of the modified surface layers produced by nitriding, carburizing and nitrocarburizing. Taking into account the most recent studies on this topic, the nature of expanded austenite was also discussed, in particular regarding the ordering of N atoms in the supersaturated lattice.

Experimental contributions have further improved our knowledge of expanded phases and their properties. The characteristics of expanded austenite formed in 316Ti austenitic stainless steel and 904L super-austenitic stainless steel using nitrogen low-energy ion implantation (LEII) were investigated by Mändl and Manova (Contribution 2). Using in situ X-ray diffraction analysis, they studied the expansion of the austenite lattice as N was solubilized, and thus the modified layer growth and N diffusion. Strong variations in lattice expansion were observed as a function of the steel composition, grain orientation, and nitriding conditions, even if N uptake and diffusion were similar for the two steels. The formation of a double layer with two distinct lattice expansions (higher near the surface and lower near the interface with the bulk) was detected for some nitriding conditions. The stability of the expanded austenite was also assessed, highlighting different behaviors regarding CrN precipitate formation for 316Ti and 904L.

Expanded austenite can form not only on massive alloys but also on additively manufactured layers or on powders to be used for thermal spraying.

The effect of low-temperature plasma nitriding on the characteristics of AISI 316L stainless steel-based WC composite layers produced by laser cladding was studied by Adachi et al. (Contribution 3). The WC particles acted as a C source, promoting the formation of mixed carbides and, presumably, C solubilization in the austenite matrix. When low-temperature nitriding was carried out, a N-rich expanded austenite formed at the surface, while a C-rich expanded austenite was present in the inner part, and the modified surface layers were similar to those produced by combining nitriding and carburizing simultaneously (i.e., nitrocarburizing) or in sequence. The Vickers microhardness values at the surface ranged from 1200 to 1400 HV and the hardness profiles were smooth. The nitriding treatment also improved the corrosion resistance in a 3.5 wt.% NaCl solution.

AISI 316L powder surface hardening using low-temperature gas nitrocarburizing and the use of this hardened powder for cold gas spraying were investigated by Lindner et al. (Contribution 4). Expanded austenite was able to form on the particles, even if this was limited to the surface area in most of the powder. Using this hardened powder, a coating was produced successfully using cold gas spraying. The phase fraction of the expanded austenite in the coating depended on the spraying conditions and decreased with the impact velocity of the particles. The spraying conditions also influenced the hardness values, the wear caused by the scratch test and the corrosion resistance.

The influence of different treatment processes on expanded austenite formation and on the properties of the modified surface layers was also explored. Schibicheski Kurelo et al. (Contribution 5) compared two nitriding techniques, glow discharge plasma nitriding (GDPN) and cathodic cage plasma nitriding (CCPN), regarding the mechanical and tribological behavior of the modified surface layers produced on AISI 316 stainless steel. Expanded austenite was obtained with both nitriding techniques, but the process type influenced nitride formation and the thickness of the modified layers. In particular, by using the same treatment temperature, the GDPN technique produced thicker and harder modified layers, having better scratch resistance. However, lower roughness and the absence of Cr depletion were observed in the CCPN-treated samples.

A low-temperature nitrocarburizing process based on the pyrolysis of a solid reagent (urea) was investigated by Illing et al. (Contribution 6). AISI 316L ferrules were used as samples, and they were introduced together with urea powder in a vessel, which was then evacuated and sealed. The heating of the vessel up to 450 °C using a two-step process activated the surface by removing the passive film, and enabled the fast diffusion of interstitial atoms (C, N) into the alloy, so that N- and C-rich expanded austenite was formed. The modified surface layers obtained with this technique were comparable to those produced by a gas- or plasma-based method, but with a significantly reduced processing time. The pyrolysis products of the reagent were also studied and their role in nitrocarburizing was considered, highlighting that the interaction of gas molecules with the alloy surface could have an auto-catalytic effect by altering the gas composition and condensed species.

The effect of plastic deformation on AISI 316L steel subjected to low-temperature vacuum carburizing without surface activation was studied by Cheon et al. (Contribution 7). Specimens with different deformation states were obtained, starting from cold-worked AISI 316L, on which solution or stress-relieving heat treatments were performed. When carburizing was carried out without surface activation, thinner modified surface layers, with a C-rich expanded austenite exhibiting a smaller lattice expansion, were observed in the specimens with high plastic deformation, while thicker and more uniform modified layers were obtained on the solution-treated specimens. These results could be correlated to the thickness of the oxide layer, which was higher for the plastic-deformed specimens. Therefore, the oxide layer appeared to be the most influential factor for the carburizing efficiency when surface activation was not performed.

The formation of expanded phases was also investigated on ferritic stainless steels. While the formation of expanded ferrite has been assessed by many authors, whether expanded austenite can also form is still a matter of debate. The formation and decomposition of expanded austenite produced by nitrogen plasma immersion ion implantation (PIII) on AISI 444 ferritic stainless steel was studied by Schibicheski Kurelo et al. using in situ synchrotron X-ray diffraction analysis (Contribution 8). It was observed that, even if expanded ferrite was the main phase that formed by nitriding at 300 and 400 °C, when the ion fluences were higher, as occurred when the treatment duration or temperature increased, expanded austenite was also formed. The decomposition of the expanded phases was assessed both during isothermal heating at 450 °C, as well as during the heating period. Nitrogen diffusion caused a reduction in the lattice parameters of the expanded phases. Moreover, in the samples nitrided with the highest fluences, expanded austenite tended to transform into α-ferrite, and CrN and oxides formed.

Low-temperature nitriding treatment has also been performed on non-ferrous alloys and expanded austenite has been successfully produced. De La Rosa et al. (Contribution 9) studied the effects of plasma nitriding on the Inconel 718 Ni-based alloy. Expanded austenite was obtained together with CrN using a 400 °C nitriding temperature. However, when nitriding was carried out at 450 °C, only austenite and CrN were detectable. Surface hardening and a higher scratch resistance were observed in the nitrided specimens. The corrosion behavior of the samples was tested in a 3.56 wt.% NaCl solution using potentiodynamic and potentiostatic tests. Pitting corrosion was detected in all the samples. It was observed that, even with CrN precipitation, the presence of expanded austenite in 440 °C nitrided specimens led to a better performance regarding crevice corrosion.

Thermochemical treatments have been applied to ferrous and non-ferrous alloys using treatment temperatures lower than those usually employed. De Abreu et al. (Contribution 10) investigated the effects of a duplex treatment on the surface properties of AISI M2 tool steel. The authors performed plasma nitriding at 400 °C followed by a cathodic cage deposition (CCPD) at 400 or 450 °C with a titanium cage in a nitrogen-containing atmosphere. As a result, Fe-based nitrides, γ'-Fe$_4$N and ε-Fe$_{2-3}$N, and titanium nitride, TiN, were produced. The duplex treatments effectively improved the surface hardness and thicker hardened layers with a smooth hardness gradient were obtained. Moreover, these treatments significantly reduced the wear rate and friction coefficient as evaluated using ball-on-ring tests.

A two-step hollow cathode treatment, which consisted of a nitriding process followed by an oxynitriding treatment in the range of 500–540 °C, was applied by Yan et al. to titanium TC4 (Ti-6Al-4V) (Contribution 11). Modified surface layers consisting of TiO$_2$, TiN, Ti$_2$N and a N-rich α-Ti phase were obtained. The two-step process was effective in producing hardened layers, which had a significantly improved corrosion resistance in the artificial saliva solutions.

Other studies have focused on different aspects, which can clarify the properties of modified surface layers in which expanded phases form.

In Fe-N systems, the different phases (solid solutions, expanded phases, nitrides) have different magnetic properties, which may also change as a function of the N content. Kaczmarek et al. (Contribution 12) studied the changes in the phase composition and magnetic properties of AISI 1010 and AISI 1085 steels, which were nitrided and then annealed. The nitriding treatment produced a compound layer consisting of ε-Fe$_{2-3}$N and γ'-Fe$_4$N nitrides. During annealing, ε-nitride decomposed first, and the γ' content significantly increased. The presence of nitrides and their change in composition influenced the magnetic properties of the steels.

The precipitation of Mn nitrides may occur in the nitriding of Mn-rich stainless steels, such as AISI 200 series or Ni-free austenitic stainless steel, affecting the properties of the modified surface layers. The electronic structure and hardness of Mn$_3$N$_2$ was investigated by Zhang et al. (Contribution 13). Bulk Mn$_3$N$_2$ samples were synthesized and their Vickers hardness was measured as 9.9 GPa. This experimental result was useful for revising the theoretical predictions. By combining first-principle simulations and experimental analyses, the authors highlighted the importance of metal bonds in determining the hardness of this nitride.

3. Conclusions and Outlook

Low-temperature thermochemical treatment with N and/or C represents a lively research field, and it allows for increasingly tailored surface modifications to be achieved. The studies collected in this Special Issue report the latest scientific achievements on expanded phases, the characteristics of modified surfaces, and process techniques, and they are valuable for scientists and engineers engaged in the research and development of surface engineering processes for application to stainless steels and other metallic materials.

Acknowledgments: The Guest Editors wish to thank all the authors for their contributions, as well as all the reviewers for their work and efforts to improve the quality of the articles. The Editors appreciate the support from the *Metals* Editorial Office throughout the publication process.

Conflicts of Interest: The authors declare no conflicts of interest.

List of Contributions:

1. Borgioli, F. The "Expanded" Phases in the Low-Temperature Treated Stainless Steels: A Review. *Metals* **2022**, *12*, 331. https://doi.org/10.3390/met12020331

2. Mändl, S.; Manova, D. Comparison of Nitriding Behavior for Austenitic Stainless Steel 316Ti and Super Austenitic Stainless Steel 904L. *Metals* **2024**, *14*, 659. https://doi.org/10.3390/met14060659

3. Adachi, S.; Yamaguchi, T.; Ueda, N. Formation and Properties of Nitrocarburizing S-Phase on AISI 316L Stainless Steel-Based WC Composite Layers by Low-Temperature Plasma Nitriding. *Metals* **2021**, *11*, 1538. https://doi.org/10.3390/met11101538

4. Lindner, T.; Löbel, M.; Grimm, M.; Fiebig, J. Cold Gas Spraying of Solution-Hardened 316L Grade Stainless Steel Powder. *Metals* **2022**, *12*, 30. https://doi.org/10.3390/met12010030

5. Schibicheski Kurelo, B.C.E.; De Souza, G.B.; Da Silva, S.L.R.; Lepienski, C.M.; Alves Júnior, C.; Chuproski, R.F.; Pintaúde, G. Tribo-Mechanical Behavior of Films and Modified Layers Produced by Cathodic Cage and Glow Discharge Plasma Nitriding Techniques. *Metals* **2023**, *13*, 430. https://doi.org/10.3390/met13020430

6. Illing, C.; Ren, Z.; Agaponova, A.; Heuer, A.; Ernst, F. Rapid Alloy Surface Engineering through Closed-Vessel Reagent Pyrolysis. *Metals* **2021**, *11*, 1764. https://doi.org/10.3390/met11111764

7. Cheon, H.; Kim, K.-S.; Kim, S.; Heo, S.-B.; Lim, J.-H.; Kim, J.-H.; Yoon, S.-Y. Effect of Deformation Structure of AISI 316L in Low-Temperature Vacuum Carburizing. *Metals* **2021**, *11*, 1762. https://doi.org/10.3390/met11111762

8. Schibicheski Kurelo, B.C.E.; Lepienski, C.M.; de Oliveira, W.R.; de Souza, G.B.; Serbena, F.C.; Cardoso, R.P.; das Neves, J.C.K.; Borges, P.C. Identification of Expanded Austenite in Nitrogen-Implanted Ferritic Steel through In Situ Synchrotron X-ray Diffraction Analyses. *Metals* **2023**, *13*, 1744. https://doi.org/10.3390/met13101744

9. Nuñez de la Rosa, Y.; Palma Calabokis, O.; Ballesteros-Ballesteros, V.; Tafur, C.L.; Borges, P.C. Assessment of the Pitting, Crevice Corrosion, and Mechanical Properties of Low-Temperature Plasma-Nitrided Inconel Alloy 718. *Metals* **2023**, *13*, 1172. https://doi.org/10.3390/met13071172

10. de Abreu, L.H.P.; Naeem, M.; Monção, R.M.; Costa, T.H.C.; Díaz-Guillén, J.C.; Iqbal, J.; Sousa, R.R.M. The Effect of Cathodic Cage Plasma TiN Deposition on Surface Properties of Conventional Plasma Nitrided AISI-M2 Steel. *Metals* **2022**, *12*, 961. https://doi.org/10.3390/met12060961

11. Yan, J.; Shao, M.; Zhou, Z.; Zhang, Z.; Yi, X.; Wang, M.; Wang, C.; Fang, D.; Wang, M.; Xie, B.; He, Y.; Li, Y. Electrochemical Corrosion Behavior of Ti-N-O Modified Layer on the TC4 Titanium Alloy Prepared by Hollow Cathodic Plasma Source Oxynitriding. *Metals* **2023**, *13*, 1083. https://doi.org/10.3390/met13061083

12. Kaczmarek, S.M.; Michalski, J.; Frączek, T.; Dudek, A.; Fuks, H.; Leniec, G. Mechanical and Magnetic Investigations of Balls Made of AISI 1010 and AISI 1085 Steels after Nitriding and Annealing. *Metals* **2023**, *13*, 1060. https://doi.org/10.3390/met13061060

13. Zhang, S.; Zhou, C.; Sun, G.; Wang, X.; Bao, K.; Zhu, P.; Zhu, J.; Wang, Z.; Zhao, X.; Tao, Q.; Ge, Y.; Cui, T. Electronic Structure and Hardness of Mn_3N_2 Synthesized under High Temperature and High Pressure. *Metals* **2022**, *12*, 2164. https://doi.org/10.3390/met12122164

References

1. Czerwinski, F. Thermochemical Treatment of Metals. In *Heat Treatment–Conventional and Novel Applications*; Czerwinski, F., Ed.; IntechOpen: Rijeka, Croatia, 2012; pp. 73–112. [CrossRef]

2. Ramezani, M.; Mohd Ripin, Z.; Pasang, T.; Jiang, C.-P. Surface Engineering of Metals: Techniques, Characterizations and Applications. *Metals* **2023**, *13*, 1299. [CrossRef]

3. Lo, K.H.; Shek, C.H.; Lai, J.K.L. Recent developments in stainless steels. *Mater. Sci. Eng. R Rep.* **2009**, *65*, 39–104. [CrossRef]

4. Cardoso, R.P.; Mafra, M.; Brunatto, S.F. Low-temperature Thermochemical Treatments of Stainless Steels—An Introduction. In *Plasma Science and Technology-Progress in Physical States and Chemical Reactions*; Mieso, T., Ed.; InTech: Rijeka, Croatia, 2016; pp. 107–130, ISBN 978-953-51-2280-7. [CrossRef]

5. Dong, H. S-phase surface engineering of Fe-Cr, Co-Cr and Ni-Cr alloys. *Int. Mater. Rev.* **2010**, *55*, 65–98. [CrossRef]

6. Peng, Y.; Duan, H.; Feng, Y.; Gong, J.; Somers, M.A.J. Surface hardening of $Al_{0.1}CoCrFeNi$ and $Al_{0.5}CoCrFeNi$ high-entropy alloys by low-temperature gaseous carburization. *Intermetallics* **2023**, *160*, 107943. [CrossRef]

7. Borowski, T. Enhancing the Corrosion Resistance of Austenitic Steel Using Active Screen Plasma Nitriding and Nitrocarburising. *Materials* **2021**, *14*, 3320. [CrossRef]

8. Sampaio, W.R.V.; Serra, P.L.C.; Monção, R.M.; de Sousa, E.M.; Silva, L.G.L.; da Silva, F.L.F.; Rossino, L.S.; Bandeira, R.M.; Feitor, M.C.; de Sousa, R.R.M.; et al. Influence of using different titanium cathodic cage plasma deposition configurations on the mechanical, tribological, and corrosion properties of AISI 304 stainless steel. *Surf. Coat. Technol.* **2023**, *475*, 130149. [CrossRef]

Review

The "Expanded" Phases in the Low-Temperature Treated Stainless Steels: A Review

Francesca Borgioli

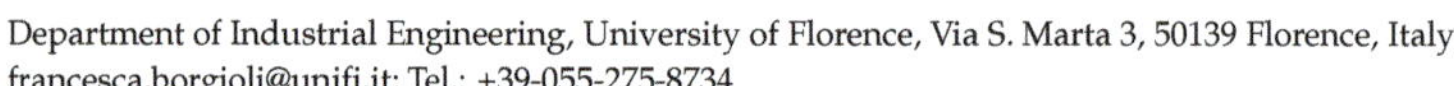

Department of Industrial Engineering, University of Florence, Via S. Marta 3, 50139 Florence, Italy; francesca.borgioli@unifi.it; Tel.: +39-055-275-8734

Abstract: Low-temperature treatments have become a valuable method for improving the surface hardness of stainless steels, and thus their tribological properties, without impairing their corrosion resistance. By using treatment temperatures lower than those usually employed for nitriding or carburizing of low alloy steels or tool steels, it is possible to obtain a fairly fast (interstitial) diffusion of nitrogen and/or carbon atoms; on the contrary, the diffusion of substitutional atoms, as chromium atoms, has significantly slowed down, therefore the formation of chromium compounds is hindered, and corrosion resistance can be maintained. As a consequence, nitrogen and carbon atoms can be retained in solid solutions in an iron lattice well beyond their maximum solubility, and supersaturated solid solutions are produced. Depending on the iron lattice structure present in the stainless steel, the so-called "expanded austenite" or "S-phase", "expanded ferrite", and "expanded martensite" have been reported to be formed. This review summarizes the main studies on the characteristics and properties of these "expanded" phases and of the modified surface layers in which these phases form by using low-temperature treatments. A particular focus is on expanded martensite and expanded ferrite. Expanded austenite–S-phase is also discussed, with particular reference to the most recent studies.

Keywords: stainless steels; low-temperature treatments; nitriding; carburizing; expanded austenite; S-phase; expanded martensite; expanded ferrite; corrosion resistance; wear resistance

check for **updates**

Citation: Borgioli, F. The "Expanded" Phases in the Low-Temperature Treated Stainless Steels: A Review. *Metals* **2022**, *12*, 331. https://doi.org/10.3390/met12020331

Academic Editor: Andre Paulo Tschiptschin

Received: 30 December 2021
Accepted: 11 February 2022
Published: 14 February 2022

Publisher's Note: MDPI stays neutral with regard to jurisdictional claims in published maps and institutional affiliations.

1. Introduction

Stainless steels are used for a wide variety of applications, from components for the chemical and power engineering industries to household utensils and kitchenware [1,2]. Their success is due to their resistance to general corrosion in many environments, owing to the thin, adherent, and self-healing chromium-rich oxide film, which forms on the steel surface in the presence of oxygen. A chromium (Cr) content of at least 10.5 wt.% is required for the formation of this protective oxide film [2]. The addition of Cr stabilizes the body-centered cubic (b.c.c.) structure of ferrite, so that the phase transformation of ferrite into the face-centered cubic (f.c.c.) structure of austenite does not occur when the Cr content is higher than ~14 wt.% [3]. When nickel (Ni) is added, the austenite phase field, the so-called γ-loop, is expanded, and the austenite becomes stable in a wide temperature range [4]. By changing the amounts of Cr and Ni, as well as the alloy elements able to stabilize ferrite, such as molybdenum (Mo), silicon (Si), niobium (Nb), titanium (Ti), or those able to stabilize austenite, such as carbon (C), nitrogen (N), manganese (Mn), ferrite, martensite, or austenite lattices and their related microstructures can be obtained. Due to the important effect of microstructure on their properties, stainless steels are usually classified according to their microstructure into five groups: ferritic, martensitic, austenitic, duplex (ferritic-austenitic), and precipitation hardening stainless steels [1,2].

Ferritic stainless steels are nonhardenable, ferromagnetic FeCr alloys, with a Cr content typically in the range of 10.5–30 wt.%, C up to ~0.2 wt.%, and with no or very little Ni. They are cheaper than austenitic CrNi grades, can have good ductility and formability, and are highly resistant to stress-corrosion-cracking, but they are subjected to ductile-to-brittle transition.

Martensitic stainless steels have Cr and C contents that are balanced so that the γ-loop is expanded, and austenite-to-martensite transformation can occur through quenching. Carbides may also be present in order to increase wear resistance. Cr is usually in the range of 10.5–18 wt.%, and C content may exceed 1.2 wt.%. High strength and hardness, as well as good toughness, can be attained through quenching and tempering heat treatments. A lower corrosion resistance is observed, in comparison with austenitic and ferritic grades.

Austenitic stainless steels have a fairly high content of austenite-stabilizing elements, such as Ni, Mn and N. Ni, with a content of 8 wt.% or higher, is the basis of CrNi AISI 300 series austenitic stainless steels, while it is partly substituted with Mn and N in CrMn AISI 200 series stainless steels. A negligible Ni amount, lower than 0.3 wt.%, is present in the so-called Ni-free stainless steels, in which Mn (9–24 wt.%) and N (0.45–1.1 wt.%) stabilize the austenite structure [5]. These steels are nonmagnetic, have excellent corrosion resistance, as well as ease of formability and weldability, and they can be employed from cryogenic to elevated temperatures.

Duplex stainless steels have a microstructure containing both ferrite and austenite. The amount of each phase is a function of the alloy element content, mainly of Cr and Ni, and heat treatment. These steels have a corrosion resistance comparable to that of austenitic stainless steels with a similar alloy element content, but they have a higher yield strength and improved resistance to stress–corrosion cracking and corrosion fatigue. Their toughness is between that of ferritic and austenitic stainless steels, but, as ferritic grades, they lack cryogenic toughness.

Precipitation hardening (PH) stainless steels contain, besides Cr and Ni, alloy elements, such as copper (Cu), aluminum (Al), and Ti, which allow them to harden by means of a solution and aging treatment. They are classified as martensitic, austenitic, and semi-austenitic, according to their microstructure after the solution-annealing heat treatment. Semi-austenitic steels transform into martensite during subsequent heat treatment. The formation of precipitates allows for high strength and maintaining a good corrosion resistance.

Modification of the surface characteristics of stainless steels may be needed in order to improve surface hardness, tribological properties, fatigue resistance, or resistance to localized corrosion in specific environments. Surface engineering techniques allow for overcoming the limitation of the different stainless steels grades by means of either coatings with hard compounds, obtained with processes as chemical vapor deposition (CVD), physical vapor deposition (PVD) and plasma spray, or through modifying the surface layers using diffusion processes [6]. Thermochemical processes, in which N and/or C surface alloying is performed by means of diffusion, are well known for the surface modification of low alloy and tool steels, and they have become an effective industrial practice. For low alloy and tool steels, nitriding is usually performed at temperatures in the range of 500–580 °C, and it produces modified surface layers consisting of an outer compound layer, in which Fe-based nitrides, γ'-Fe$_4$N, and ε-Fe$_{2-3}$N, are present, and an inner diffusion layer, in which alloy element nitrides precipitates form in a N-rich ferrite matrix [7]. Carburizing is usually carried out at temperatures in the range of 850–950 °C, at which austenite is stable. After the carburizing step, quenching and tempering are performed, so that a C-rich martensite case can be produced [6].

When nitriding or carburizing treatments are applied to stainless steels using the same temperatures employed for low alloy and tool steels, hard Cr compounds (nitrides, carbides) are able to form, and the Cr-depleted matrix cannot maintain a uniform and protective passive film, therefore corrosion can easily occur [8,9]. The uptake of interstitial elements from the treatment atmosphere to the substrate is hindered by the passive layer, which acts as a diffusion barrier, so that a uniform treatment can be obtained only by previously removing or reducing this oxide layer [10–12]. These drawbacks of nitriding and carburizing can be overcome using the so-called low-temperature diffusion treatments. In fact, when the treatment temperature of nitriding or carburizing process is decreased, so that significant Cr (substitutional) diffusion is avoided while interstitial atoms can diffuse, in the so-called para-equilibrium conditions [13], it is possible to inhibit the formation of Cr nitrides and carbides precipitates and obtain supersaturated solid solutions of the diffused interstitial atoms in austenite, ferrite, or martensite, which allow for a surface hardness increase without impairing the corrosion resistance.

The possibility of obtaining "expanded" phases by means of thermochemical treatments, performed at temperatures lower than those usually employed for nitriding and carburizing, was not recognized for many years. For austenitic stainless steels, the formation of a N-rich phase, which is now known as "expanded austenite" or "S-phase", was reported in the mid-1980s, more or less in the same period, by different groups (Zhang and Bell [14,15], Ichii et al. [16,17], De Benedetti et al. [18–20], and Yasumaru and Kamachi [21], as recently reviewed in [11]), even if studies on nitriding at low temperatures date back to the 1970s [22,23]. Low-temperature carburizing dates back to 1983, when Kolster reported a process which was able to increase the fatigue and wear resistance of stainless steels, while maintaining the corrosion resistance [24]. This process was the basis of the proprietary treatment known as Kolsterizing. The possibility of applying low-temperature nitriding to martensitic and ferritic stainless steels has been neglected in consideration. In the same period, many studies reported nitriding treatments carried out at temperatures comparable to those used for low alloy steels, i.e., at temperature of 500 °C or higher [25–32]. Lower temperatures were used in a few studies, but the formation of "expanded" phases was hardly recognized [25,29,33]. The term "expanded martensite" was used in 2000 by Kim et al. [34] for the expanded phase observed in a nitrided AISI 420 martensitic stainless steel, while "expanded ferrite" was used later for the phase formed in treated martensitic [35], duplex [36], and ferritic stainless steels [37].

From the few studies reported in the 1980s, the interest for the characteristics of these expanded phases has grown, since the formation of these phases allows for high surface hardness, improved wear and fatigue resistance, and high corrosion resistance in chloride-ion (Cl^-) containing solutions. Until now, more than a thousand of papers have been published in academic journals, conference proceedings, and professional reports. Most part of the research papers, as well as of the brief overviews [38–45] and extended reviews [10–12,46–48], regarded the low-temperature treatments of austenitic stainless steels. Less attention has been paid to the other stainless steel types. Only few reviews reported the studies on expanded martensite and expanded ferrite [10,46,49]; however, an in-depth discussion of the studies on all the expanded phases is lacking.

The aim of the present review is to report the main studies on the formation, characteristics, and properties of the "expanded" phases obtained in stainless steels by using low-temperature treatments, and on the characteristics of the modified layers in which they form, with a particular focus on expanded martensite and expanded ferrite. The review is organized in four main sections. In the first section, the solid solutions of interstitial atoms in Fe phases and the formation of "expanded phases" are discussed. The second section regards expanded martensite, formed in martensitic and martensitic PH stainless steels, while the third section is devoted to expanded ferrite, formed in ferritic and duplex stainless steels. Finally, in the fourth section, expanded austenite–S-phase, formed in austenitic stainless steels, is discussed, with particular reference to the most recent studies.

2. N and C Solid Solutions in Fe Phases and Formation of "Expanded" Phases

N and C atoms are recognized as the most effective strengtheners of Fe-based solid solutions. According to phase diagrams, for pure Fe, the maximum solubility of N in b.c.c. α-Fe is 0.4 at.% at 592 °C, and in f.c.c., γ-Fe 10.3 at.% at 650 °C [50], while the maximum solubility of C in α-Fe is 0.104 at.% at 727 °C, and in γ-Fe 9.23 at.% at 1147 °C [51].

As reported by Gavriljuk [52], the N and C alloying in the f.c.c. phase influences the density of the electron states at Fermi level, causing different effects. When N atoms are in a solid solution in a f.c.c. Fe-based lattice, they increase the concentration of free electrons, having higher energy, and hence enhance the metallic character of the interatomic bonds, while C atoms promote the localization of electrons on the atomic sites. As a consequence, even if the atomic radius of N is smaller than that of C, the effective atom size of N atoms in solid solution is larger, and thus a larger distortion of the f.c.c. lattice is produced. It should be noted that a similar effect on electron states was also observed for substitutional alloy elements, with elements located to the right of Fe in the periodic table, such as Ni, Cu, Si, and Al, which increase the concentration of free electrons in austenitic steels, while elements located to the left of Fe, such as Mn, Cr, and Mo, decrease it [53].

The influence of N and C atoms on the interatomic bonds leads to different atomic distribution in solid solutions. Using Mössbauer spectroscopy, it was shown that in the Fe–N solid solution, there is a tendency towards short-range ordering of N atoms, with the N atoms occupying two adjacent octahedral sites in 180° N–N pairs, as in the superstructure of γ'-Fe$_4$N, while in the Fe–C solid solution, there is a tendency towards clustering of C atoms [52]. N and C atoms also affect the distribution of substitutional solutes, with N promoting the distribution of substitutional atoms in austenite and C assisting the clustering of substitutional atoms [52]. Thus, it has been concluded that, in Fe-based solid solutions, alloy elements increasing the concentration of free electrons, such as N, Ni, Si, and Al, cause short range atomic ordering or a more homogeneous distribution of solute atoms, while atoms which promote the localization of electrons on atomic sites, such as C, Cr, Mn, and Mo, cause atomic clustering [52].

When martensite structure is taken into account, a similar behavior is observed, with N atoms having a tendency towards ordering and C atoms towards clustering [52].

The easy precipitation of Cr carbides can be considered a consequence of the clustering of C atoms and C-assisted clustering of Cr atoms. On the contrary, N atoms tend to shift the curve of the sensitization in the time-temperature-transformation diagram to the right on the time scale, and in the tempering of martensite they hinder the clustering of Cr atoms and precipitation of Cr nitrides. Thus, in an austenitic stainless steel, such as AISI 316, the C content is lower than 0.015 at.% in order to avoid carbide precipitation [54], while the equilibrium solubility of N in austenite is larger, even if fairly low, measuring less than 0.65 at.% [46].

In the "expanded" phases, the interstitial content can be several hundred times higher than the equilibrium solubility, causing a significant lattice expansion. According to phase diagrams, nitrides and carbides should be expected instead of interstitial solid solutions. From a thermodynamic perspective, the phase transition towards the corresponding nitrides and carbides is favorable enough for transition metals of lower groups, such as Ti, while for Fe nitrides and carbides, there is not a strong driving force towards compound formation [55,56], so N and C atoms can remain in solid solution in a Fe phase. Regarding Cr, it has a higher affinity to N and C in comparison with Fe, but the formation of Cr nitrides and carbides can be hindered by the use of low-temperature treatments. In fact, under these conditions, the diffusion of interstitial atoms such as N and C is several orders of magnitude higher than that of substitutional atoms (Cr, Mn, Mo, Ni), which are relatively "immobile" in the lattice [57], therefore, interstitial atoms tend to be retained in solid solutions, and the so-called "colossal" supersaturation [54,58] can occur.

As has been observed for interstitial solid solutions, the maximum N content in the expanded phases is higher than that of C. In the expanded austenite–S-phase, the maximum observed N content is ~38 at.% [58], also significantly beyond that of γ'-Fe$_4$N (19.3–20 at. % [50]), having a f.c.c. structure of metal atoms, while the maximum C content is ~19 at.% [59]. In expanded ferrite, obtained in ferritic stainless steels, a maximum N content of ~24 at.% was observed [60], and the reported maximum C content was ~10 at.% [61]. Similar values were detected in expanded martensite, with a maximum N content of ~22 at.% [62] and a maximum C content of ~10 at.% [63].

It should be noted that the large octahedral interstitial sites of the f.c.c. austenite phase allow for the accommodation of much more solute atoms (N and C) than the interstitial sites of the b.c.c. ferrite phase, therefore lower N and C contents were observed in expanded ferrite, and hence in expanded martensite. Thus, nitrides tend to precipitate more easily in ferrite and martensite than in austenite [46].

3. Expanded Martensite

The term "expanded martensite" can be ascribed to Kim et al. [34,64], who observed the presence of a martensitic phase with larger lattice parameters, in comparison with those of the bulk, in AISI 420 martensitic stainless steel samples nitrided at 350 and 400 °C. They named this phase by analogy with the designation "expanded austenite" for the phase observed in low-temperature nitrided austenitic stainless steels. Even if the term may seem odd, since martensite is already an expanded phase, and "expanded ferrite" has been observed to be a more appropriate name [65], it is widely used to indicate the phase that forms as an expansion of the original martensite phase in martensitic and martensitic PH stainless steels which have been nitrided and/or carburized at low temperatures. Similarly, the symbols α_N' and α_C' are used to indicate N- or C-rich expanded martensite phases, respectively.

The main characteristics of the modified surface layers containing expanded martensite, obtained by the nitriding or carburizing of martensitic and martensitic PH stainless steels, are summarized in Table 1.

Table 1. Main characteristics of the modified surface layers containing expanded martensite, obtained by the nitriding or carburizing of martensitic and martensitic PH stainless steels.

Properties		Nitriding	Carburizing
Formation temperature (without formation of Cr compounds) (°C)		200–350	200–450
Max. interstitial content in expanded martensite (at. %)		22	10
Max. surface hardness ($\mathrm{kg_f\ mm^{-2}}$)	- expanded martensite only - with Fe-based compounds	1120 1560	970 1530
Hardness profile	- expanded martensite only - with Fe-based compounds	gradual change abrupt change	gradual change gradual change
Load bearing capacity	- expanded martensite only - with Fe-based compounds	high low	high high
Dry sliding wear resistance		very good	good
Localized corrosion resistance	- expanded martensite only - with Fe-based compounds	good variable [1] / good [2]	good good

[1] Nitrides dispersed in expanded martensite. [2] Nitrides forming a continuous outer layer.

3.1. Formation of Expanded Martensite

Expanded martensite has been reported to form in stainless steels which have a martensitic structure, i.e., martensitic [33,49,64–67] and martensitic PH [62,65,68–71] stainless steels.

In a martensite lattice, the diffusion of both interstitial and substitutional atoms is faster than that of the same atoms in f.c.c. austenite. As a consequence, modified surface layers are usually thicker than those produced in austenitic stainless steels using the same treatment conditions. Moreover, due to the low content or absence of Ni, which does not form nitrides and promotes the formation of a solid solution, the maximum treatment temperature to be used for avoiding the precipitation of nitrides and carbides, particularly Cr-based ones, is lower, and the treatment time is shorter than those used for austenitic stainless steels, especially for CrNi-based ones.

The phases present in the modified surface layers markedly depend on the treatment conditions, and, in particular, on the treatment temperature. For AISI 420 martensitic stainless steel, modified layers consisting of expanded martensite only were observed on samples which had been nitrided for 4 h at ultra-low temperatures, i.e., in the range of

200–300 °C [49]. When the treatment temperature was increased up to 400 °C, the h.c.p. ε-Fe$_{2-3}$N nitride was also formed, and when using temperatures higher than 400 °C, Cr nitride precipitates were present [49]. The early precipitation of Cr nitrides, such as that observed in Sandvik Nanoflex$^{®}$ (C 0.02 wt.%, Cr 12 wt.%, Ni 9 wt.%, Mo 4 wt.%, Cu 2 wt.%) PH stainless steel gas nitrided at 420 °C for 20 h, was ascribed to an energetically favorable orientation relationship between the expanded body centered tetragonal (b.c.t.) martensite lattice and NaCl-type CrN [69], analogous to the Baker–Nutting orientation relationship between the expanded α-Fe b.c.c. lattice and CrN [72], which lowered the nucleation barrier.

Nitriding at 450 °C or at higher temperatures caused the formation of an outer compound layer, in which both Cr and Fe nitrides were present [49,71,73,74]. Alphonsa et al. [30] observed the formation of an outer thin (2–5 μm) layer consisting mainly of ε-Fe$_{2-3}$N with dispersed CrN on AISI 420 samples, nitrided at 530 °C for 20 h, while Dong et al. [71] reported the presence of CrN and γ'-Fe$_4$N in the modified layer formed on 17–4 PH stainless steel samples, which were plasma nitrided at temperatures in the range of 460–500 °C for 10 and 30 h.

The treatment atmosphere, and, in particular, the N$_2$ volume fraction and the addition of carburizing gas, also influences the phases that form in the modified layer. By increasing the N$_2$ content, the formation of nitrides was promoted, and the addition of CH$_4$ tended to favor the formation of ε-Fe$_{2-3}$(N,C) carbonitride [75]. For the plasma immersion ion implantation (PIII) process, the pulse voltage, and hence the energy of ion implantation, influenced the phases formed in the modified surface layers. On AISI 630 martensitic stainless steel, PIII nitriding caused the formation, in addition to the expanded martensite, of Fe$_3$N when high energy ion implantation at 25 kV was employed, and of Fe$_4$N when low energy, with 1.5 kV voltage, was used [76].

Regarding the carburizing treatment, when quenched AISI 420 martensitic stainless steel was carburized at 420 and 450 °C for 4 h [77], and at 400 °C for 8 h [78], Fe$_3$C was observed, together with C-rich expanded martensite.

Since the formation of Cr nitrides and carbides is related to diffusion phenomena, both treatment temperature and time are the process parameters which must be carefully controlled in order to avoid the formation of these precipitates. For martensitic stainless steels, the limit temperature for avoiding Cr nitride/carbide precipitation is considered to be ~350 °C for nitriding [10,79], ~ 400 °C for nitrocarburizing [79,80], and ~450 °C for carburizing [79,81], with a treatment duration of up to 8 h. When a treatment temperature of 400 °C was employed for AISI 420 martensitic stainless steel, 4 h was reported to be the threshold for sensitization, i.e., the precipitation of Cr compounds, for nitriding treatments [64,78], 6 h for nitrocarburizing [80], and more than 36 h for carburizing [82].

It was observed that, with nitrocarburizing treatments, thinner hardened layers were obtained as compared to those produced with nitriding at the same temperature and for the same treatment time [78,79,83], in contrast with the fact that C atoms diffuse faster than N atoms. As such, thicker hardened layers were expected for the nitrocarburized samples, as observed in austenitic stainless steels [84]. It has to be noted that, for the treatment conditions reported by Scheuer et al. [78,79] for the treatment of AISI 420 steel, in the modified layer, besides expanded martensite, ε-Fe$_{2-3}$N and ε-Fe$_{2-3}$(N,C) were also able to form in the outer part of the layer of nitrided and nitrocarburized samples, respectively. By comparing X-ray diffraction patterns, the volume fraction of ε-Fe$_{2-3}$(N,C) seemed higher than that of ε-Fe$_{2-3}$N, therefore it may be hypothesized that a slight thicker layer of carbonitride formed in nitrocarburized samples, in comparison with the nitride layer formed in nitrided steel. Taking into account that the diffusion rate of interstitial atoms is lower in nitride and nitrocarbide than in b.c.c. α-Fe [78], the apparently odd difference in the thickness of the hardened layers for the two treatments might be explained.

Taking advantage of the different diffusion rates of N and C atoms, and their possible interaction during the diffusion process, it was possible to delay the precipitation of Cr nitrides and/or carbides, and obtain modified layers consisting of expanded martensite and Fe-based compounds by combining sequentially nitriding, nitrocarburizing, or carburizing [78]. The microstructure of the modified layer and hardness profile depended on the succession of secondary phases formation, which, in turn, depended on the sequence of thermochemical treatments.

Alloy composition influences the phases formed in the modified layer, and, in particular, the precipitation of CrN, and the thickness of modified layers, especially when low nitriding temperatures were used [60,85], therefore the maximum treatment temperature which allows the prevention of nitride and carbide precipitation depends on steel type [62].

The characteristics of the modified surface layers also depend on the previous heat treatment history of the steel. For martensitic stainless steels, low-temperature treatments are performed after quenching. Therefore, both tempering and surface treatment can be carried out in one step, or the surface treatment can be performed after the tempering step. Similar strategies can be employed for martensitic PH stainless steels, with the surface treatment and aging carried out as one step, or the surface treatment performed after aging.

The effect of previous heat treatment was reported by Figueroa et al. [86]. AISI 420 martensitic stainless samples were quenched and tempered at two different temperatures, 310 and 530 °C, for 1 h, and then they were PIII nitrided in the range of 340–500 °C. It was observed that the tempering temperature influenced the volume fraction of retained austenite, and that this phase, in turn, influenced the phases which formed during nitriding. Expanded austenite was able to form from the large amount of retained austenite present in the samples which were tempered at 310 °C, and by increasing the nitriding temperature up to 400 °C, it tended to become the prevailing phase, as compared to ε-nitride. In the samples which were tempered at 530 °C, containing a lower amount of retained austenite, ε-nitride formed when nitriding was performed at 360 °C. By increasing the treatment temperature to 430 °C, faster N diffusion tended to prevent the formation of ε-nitride, and expanded martensite, together with a low amount of CrN, precipitated at grain boundaries, was observed. It is interesting to note that, when nitriding was carried out at 475–500 °C, the modified layers of both sample types consisted of CrN precipitates in the ferrite matrix.

Nitriding and nitrocarburizing treatments allowed for the simultaneous aging step for PH steels [62,69,70,87]. Bottoli et al. [69] evaluated the effects of both the initial alloy microstructure and the low-temperature thermochemical treatment type on the surface and bulk properties of the Sandvik Nanoflex® PH stainless steel. By combining annealing and deformation treatments, it was possible to obtain a bulk material with different amounts of martensite and austenite. The characteristics of the modified surface layers obtained by means of nitriding or nitrocarburizing depended on the phases present in the alloy, as well as the hardening effect obtained by the aging of the bulk. Since N diffusivity is different in austenite and martensite, when both of these phases were present, inhomogeneous layers consisting of expanded austenite and expanded martensite were obtained. When martensite became the prevalent phase, an expanded martensite modified layer was obtained. When nitriding was performed at 420 °C (20 h), dark etched regions were present at grain boundaries, suggesting that Cr nitride precipitates were able to form, but when the nitriding temperature was reduced at 400 °C, Cr nitride precipitation was avoided. By using a nitrocarburizing atmosphere, the expanded martensite layer had a two-layer structure, with an outer N-rich zone and an inner C-rich one, as was observed for expanded austenite. When austenite was prevalent in the steel, the modified layer consisted of expanded austenite, and it was thinner than that consisting of expanded martensite obtained with the same nitriding conditions.

The possibility of performing the nitriding treatment after aging, or as a single treatment which allowed both nitriding and aging, was studied by Pinedo et al. [70] for a 17–4 PH stainless steel. Even if, for both strategies, modified surface layers consisting of expanded martensite were obtained with nitriding (400 °C, 20 h), a significantly thicker modified layer was produced in the solubilized and aged steel. It was hypothesized that the formation of the Cu-rich coherent precipitates, formed during aging, ensured that Cu did not lower the N activity on martensite matrix during nitriding, contrary to what occurred in the solubilized only steel.

The shape of concentration profile of interstitial atoms as a function of depth, observed in the modified surface layers, depends on the phases that are able to form. When expanded martensite is the main phase and nitrides or expanded austenite are absent or in negligible amounts, the shape of N concentration profile vs. depth is very close to a complementary error function [87–90], suggesting a concentration independent diffusivity. When nitrides or expanded austenite form together with expanded martensite, the concentration profile is affected by the different diffusion rate in the different phases, and a trend with a fairly sharp decrease, a more or less wide plateau followed by a smoother decrease was observed [62,76].

The thermal stability of modified surface layers containing expanded martensite was studied by Baniasadi et al. [63] for carburized martensitic PH stainless steels. For carburized (450 °C, 18 h) 17–4 PH steel, the precipitation of Fe_3C was observed, beginning with heating at 200 °C for 2 h, while for 13–8Mo PH steel, carbide precipitation occurred during carburizing.

3.2. Characteristics of the Modified Surface Layers

The micrographs of the cross-section of AISI 420 martensitic stainless steel samples, which were plasma nitrided at 250 and 350 °C, are shown in Figure 1, together with the X-ray diffraction patterns of samples, untreated and treated in the range of 200–400 °C. The surface morphology and the microstructure of the cross-section of an AISI 431 martensitic stainless steel sample, plasma nitrided at 400 °C for 5 h, together with its X-ray diffraction patterns before and after the treatment, are depicted in Figure 2.

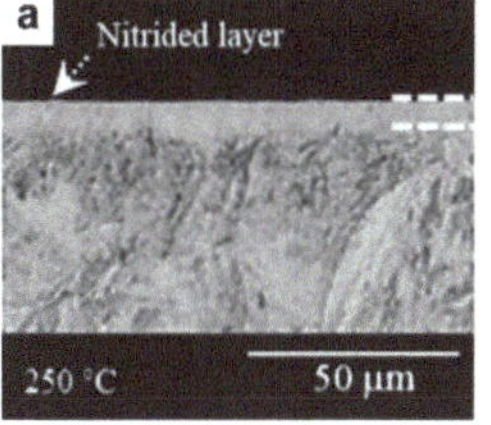

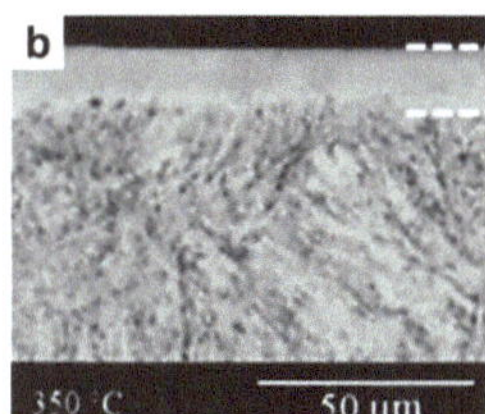

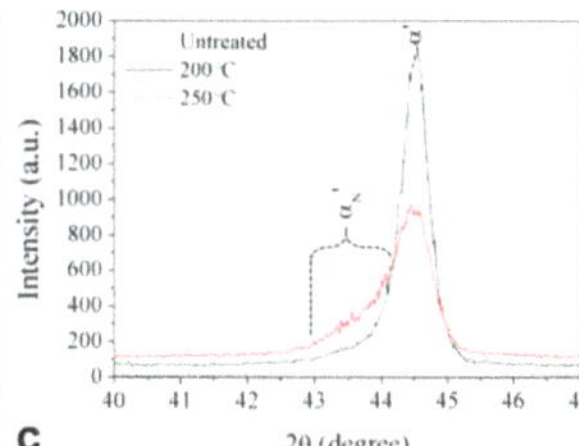

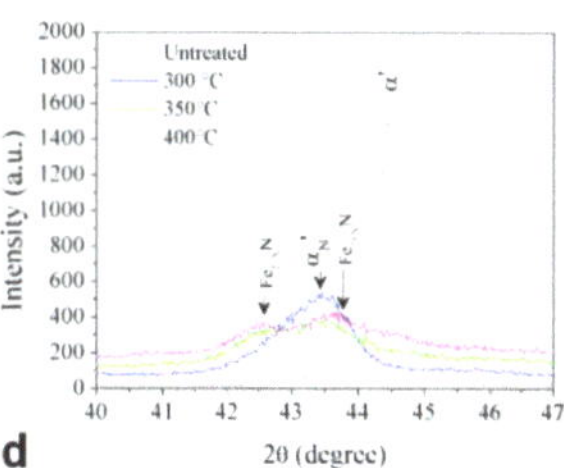

Figure 1. Micrographs of the cross-section of AISI 420 martensitic stainless steel samples plasma nitrided at 250 (**a**) and 350 (**b**) °C (etchant: Vilella's reagent), and X-ray diffraction patterns of samples, untreated and nitrided at 200 and 250 °C (**c**), and in the range of 300–400 °C (**d**). (For further details, see ref. [49]). (Reprinted from *Materials Research Express*, 6 (2019), C.J. Scheuer et al., Ultra-low—to high-temperature plasma-assisted nitriding: revisiting and going further on the martensitic stainless steel treatment, 026529, ©IOP Publishing. Reproduced with permission. All rights reserved.).

The surface of low-temperature treated stainless steels with a martensite microstructure shows an etched appearance when plasma-based treatments are used, with the grains well delineated and reliefs at grain boundaries [66,70,83,91,92], as depicted in Figure 2a.

In nitrided samples, when Cr nitrides are not able to form, the cross-section microstructure of the outer part of the modified layer is almost featureless (Figure 1a,b). At a greater depth, the inner part of the modified layer becomes delineated by the chemical etching, and it does not show a strongly etched interface with the substrate. This characteristic was observed for modified layers consisting of expanded martensite only [49,68–71,89], and layers in which both expanded martensite and ε-$Fe_{2-3}N$ nitride [49,67,93], or expanded martensite and expanded austenite [62,65,69], or expanded martensite, expanded austenite, and ε-$Fe_{2-3}N$ [83] were detected. It should be noted that this featureless appearance may be due to the chemical etchings used, such as Marble's reagent [33] or Vilella's reagent [49,65,70,71,83], which are unable to etch Fe-based nitrides as well as N-rich phases, as was observed for the expanded austenite [11]. The microstructure of the modified layer can be better delineated, highlighting the martensite structure, by using acetic glyceregia [91], as showed in Figure 2b, or Kalling's reagent [92].

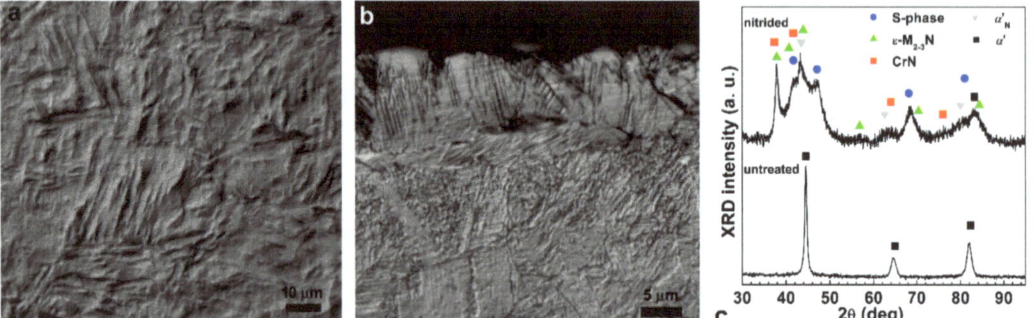

Figure 2. Surface morphology (**a**), cross-section micrograph (**b**) (etchant: glyceregia) and X-ray diffraction patterns before and after the treatment (**c**) of an AISI 431 martensitic stainless steel sample plasma nitrided at 400 °C for 5 h. (For further experimental details, see ref. [91]).

The microstructure of the modified surface layers of nitrided samples depends on the formed phases and their volume fraction When Fe-based nitrides are present in small amounts, they are not usually able to form a continuous layer. Acicular Fe-based nitrides precipitates were observed together with expanded martensite [66]. When the volume fraction of Fe nitrides increased, the formation of a double-layer structure was reported, with an outer compound layer consisting of nitrides and an inner diffusion layer, in which expanded martensite was present [94]. However, it should be noted that the featureless "white" part of the modified layer cannot be considered a compound layer, since it usually does not contain nitrides only. This part has a lower thickness in comparison with that of the whole hardened layer [78], thus it may be hypothesized that it is related to the N-richer part of the modified layer, consisting of Fe nitrides and N-rich expanded martensite and austenite, while the etched part of the layer is similar to the diffusion layer observed in low alloy steels. In nitrided samples, the "diffusion layer" is fairly thin, while it is thicker when nitrocarburizing treatment is performed [78]. In martensitic and martensitic PH stainless steels which are subjected to low-temperature nitriding, expanded austenite can form both from the retained austenite and as a consequence of the transformation of martensite in austenite due to the austenite stabilizing effect of N [83,86,95].

A fairly thin featureless layer, in which C-rich expanded martensite was present, with a thicker "diffusion layer", was observed in carburized samples [77,78,82]. As an example, for AISI 420 martensitic stainless steel samples, which were quenched and then carburized at 450 °C for 4 h, Vilella's reagent delineated a featureless "white" modified layer, which had a fairly low thickness (1.2–2.6 μm). By comparing the thickness of the modified layer with that of the hardened layer (about 55 μm), it was argued that C atoms diffused more deeply in the bulk than the etchant was able to reveal [77].

A different microstructure was observed when a large amount of expanded austenite was able to form. Using Kalling's reagent, Frandsen et al. [62] showed that the modified layers of two PH stainless steels, Uddeholm Corrax® (C 0.03 wt.%, Cr 12 wt.% Ni, 9.2 wt.%, Mo 1.4 wt.%, Al 1.6 wt.%) and Sandvik Nanoflex®, gas carburized at 508 °C, consisted of a two-layer microstructure, with an outer C-rich expanded austenite layer and an inner C-rich expanded martensite layer.

When C-rich expanded martensite formed, a martensite coarsening occurred, and plate-type martensite was present in the outer part of the modified layer [92]. This feature was observed in AISI 420 samples that were quenched and then carburized in the range of 350–450 °C for 12 h, and at 400 °C for times in the range of 12–36 h. The coarsening was related to the transformation from a predominantly lath-type martensite, present in the as-hardened condition and characteristic of low C content, to plate-type martensite, characteristic of high C content. When higher treatment temperatures or longer durations were used, Cr-based carbides were able to form, thus a decrease in the C content in martensite occurred up to ferrite formation, and platelet refinement was observed, suggesting coarsening/refinement transformations.

Only a few studies have reported transmission electron microscopy (TEM) analysis of the modified layers. TEM analysis of the N-rich expanded martensite, formed in 17–4 PH stainless steel samples nitrided at 350 °C for 10 h, evidenced the presence of twins and slip lines across martensite laths [71]. Moreover, the selected area electron diffraction (SAED) patterns showed faint diffraction spots and streaks along <111> directions, suggesting the presence of high residual stress. In the same steel, nitrided at 440 °C for 8 h, stacking faults and dislocation groups were observed [68]. An interesting feature was observed in 15–5 PH and 17–7 PH stainless steels nitrided at 347 and 397 °C for 20 h [96,97]. TEM analysis evidenced the presence of plate-like structures, which had a lenticular shape, present in the martensite laths near the surface. These plates apparently grew from the surface, richer in N atoms, toward the bulk, and they were aligned along a few distinct orientations relative to the martensite laths. The analysis of SAED patterns suggested the formation of austenite, which was hypothesized to form as a consequence of a shear transformation, thus it was indicated as "martensitic austenite".

The maximum nitrogen content in the expanded martensite depends on the treated stainless steels. In Sandvik Nanoflex® PH stainless steel nitrided at 425 °C, when using a nitriding potential K_N = 2.37 and a duration of 24 h in order to obtain a modified layer consisting of expanded martensite only, a N content as high as ~22 at.% was observed at the surface [62]. A maximum N content of ~20 at. % was registered in an AISI 431 martensitic stainless steel, subjected to ion implantation at 380 °C for 1 h at a voltage of 10 kV, [60], as well as in AISI 420, AISI 420C, and AISI 431 steels nitrided with the PIII process at 320 °C for 2 h [90]. A maximum C content of ~10 at.% was detected in 17–4 PH stainless steel carburized at 450 °C for 18 h [63].

When a high content of interstitial atoms is solubilized in martensite, as in the outer part of the modified layer, it may be hypothesized that the martensite lattice is distorted, and its tetragonality increases due to Zener ordering [98,99]. Under this condition, the interstitial atoms occupy the octahedral interstices of the ferrite lattice so that all the expanded <001> directions are parallel, forming a tetragonal structure. Evidence of the transition from a martensite having a more cubic structure, in which interstitial atoms occupy randomly all three sets of octahedral interstices in the three possible <001> directions, to an "ordered" tetragonal structure is given by the observed martensite coarsening in low-temperature carburized AISI 420 samples, from a lath-type martensite, related to the b.c.c. disordered structure, to a plate-type martensite, related to b.c.t. ordered structure [92]. A similar effect was not reported for nitrided samples, even if it may be hypothesized that it occurs, due to the similar microstructures reported for high-N and high-C martensites [100].

The evaluation of lattice tetragonality by means of X-ray diffraction analysis is complicated by the broadening of martensite peaks and their superposition with those of other phases. In the patterns of untreated martensitic or martensitic PH stainless steels, the tetragonality of the lattice is low, thus the peaks of a b.c.t. structure are usually indistinguishable from the peaks of a b.c.c. structure, even if the steel has a fairly high C content (see, for example, [101]). When N or C atoms are solubilized in the martensite structure, as a consequence of low-temperature treatment, a "tail" at lower diffraction angles is observed [33,60,69,77,94], thus an expansion of the lattice is hypothesized. The distinct shift of the peaks towards lower angles was reported as the interstitial content increases, hence the lattice expansion [64,69,70,77,102], even if it was not so marked as that observable for expanded austenite [103]. A broadening of the peaks was also registered [64,69,77,94]. The broadening of the peaks may be so marked, and Sun and Bell [89] hypothesized the formation of an amorphous phase in the modified layers of nitrided 17–4 PH steel. Frandsen et al. [62] observed a splitting of the peaks in the nitrided Uddeholm Corrax® and Sandvik Nanoflex® PH stainless steels, which was ascribed to the tetragonal distortion of the b.c.c. lattice to b.c.t., while only a broadening was registered for carburized Nanoflex® samples, which was similarly interpreted as due to a tetragonal distortion. Manova et al. [60] observed an anisotropic lattice expansion normal to the surface up to 3.5%, generally larger for the (110) planes than for the (200) planes, in martensitic and martensitic PH stainless steels which were subjected to nitrogen ion implantation at 380 °C for durations up to 3 h. On the contrary, Luiz et al. [104] observed an increase of 4.18% for the lattice parameter calculated for the (200) plane, while only of 2.10% for the lattice parameter calculated for the (110) plane, when the expanded martensite formed in nitrided UNS S41426 super-martensitic stainless steel was analyzed. On the basis of the analysis of X-ray diffraction patterns, SAED patterns and high resolution TEM (HRTEM) images of nitrided and carburized 15–5 PH stainless steel, Zangiabadi [96] reported that a high tetragonality occurred for N-rich expanded martensite, while it was smaller for C-rich expanded martensite.

3.3. Hardness of the Modified Surface Layers

The solubilization of further interstitial atoms in the martensite lattice causes an increase of surface hardness. The reported data often are also influenced by the presence of nitrides and/or carbides, that form together with expanded martensite, as well as by the hardness profiles. Regarding the expanded martensite only, values of 11 GPa ($\sim$1120 kg$_f$ mm^{-2}) were registered on nitrided samples through the nanoindentation technique [49]. The precipitation of ε-nitride in the expanded martensite caused a further increase of surface hardness, and values of 13.7 GPa ($\sim$1400 kg$_f$ mm^{-2}) were detected [49], while with a more continuous outer ε-nitride layer, values of $\sim$1560 kg$_f$ mm^{-2} were registered [104]. When a fairly high amount of Cr nitrides was able to form, values as high as 1700–2000 kg$_f$ mm^{-2} were reported [102]. Typical surface hardness for carburized samples was $\sim$970 kg$_f$ mm^{-2} [81], but values up to $\sim$15 GPa ($\sim$1530 kg$_f$ mm^{-2}), measured using nanoindentation techniques and probably also due to the strengthening effect of carbides precipitates, were registered [105]. The hardness profiles depended on the treatment conditions, and thus on the modified layer microstructure, as shown in Figure 3. When expanded martensite is able to form without relevant amounts of nitrides, as when fairly low nitriding temperatures are used (as 200–350 °C in Figure 3b), the hardness profile is smooth [49,67]. As high volume fractions of nitrides, and, in particular, Cr nitrides, are able to form (as for nitriding temperatures in the range 450–600 °C in Figure 3a), a thick hardened layer is observed, with values as high as $\sim$1400 kg$_f$ mm^{-2} [49], followed by a region having a steep hardness decrease [30,49,74,106,107]. For carburized samples, fairly smooth hardness profiles are usually observed [81,107].

3.4. Tribological Properties of the Modified Surface Layers

The tribological properties of low-temperature treated martensitic and martensitic PH stainless steels depend on the phases formed in the modified surface layers and on the test conditions.

An improvement of wear resistance was reported for AISI 420 and AISI 431 martensitic stainless steels, which were subjected to N-ion implantation at temperatures in the range of 320–380 °C so that expanded martensite modified layers were able to form, when they were tested in ball-on-disc configuration against an alumina ball [88,90]. A significant wear reduction was observed also for AISI 630 martensitic stainless steel samples, nitrided and nitrocarburized by means of PIII process at 380 °C for 3 h, when they were tested using an oscillating ball-on-disc tribometer in unlubricated conditions against a tungsten carbide ball, and a reduction of the friction coefficient from approximately 0.8, for the untreated steel, to about 0.4, for the nitrided samples, was reported [76].

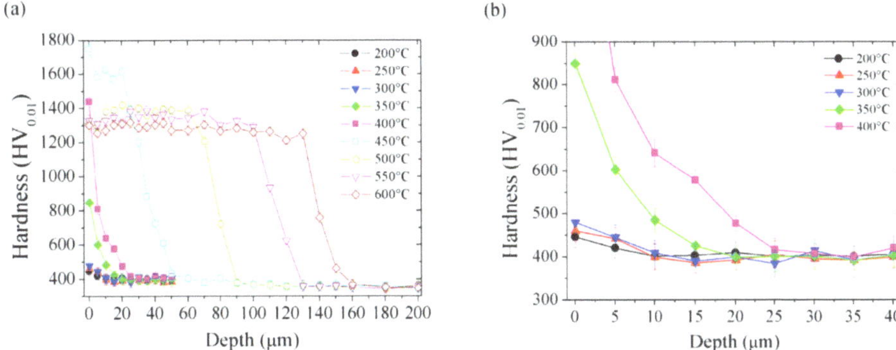

Figure 3. Microhardness profiles of AISI 420 martensitic stainless steel samples plasma nitrided in the range of 200–600 °C (**a**) and in the range of 200–400 °C (**b**). (For further details, see ref. [49]). (Reprinted from *Materials Research Express*, 6 (2019), C.J. Scheuer et al., Ultra-low—to high-temperature plasma-assisted nitriding: revisiting and going further on the martensitic stainless steel treatment, 026529, ©IOP Publishing. Reproduced with permission. All rights reserved.).

The wear resistance of martensitic 17–4 PH stainless steel, tested in unlubricated block-on-ring conditions with an AISI 52100 steel ring as counterface, was substantially improved when a modified layer consisting mainly of expanded martensite was obtained by means of plasma nitriding at 350 °C for 4 h [68].

The evaluation of the scratch resistance of a Fe-13Cr-3Ni-Mo martensitic stainless steel, subjected to nitriding and carburizing at temperatures in the range of 300–400 °C, evidenced that the scratch resistance was influenced by the different characteristics of the obtained modified surface layers [107]. In the nitrided samples, the modified layers consisted of N-rich expanded martensite and hard but brittle Fe-based nitrides, which allowed to have higher surface hardness and a low friction coefficient, but caused a micro-cutting predominance of the wear phenomena and many tensile cracks at the surface when a 15 N load was used. The carburized samples, having modified layers consisting of C-rich expanded martensite and small amounts of carbides, showed smoother microhardness profiles, which allowed a high load-bearing capacity, therefore, micro-ploughing was the predominant wear mechanism, and semi-circular cracks were observed for the 15 N load. The lower load-bearing capacity of modified layers consisting of an outer nitride-rich layer was observed also for micrometric-size protrusions, formed by sputter etching on AISI 420 martensitic stainless steel and then subjected to nitriding [108]. A significantly higher load-bearing capacity and scratch resistance of these protrusions was observed when nitriding produced modified surface layers, consisting mainly of expanded martensite and only a reduced amount of nitrides.

Similarly, the improvement of cavitation erosion resistance was reduced when Fe nitrides precipitates formed on the top of the modified surface layer of a plasma-nitrided AISI 410 martensitic stainless steel, allowing the detachment of entire grains due to shock-wave impact over the surface. However, when the removal of the outermost nitride-rich regions was performed, a marked decrease of the cavitation damage and erosion rate was observed [66]. The reduction of Fe nitrides precipitates and their decrease to nanometer size, such as those obtained using an active screen nitriding process, allowed the expanded martensite layer to markedly improve cavitation erosion resistance [67].

Modified surface layers consisting of expanded martensite and ε-nitride, such as those formed on an AISI 420 martensitic stainless steel plasma nitrided at 380 °C for 15 h, were able to improve the resistance to erosion and erosion corrosion, using neutral and acid slurries [109].

The precipitation of large volume fraction of hard nitrides is able to improve wear resistance in dry sliding conditions, in comparison with the untreated steels [27,74,110].

3.5. Corrosion Behavior of the Modified Surface Layers

When Cr is retained in solid solution, a protective surface oxide layer is maintained, and a good corrosion resistance is achieved. The further addition of N and/or C can improve the corrosion resistance, especially in Cl^--containing solutions. As a matter of fact, all of the different phases that form as a consequence of low-temperature nitriding (expanded martensite, expanded austenite, Fe-based nitrides, Cr-based nitrides) contribute to the corrosion behavior. In the modified surface layers produced on martensitic and martensitic PH stainless steels, expanded martensite may be present together with other phases (Fe nitrides, expanded austenite), thus the evaluations of the corrosion behavior of expanded martensite alone are very limited. N-rich expanded martensite with a very small amount of γ'-Fe_4N nitride showed a good corrosion resistance in NaCl solution and reduction of pitting phenomena [94] (Figure 4), as well as C-rich expanded martensite [82]. An improved corrosion resistance in NaCl solutions was obtained when ε-$Fe_{2\text{-}3}N$ nitride was able to form a continuous layer at the surface [74,86,93,106], or when ε-nitride formed together with expanded austenite [86], therefore both an increase in corrosion and pitting potential and a decrease in anodic current density were registered. When a multiphase structure was formed, such as expanded martensite with expanded austenite and/or ε-nitride, different behaviors were reported, with an increase [83,91,111] or slight decrease [65] in corrosion resistance. The study of Luiz et al. [104] on the corrosion behavior of PIII-nitrided UNS S41426 super-martensitic stainless steel showed that the modified layer consisting of expanded martensite and ε and γ' Fe-based nitrides was efficient in significantly improving the stability of the passive film for up to 30 days of immersion in 3.5% NaCl solution. When cyclic voltammetry tests were performed in 3.5% NaCl, the nitrided samples showed an enhanced resistance to passive film breakdown, in comparison with the untreated steel. However, instead of the typical localized corrosion phenomena, uniform corrosion of the nitrided layer was predominant, presumably due to the high content of Fe-based nitrides at the surface.

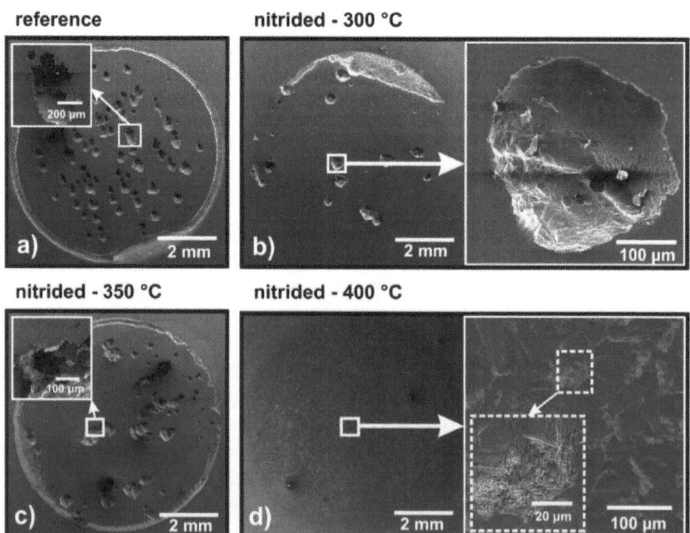

Figure 4. Surface microstructure of UNS S41426 super-martensitic stainless steel untreated (reference) (**a**) and nitrided at 300 (**b**), 350 (**c**), and 400 (**d**) °C, after corrosion tests in 3.5% NaCl. (For further details, see ref. [94]). (Reprinted from *Surface & Coatings Technology*, 351, B.C.E.S. Kurelo et al., Performance of nitrogen ion-implanted supermartensitic stainless steel in chlorine- and hydrogen-rich environments, 29–41, Copyright (2018), with permission from Elsevier).

The precipitation of fairly large volume fractions of Cr nitrides caused, as expected, a marked decrease in corrosion resistance [83,86]. Only when a thick compound layer consisting of nitrides was able to form, was higher corrosion resistance observed, such as for the AISI 410 martensitic stainless steel samples, which were plasma-nitrided in the range of 420–500 °C for 20 h and tested in 3.5% NaCl solution and 1% HCl acidic water solution [106].

4. Expanded Ferrite

The term "expanded ferrite" was initially used to indicate the phase that forms together with CrN when expanded austenite decomposes [112], or as a designation of "expanded martensite" [35,64,65]. Only since the mid-2000s was it used to indicate the interstitial-rich phase that forms in FeCr ferritic alloys [37] and duplex stainless steels [36]. The symbols α_N and α_C are also used to indicate N- or C-rich expanded ferrite phases.

Expanded ferrite has been observed to form in ferritic and duplex stainless steels subjected to low-temperature treatments.

4.1. Ferritic Stainless Steels

The main characteristics of the modified surface layers containing expanded ferrite, obtained by the nitriding or carburizing of ferritic stainless steels, are reported in Table 2.

Table 2. Main characteristics of the modified surface layers containing expanded ferrite, obtained by the nitriding or carburizing of ferritic stainless steels.

Properties		Nitriding	Carburizing
Formation temperature (without formation of Cr compounds) (°C)		<380	<470
Max. interstitial content in expanded ferrite (at. %)		24	10
Max. surface hardness (kg_f mm^{-2})	- expanded ferrite only - with Fe-based compounds	1090 1200	1000 -
Hardness profile	- expanded ferrite only - with Fe-based compounds	gradual change abrupt change	gradual change -
Dry sliding wear resistance		good	-
Localized corrosion resistance	- expanded ferrite only - with Fe-based compounds	good variable [1] / good [2]	good -

[1] Nitrides dispersed in expanded ferrite. [2] Nitrides forming a continuous outer layer.

4.1.1. Formation of Expanded Ferrite

In the ferritic stainless steels, the formation of a supersaturated solid solution of interstitial atoms in the b.c.c. lattice is reported [113–115], but it is competitive with that of the ε-$Fe_{2-3}N$ nitride, which tends to form a continuous outer layer [116–120]. It is interesting to note that at low treatment temperatures the presence of γ'-Fe_4N is not usually reported, as well as that of a more or less expanded austenite, which should be stabilized by N or C atoms. The formation of expanded austenite was observed in a super-ferritic stainless steel, containing 28.12 wt.% Cr, 3.91 wt.% Ni, 2.44 wt.% Mo, which was subjected to N ion implantation at 100 °C at a lower implantation dose, whereas using higher implantation dose and energy N-rich martensite, together with Fe and Cr nitrides, formed [121]. It may be hypothesized that expanded austenite formation was promoted by the presence of Ni in the steel. The presence of expanded austenite was also reported in X10CrAl18 steel, which was nitrided with the PIII process at 300 °C for 3 h [122]. The Fe-based nitrides ε-$Fe_{2-3}N$ and γ'-Fe_4N were observed, together with CrN, in ferritic stainless steels nitrided at temperatures typical of the nitriding of low alloy steels [31,114,116,123].

A continuous layer of expanded ferrite was observed by Kurelo et al. [113] on a UNS S44400 super-ferritic stainless steel subjected to nitriding by means of PIII process at temperatures in the range of 300–400 °C for 3 h. The presence of Fe nitride precipitates was assumed for the samples nitrided at 350 and 400 °C, even if clear peaks of these phases were not observed in X-ray diffraction patterns. Alphonsa et al. [114] reported the formation of expanded ferrite when AISI 430F steel was plasma nitrided at 380 °C (4 h, 80% N_2 + 20% H_2), while plasma nitrocarburizing (4 h, 78% N_2 + 20% H_2 + 2% C_2H_2) at the same temperature caused the formation of Fe_3N (or, more likely, the ε-$Fe_{2\text{-}3}(N,C)$ carbonitride), together with a N- and C-rich expanded ferrite. With a treatment temperature of 400 °C, Fe_3N was present in both nitrided and nitrocarburized samples (probably, $Fe_{2\text{-}3}(N,C)$ in nitrocarburized samples), and a further increase of the treatment temperature caused the formation of Cr nitrides in all the samples and of Fe and Cr carbides in the nitrocarburized ones. Modified surface layers of nitrided samples were thicker than those of samples nitrocarburized at the same temperature, similar to what was observed for martensitic stainless steels. Again, X-ray diffraction analysis suggests that the volume fraction of nitride in nitrided samples was lower than that of carbonitride in nitrocarburized samples, so that it may be hypothesized that the thickness of the modified layers was influenced by the slower diffusion of interstitial atoms in nitride/carbonitride [78]. De Sousa et al. [115] observed the formation of a thin, uniform layer consisting mainly of expanded ferrite on AISI 409 steel subjected to cathodic-cage nitrocarburizing at 400 °C for 5 h with a 95% N_2 + 5% CH_4 gas mixture, while for a plasma nitrocarburizing treatment, performed at the same temperature but for 10 h with a 78% H_2 + 20% N_2 + 2% CH_4 gas mixture, the formation of a significant amount of carbides and nitrides could also be inferred. Regarding carburizing treatments, C-rich expanded ferrite layer with a small volume fraction of carbides was reported to form on a E-BRITE stainless steel (Cr 25.8 wt.%, Ni 0.11 wt.%, Mo 0.99 wt.%) using a treatment temperature of 470 °C [61].

The reported studies put in light that the formation of expanded ferrite with none or a small amount of Fe-based nitrides is dependent on both treatment conditions and steel composition. Treatment temperature should be low enough to reduce the diffusivity of substitutional atoms, in particular of Cr atoms. N atoms inlet in the material should be reduced to prevent or reduce the formation of nitrides, but, at the same time, taking into account the high N atom diffusivity in ferrite, the amount of N inlet cannot be too low, in order to allow to form expanded ferrite. Since the formation of Cr-based compounds is both a chemical and diffusion driven process, not only the treatment temperature but also the duration must be taken into account. For nitriding and nitrocarburizing treatments, temperatures lower than 450 °C should be employed, in order to avoid Cr nitride precipitates [114]. However, by nitriding at 400 °C, the formation of Cr nitrides was avoided when the treatment time lasted for up to 4 h [113,114], but the formation occurred when a 20-h duration was used [124]. Similarly, with a 380C nitriding temperature Cr nitride was not detected performing the treatment for 28 h, while the presence of this compound could not be excluded when the treatment time was 40 h [116]. Regarding carburizing, according to the study of Michal et al. [61], temperatures lower than 470 °C should be used to avoid carbide precipitation.

The shape of the concentration profile for N- and C-rich expanded ferrite was similar [60,61], and it was approximately that of a complementary error function, thus it was hypothesized that the diffusion coefficient of interstitial atoms in ferrite has a minimum dependence on concentration [61].

4.1.2. Characteristics of the Modified Surface Layers

The surface of low-temperature nitrided ferritic stainless steels has a slightly etched appearance, with grains well delineated, and small reliefs present at grain boundaries [125].

The cross-section microstructure of the modified surface layer containing expanded ferrite is almost featureless, as a consequence of the higher corrosion resistance of this layer in comparison with that of the substrate [61,113,114,125]. Below this layer, in the region in which the N concentration decreased considerably, Kurelo et al. [113] observed needle-like structures, which were supposed to be related to the precipitation of γ'-Fe$_4$N nitride (Figure 5). Larisch et al. [116] reported that, when ε-Fe$_{2\text{-}3}$N nitride formed together with expanded ferrite, no sharp boundary with the substrate was delineated by chemical etching, and a peculiar microstructure was observed, with a "white" unetched outer zone which became progressively less "white" and containing needle-like structures, interpreted as deformation bands instead of γ'-Fe$_4$N precipitates. The comparison of the thickness of the unetched layer with that of the hardened layer suggests that diffusion of N atoms extends far beyond the fairly thin unetched zone [113], as was observed for martensitic stainless steels. Thus, it may be hypothesized that, in the unetched layer, N-rich expanded ferrite is present, eventually coming together with Fe-based nitrides, while a "diffusion" zone with a lower N-content extends below this layer. On the contrary, when carburizing was performed, comparable thicknesses were reported for the unetched layer and the hardened one [61].

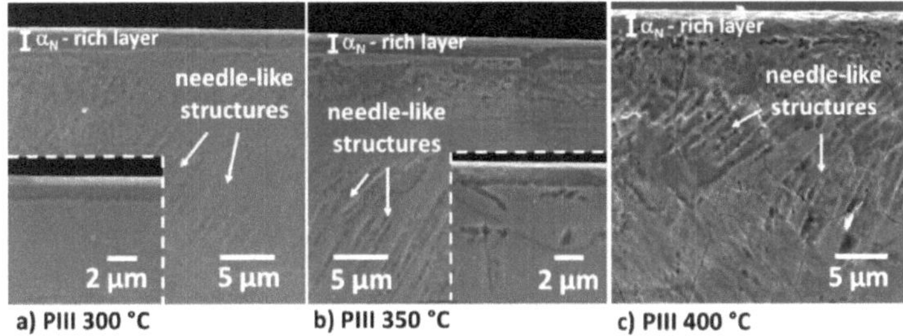

Figure 5. SEM micrographs of the cross-sections of UNS S44400 super-ferritic stainless steel samples PIII nitrided at 300 (**a**), 350 (**b**) and 400 (**c**) °C (etchant: Murakami's reagent) (α_N: expanded ferrite). (For further details, see ref. [113]). (Reprinted from *Surface & Coatings Technology*, 430, B.C.E.S. Kurelo et al., Mechanical properties and corrosion resistance of α_N-rich layers produced by PIII on a super ferritic stainless steel, 126388, Copyright (2020), with permission from Elsevier).

N-content far beyond the solubility limit was detected in the expanded ferrite layer: a N content of ~12 at.% was present in UNS S44400 super-ferritic stainless steel which was PIII-nitrided at 400 °C for 3 h [113], while a value as high as ~24 at.% was detected in an AISI 430F ferritic stainless steel which subjected to ion implantation at 380 °C for 1 h with a voltage of 10 kV [60]. X-ray diffraction analysis suggested that the expansion of the ferrite b.c.c. lattice was not isotropic, with a higher value for the (200) plane than that for the (110) one [60,104,113] (Figure 6). Variations up to 6.62% for the lattice parameter calculated from the (200) peak and 5.24% for the lattice parameter calculated from the (110) peak were observed [104]. It was hypothesized that this anisotropic expansion was due to the anisotropy of the elastic modulus, with lower elastic modulus for the <200> direction in comparison with that of the <110> one [104].

Michal et al. [61] observed, on E-BRITE stainless steel subjected to carburizing treatment at 470 °C, the formation of C-rich expanded ferrite layer with a maximum C content of ~10 at.%, which caused a lattice expansion of ~0.3%.

4.1.3. Hardness of the Modified Surface Layers

The formation of the expanded ferrite allows for increased surface hardness, due to solid solution strengthening effect. The precipitation of nitrides and carbides increased the microhardness further on, and higher values were observed as the volumetric fraction of nitrides and carbides increased, or when a continuous nitride layer was able to form. For N-rich expanded ferrite layer, microhardness values up to ~1090 kg$_f$ mm^{-2} were registered [113]. Microhardness values higher than 1000 kg$_f$ mm^{-2} were observed on carburized samples [61]. The formation of Fe-based nitrides increased surface microhardness, and values up to ~1200 kg$_f$ mm^{-2} were registered [104]. A further increase, with values as high as 1550 kg$_f$ mm^{-2}, was observed when a large volume fraction of Fe- and Cr-based nitrides formed [114].

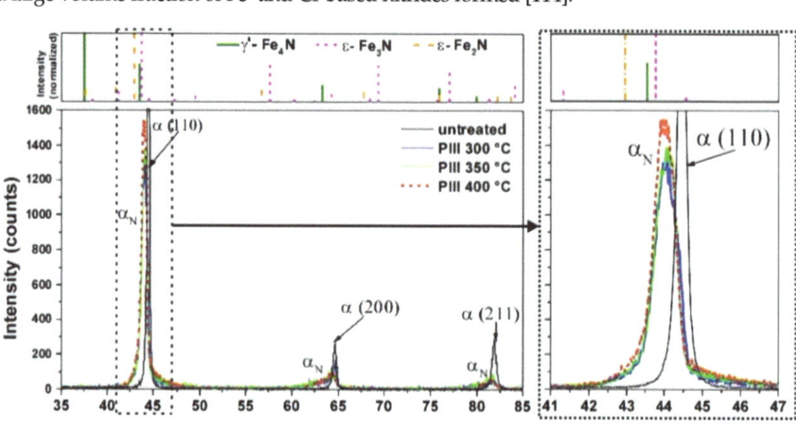

Figure 6. X-ray diffraction patterns of UNS S44400 super-ferritic stainless steel samples untreated and PIII nitrided as indicated (α: ferrite, α$_N$: expanded ferrite). (For further details, see ref. [113]). (Reprinted from *Surface & Coatings Technology*, 430, B.C.E.S. Kurelo et al., Mechanical properties and corrosion resistance of α$_N$-rich layers produced by PIII on a super ferritic stainless steel, 126388, Copyright (2020), with permission from Elsevier).

When expanded ferrite was able to form without nitrides or carbides or with a negligible amount of Fe-based compounds, fairly smooth hardness profiles were observed [61,113]. When fairly high amounts of nitrides or carbonitrides form, the microhardness profile shows the presence of high hardness values at the surface, forming a more or less extended plateau, and then a steeper decrease to matrix values [115,124].

4.1.4. Tribological Properties of the Modified Surface Layers

Owing to the increase of surface hardness, the presence of an interstitial-rich expanded ferrite layer should improve the wear resistance of ferritic stainless steels. However, in international literature, the main studies reported the wear behavior of ferritic stainless steels subjected to treatments that produced more or less large amounts of nitrides, which do not allow for an evaluation of the effect of the expanded phase only. Improvement of wear resistance was observed for samples in which Fe-based nitrides were formed, tested in ball-on-disk configuration with an alumina ball as a counterpart [90], and for specimens having both Fe- and Cr-based nitrides, subjected to micro-abrasive wear tests [123] and sliding wear tests using an Amsler machine [124].

4.1.5. Corrosion Resistance of the Modified Surface Layers

The formation of a continuous N-rich expanded ferrite layer allowed for an increase in the corrosion resistance in NaCl solutions. Polarization curves usually showed an increase in corrosion potential and pitting potential, and a decrease in anodic current density in the passive branch [113,114,120]. The precipitation of Fe-based nitrides, which are nobler than the expanded phase, can cause a microgalvanic effect, thus a worsening of the corrosion resistance may occur [113,114,123]. Kurelo et al. [113] observed an increase of the passive current density values for UNS S44400 super-ferritic stainless steel PIII nitrided at 400 °C, in comparison to those of samples nitrided at 300 °C, and assumed that this fact was due to the precipitation of Fe-based ε-nitride, and possibly of Cr nitrides. Similar nitrided samples, having a higher nitride content, were tested by Luiz et al. [104] in a 3.5% NaCl solution for up to 30 days of immersion. For both nitrided and untreated samples, an increase of the open circuit potential values for up to 15 days was observed, followed by a slight decrease.

However, the potential values of nitrided samples were nobler than those of the untreated alloy, suggesting an improvement of passive film stability after nitriding.

A continuous layer of ε-nitride may increase the corrosion resistance in Cl^--containing solutions further on [126]. Increasing the treatment temperature and/or the treatment duration may promote the precipitation of Cr nitrides, which causes a marked worsening of the corrosion resistance. When AISI 430 was nitrided at 420 °C, the pitting potential increased with respect to that of the untreated steel in a 0.5 M NaCl solution, but when the treatment was prolonged up to 20 h, a marked decrease in corrosion and pitting potentials was observed [120]. When a large volume fraction of Cr nitrides was able to form, a marked decrease in corrosion resistance was reported [114,123,127].

Different corrosion behaviors were observed when low-temperature treated ferritic stainless steels were tested in Cl^--free solutions. Spies [120] reported a significant increase of corrosion potential and decrease in the corrosion rate for AISI 430 steel samples nitrided at 300 °C for 60 h or at 350 °C for 26 h and tested in a 0.05 M H_2SO_4 solution, in comparison with the untreated alloy. A similar behavior was observed for AISI 430 samples nitrided at 250 °C for 39 h or 420 °C for 36 h [127]. On the contrary, Gontijo et al. [128] reported only a slight increase in corrosion potential and a marked increase in anodic current density for AISI 409 samples nitrided in the range of 350–500 °C and tested in a 0.1 M H_2SO_4 solution, in comparison with the untreated steel.

Regarding carburizing treatment, for the C-rich expanded ferrite an improved resistance to localized corrosion can be inferred on the basis of the high resistance to chemical etching of the modified layer [61].

4.2. Duplex Stainless Steels

Table 3 summarizes the main characteristics of the modified surface layers, in which expanded ferrite forms, obtained by the nitriding or carburizing of duplex stainless steels.

Table 3. Main characteristics of the modified surface layers containing expanded ferrite, obtained by the nitriding or carburizing of duplex stainless steels.

Properties	Nitriding	Carburizing
Formation temperature (without formation of Cr compounds) (°C)	<430	<480
Max. interstitial content in expanded ferrite (at. %)	25	18
Max. surface hardness (kg_f mm^{-2})	1510	1500
Hardness profile	abrupt change	abrupt change
Dry sliding wear resistance	good	-
Localized corrosion resistance	very good	good

4.2.1. Formation of Expanded Ferrite

The basic microstructure of duplex stainless steels consists of austenite grains in a ferritic matrix. The ferrite and austenite regions have different compositions, due to the partitioning of alloy elements during cooling. When a fairly slow cooling is performed, ferrite is richer in ferrite-stabilizing elements, such as Cr and Mo, while austenite is rich in austenite-stabilizing elements, such as Ni and Mn. When a low-temperature treatment is performed on a duplex stainless steel, the interstitial atoms have a different diffusion rate in austenite and ferrite. In fact, N and C atoms diffuse faster in the b.c.c. ferrite grains than in f.c.c. austenite ones, therefore different thickness values of the modified layers were observed in the two phases [116]. Moreover, depending on treatment conditions and alloy composition, ferrite may solubilize a "colossal" amount of interstitial atoms, thus expanded ferrite is observed [129–133], or transforms into expanded austenite [116,132–136]. The original ferrite grains have a greater tendency to form nitride precipitates, due to the presence of a fairly larger amount of nitride-forming elements, such as Cr and Mo. Bielawski and Baranowska [129] observed the formation of some strongly etched regions in a part of the modified layer formed on ferritic substrate of a X2CrNiMo 22–5–3 duplex stainless steel which was gas nitrided at 430 °C with high N potential, while the modified layer formed on austenite was unetched. This effect became more marked by increasing the treatment temperature up to 475 °C. Cr nitride was observed in UNS S31803 steel samples subjected to active screen plasma nitriding at 435 °C for 10 h [131]. The Cr-based carbide, $M_{23}C_6$, was observed in F51 steel samples which were carburized at 480 °C for 12 h [137]. Thus, the formation of Cr nitride/carbide precipitates can be avoided using treatment temperatures lower than those usually used for the treatment of austenitic stainless steels. However, by decreasing the treatment temperature further on, the formation of Fe-based nitrides may occur. In fact, ε-nitride precipitates were reported to form in the expanded ferrite regions of a UNS S31803 steel, plasma nitrided with the active screen process at 400 °C for 20 h [132]. Fe-based nitride precipitates were detected by X-ray diffraction analysis, and they were hypothesized to form in the expanded ferrite grains of UNS S32750 super-duplex stainless steel samples, subjected to balanced plasma nitriding at temperatures as low as 292–355 °C, due to the their brittleness under indentation [133].

4.2.2. Characteristics of the Modified Surface Layers

When both expanded ferrite and expanded austenite are able to form, the surface morphology of low-temperature treated duplex stainless steels has features similar to those observed for austenitic stainless steels, with delineated grain boundaries, reliefs, and shear lines present in austenite grains, while the ferritic part remains fairly smooth [125].

Differences in the austenite and ferrite grains are also present when the cross-section microstructure is examined (Figure 7). As recounted previously, when duplex stainless steels are treated at low temperatures, the part of the modified surface layer containing ferrite is usually thicker than that containing austenite [125,130,132,138]. In the austenitic part, the layer, delineated by chemical etching, has a characteristic arc-like shape, and it is thicker close to the ferrite boundary and thinner in the central part of the austenite grain [125,130] (Figure 7a,b). This fact can be ascribed to grain boundary diffusion and lateral flow of N atoms from ferrite to austenite [139]. Apart from the differences in thickness, the whole modified surface layer was fairly smooth, and both the expanded ferrite and expanded austenite grain types were not significantly etched by chemical etching [125]. Deformation bands were present in the expanded ferrite regions (Figure 7d), due to intense compressive stresses developed during nitriding, as well as slip lines in the expanded austenite regions (Figure 7c), when the cross-section of a F51 steel, nitrided at 400 °C for 20 h and etched with oxalic acid (10%), was analyzed [130]. A similar microstructure with deformation bands was observed in UNS S31803 steel samples subjected to active screen plasma nitriding at 400 °C for 20 h [132]. Moreover, TEM analysis highlighted the presence of needle-like coherent ε-nitride precipitates having an orientation relationship of [111] α_N// [120] ε-Fe$_3$N.

Recently, Dalton et al. [140] performed a high-resolution spatially-resolved compositional and structural analysis of the δ-ferrite grains of 2205 duplex stainless steel samples nitrided in the range of 325–400 °C, and observed two competitive responses to nitriding. Some regions had a N content up to ~25 at.% and maintained a b.c.c. lattice without evident distortions. In these expanded ferrite grains, a spinodal decomposition occurred, with Fe-rich nanocrystals and nanometer Cr-rich domains having an apparently amorphous structure, and a preferential N segregation to the Cr-rich regions was observed. In other regions of the modified layer, austenite formed with a diffusionless transformation from ferrite, maintaining the composition of the expanded ferrite. The plates of this austenite contained a significant concentration of faults and/or twins, and tended to grow with a preferred orientation relationship from the parent ferrite, suggesting that their growth was due to a shear mechanism characteristic of martensitic reactions.

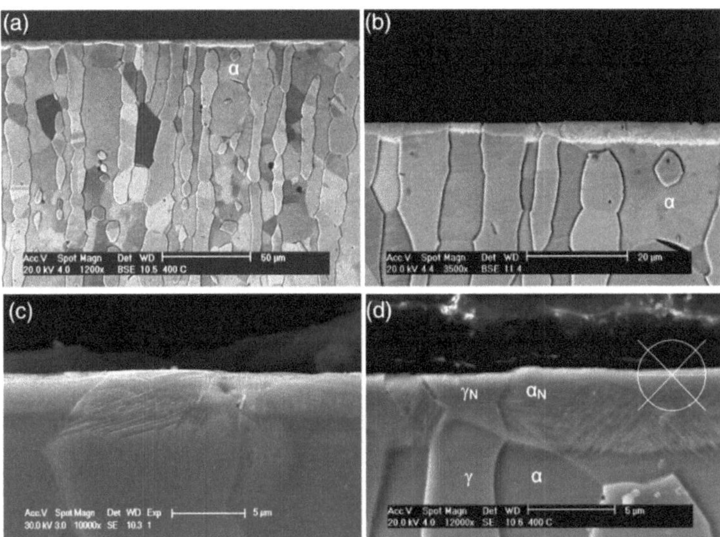

Figure 7. Micrographs of the cross-section of an AISI F51 duplex stainless steel sample plasma nitrided at 400 °C for 20 h: different thicknesses in ferrite and austenite grains (**a**,**b**), slip lines in austenite grains (**c**), deformation bands in ferrite grains (**d**) (etchant: oxalic acid (10%)) (α_N: expanded ferrite, γ_N: expanded austenite). (For further details, see ref. [130]). (Reprinted from *Surface & Coatings Technology*, 232, C.E. Pinedo et al., Low-temperature plasma nitriding of AISI F51 duplex stainless steel, 839–843, Copyright (2013), with permission from Elsevier).

A study [141] on the characteristics of the δ-ferrite present in the modified surface layers produced in 2205 duplex stainless steel and 17–7 semi-austenitic PH stainless steel subjected to carburizing at 380 °C for 150 h showed the presence of a high density of dislocations in the C-rich ferrite grains. A spinodal-like decomposition produced Fe-rich and Cr-rich regions, and the Cr-rich regions were enriched in C. Even if the C content in these modified grains was very high, with an average value as high as 18 at.%, no tetragonality was observed in the ferrite b.c.c. lattice, and carbides were not detected. It was hypothesized that C atoms were segregated to the dislocation cores, thus carbide precipitation was prevented. It was also suggested that the high dislocation density was due to the fairly large stresses applied by the neighboring austenite grains due to their volume expansion. According to Sasidhar and Meka [142], the observed high C content was in accordance with a paraequilibrium that occurred between the carburization atmosphere and the spinodally-decomposed ferrite.

Although there is a general consensus on the fact that the interstitial content is different in the ferritic part of the modified layer as compared with that of the austenitic one, conflicting results have been reported. Bielawski and Baranowska [129] detected a higher N content in the austenite part of the modified layer than in the ferritic one for X2CrNiMo 22–5–3 duplex stainless steel samples which were gas nitrided in the range of 400–475 °C. On the contrary, Pinedo et al. [130] reported a higher N content in expanded ferrite grains than in expanded austenite ones of a F51 steel which was plasma nitrided at 400 °C for 20 h. A similar result was found by Chiu et al. [131] in UNS S31803 steel samples subjected to active screen plasma nitriding for 10 h, and by Alphonsa et al. [138] in 2205 duplex grade samples nitrided at 400 °C for 4 h. When nitrocarburizing of 2205 duplex stainless steel was performed at 400 °C for 4 h, a higher N and C content was detected in austenite grains than in ferrite grains, while the thickness of the modified layer in the two grain types was comparable [138]. De Oliveira et al. [133] observed a larger N content and higher thickness in expanded austenite grains than in expanded ferrite ones for balanced plasma-nitrided

UNS S32750 super-duplex stainless steel samples. These authors ascribed this behavior to the higher Cr content of the super-duplex steel (~26 wt.%) in comparison with duplex stainless steels (~22 wt.%), and they hypothesized that, owing to the trapping mechanism in Cr sites, the ferrite may present a slower N diffusion in super-duplex stainless steels than in duplex stainless steels.Accordingly, better conditions for the N-saturation may occur, together with an earlier precipitation of nitrides, and thus the formation of thinner modified layers.

Higher C content was observed in the ferritic grains of a 2205 duplex stainless steel carburized at 380 °C for 150 h, in comparison with austenitic grains [141].

4.2.3. Hardness and Tribological Properties of the Modified Surface Layers

Low-temperature treatments improve the surface hardness of duplex stainless steels. However, alloy composition and treatment conditions influence both the interstitial atom content in the ferrite and austenite grains, present in the modified layers, and the eventual transformation of ferrite into expanded austenite and precipitation of nitrides and carbides, therefore, different hardness values in the different grain types were detected. Thus, it is not surprising that, for the same steel (2205), ferrite grains were observed to have higher hardness than austenite grains [138], or lower hardness [143]. In X2CrNiMo 22–5–3 duplex steel, the layer formed on ferrite was slightly (about 100–150 kg_f mm^{-2}) harder than that formed on austenite [129]. Values up to ~1510 kg_f mm^{-2} were registered in nitrided samples [132], and of ~1500 kg_f mm^{-2} in carburized specimens [144]. Moreover, the formation of nitrides or carbides precipitates increased the hardness further on, and values as high as 1800 kg_f mm^{-2} were measured [129,137]. The microhardness profiles of both austenite and ferrite grains showed high values at the surface and then a fairly steep decrease to matrix values [130,144].

Tribological properties were improved by the presence of expanded ferrite and/or expanded austenite. Higher scratch resistance and a lower friction coefficient were observed for nitrided SAF 2507 and SAF 2205 duplex stainless steels in comparison with the untreated alloys [143]. A lower mass loss was registered for UNS S31803 steel which was subjected to active screen plasma nitriding in the range 420–450 °C and tested in unlubricated ring-on-disk ITO type wear tests, as compared to the untreated steel [131].

4.2.4. Corrosion Resistance Properties of the Modified Surface Layers

Although low-temperature treated duplex stainless steels have modified surface layers that may contain both expanded ferrite and expanded austenite, consisting of different interstitials and alloy elements content, this bi-phase structure does not impair corrosion resistance in Cl^--containing solutions.

Increase of corrosion potential and decrease of anodic current density in the passive branch were observed for UNS S31803 steel samples subjected to active screen plasma nitriding at 420 °C for 10 h, and tested in 3.5% NaCl solution [131]. By using longer treatment times, the anodic current density values slightly increased, due to the precipitation of a small volume fraction of nitrides. Higher corrosion resistance in a 3.5% NaCl solution was observed for a 2205 duplex stainless steel nitrided and nitrocarburized at 350 and 400 °C for 4 h [138], but it is not clear whether this improvement was due to the presence of both expanded ferrite and austenite or to the transformation of ferrite into expanded austenite. Corrosion resistance comparable to that of the untreated alloy was observed for 2205 stainless steel specimens, carburized at 350 and 380 °C for 150 h, and tested in a 0.6 M NaCl solution [144]. An increase of corrosion resistance in NaCl solutions was observed when the modified layers consisted mainly of expanded austenite [104,136,143].

5. Expanded Austenite–S-Phase

As mentioned previously, the austenite-based expanded phase was the first super-saturated solid solution of interstitial atoms in a Fe-based lattice recognized to form as a consequence of low-temperature treatments of stainless steels. Today, the most used names of this metastable phase are "expanded austenite", "S-phase", γ_N (for the N-rich form), and γ_C (for the C-rich form). In particular, "expanded austenite" and "S-phase" are mainly related to two features of this phase: the solubilization of interstitial atoms far beyond the solubility limit, thus a "colossal" supersaturation occurs [54,58], and an expansion of austenite lattice is observed [145], and the peculiar X-ray diffraction pattern of this phase, with the peaks markedly shifted towards lower angles with respect to those of austenite. Ichii et al. [17], indicated these peaks, that were not indexed in the ASTM index, as S1-S5 and the new "phase" as "S-phase", and they assumed, also taking into account its high resistance to chemical etching and its ferromagnetism, that it was a M_4N-type nitride.

As a matter of fact, it is now recognized that expanded austenite–S-phase is not a "new" equilibrium phase, and that it derives from austenite. However, many studies have highlighted that the characteristics of this phase are beyond a simple expansion, or a "colossal" supersaturation, of the austenite lattice, even if some of them can be related to it.

Expanded austenite was observed to form not only in austenitic stainless steels, both in massive specimens [8,11,46,47,146] and coatings obtained with various techniques [147–154], but also in stainless steels, in which the high N solubilization might induce the transformation of ferrite or martensite into austenite, as duplex [104,116,132,134–136], martensitic [65,91,95], ferritic [121,122], and precipitation hardening [65,71,155,156] stainless steels.

Hereafter, the main characteristics and properties of the modified surface layers containing expanded austenite are summarized. The main characteristics of the modified surface layers obtained with nitriding or carburizing of austenitic stainless steels are reported in Table 4. Further information can be found in the reviews on this topic [11,12,46,47].

Table 4. Main characteristics of the modified surface layers containing expanded austenite, obtained by the nitriding or carburizing of austenitic stainless steels.

Properties	Nitriding	Carburizing
Formation temperature (without formation of Cr compounds) (°C)	300–450	300–550
Max. interstitial content in expanded austenite (at. %)	38	19
Max. surface hardness ($kg_f\ mm^{-2}$)	1450	1000
Hardness profile	abrupt change	gradual change
Load bearing capacity	low	high
Dry sliding wear resistance	very good	good
Localized corrosion resistance	very good	good

5.1. Formation of Expanded Austenite

When low-temperature treatments are carried out on austenitic stainless steels, a supersaturated solid solution of interstitial atoms in the austenite lattice is easily formed, and Fe-based nitrides are not usually observed, contrary to what occurs in martensitic and ferritic stainless steels. As for the low-temperature treatments, treatment temperature is the main parameter to be controlled in order to hinder Cr diffusion and prevent the formation of large amounts of Cr nitrides and carbides. Nitriding is usually performed at temperatures $\leq$450 °C [8,38], while carburizing is carried out at temperatures $\leq$550 °C [38]. Treatment duration is also highly important, and the transition from the solubilization of interstitial atoms to the formation of nitrides and carbides in the austenite matrix was depicted in time–temperature–transformation diagrams, evidencing that for a specific treatment temperature exists a critical time beyond which the formation of Cr nitrides or carbides occurs [38,42]. The feeding of interstitial atoms is a further important factor for obtaining expanded austenite, as observed in plasma nitriding treatments. By decreasing the treatment pressure, an enhancement of nitriding efficiency occurs, and a greater N atoms inlet is produced [103,157,158]. As a consequence, nitride precipitates are able to form at temperatures at which their formation should typically be delayed [103].

The formation of nitrides and carbides is also influenced by the alloy elements of the stainless steels. Cr is not only able to form nitrides or carbides, but it also promotes the formation of expanded austenite. Ni, which is the main austenite-stabilizing element of CrNi-based AISI 300 series austenitic stainless steels, is not a nitride-/carbide-forming element, therefore it tends to promote the formation of a nitride-free expanded austenite structure up to 450 °C. When Ni is substituted, in whole or in part, by Mn, as in the CrMn-based AISI 200 series or Ni-free austenitic stainless steels, an additional nitride-forming element is introduced in the austenite lattice. As a consequence, the diffusion of substitutional atoms should be decreased further, and lower treatment temperatures and shorter treatment durations should be employed, in comparison with those used for AISI 300 series stainless steels [91,159–161]. As an example, plasma nitriding at 400 °C, 500 Pa for 3 and 5 h, using a 80 vol.% N_2 + 20 vol.% H_2 gas mixture, allowed for the production of modified surface layers consisting mainly of expanded austenite on AISI 316L, while on AISI 202, sensible amounts of nitrides were detected [91], as well as on a Ni-free stainless steel [161]. Similarly, a modified layer mainly consisting of expanded austenite was produced on the high-Ni RA330® (Fe-19Cr-35Ni-1.2Si), triode plasma nitrided at 450 °C for 20 h, while by using the same treatment conditions, large volume fractions of nitride precipitates were observed on both high-Mn Staballoy® AG17 (Fe-17Cr-20Mn-0.5N) and AISI 304 [162].

When low-temperature treatments are performed on austenitic stainless steels, N and C diffusion in f.c.c. austenite results enhanced. Williamson et al. [57] reported that, when N-rich expanded austenite layers formed, N diffusion was four to five orders of magnitude faster than that expected for N in f.c.c. austenite and the activation energy was significantly reduced. Regarding C-rich expanded austenite, Ernst et al. [163] observed that C diffusion coefficient was more than two orders of magnitude larger for high C levels, in comparison to that of a dilute solution. The concentration profile of interstitial atoms in low-temperature treated austenitic stainless steels cannot be described by the complementary error function, which is characteristic of concentration independent diffusion. In nitrided samples, N concentration vs. depth profile has high values at the surface, a steep decrease to lower values followed by a nearly constant plateau, then a steep decrease to matrix values [11,45,46,149,159]. In carburized samples, a steep decrease from the high values at the surface, followed by a smoother decrease to matrix values, is present [11,45,46,149,159,163]. When nitrocarburizing is carried out [45,46,147,149,159], or sequential treatments are performed, i.e., nitriding followed by carburizing [164] or carburizing followed by nitriding [134,147,149,164], an outer N-rich expanded austenite layer and an inner C-rich expanded austenite layer are produced, which together allow a smoother profile to be formed, in comparison with that of nitrided steels [11,45,147,149,159,164].

Many effects may contribute to the enhanced diffusion and "colossal" supersaturation of the expanded austenite. Diffusion of interstitial atoms seems to depend on their concentration, and not on a composition gradient, with higher values for fairly high N and C content and lower values for low concentration [163,165]. Cr atoms are hypothesized to have a "trapping effect" on N atoms in the octahedral sites nearby, causing short-range ordering, while additional N atoms can be "detrapped" and diffuse rapidly through the expanded austenite layer [166–168]. A further phenomenon, which may influence the diffusion of interstitial atoms, is the high compressive residual stresses which are produced in the modified layer as a consequence of the expansion of austenite lattice, and may become an additional component of the driving force for the interstitial atoms transport [169–171].

As a metastable phase, expanded austenite tends to decompose when the service temperature is too high. The incubation time for the transformation of N-rich expanded austenite into CrN and α-Fe ranges from minutes at 500 °C to thousands of hours at 300 °C [172], while for C-rich expanded austenite decomposition and formation of carbides occur in thousands of hours at 350 °C [173].

5.2. Characteristics of the Modified Surface Layers

Figure 8 depicts the surface morphology and the cross-section microstructure of an AISI 202 sample which was plasma-nitrided at 380 °C, 340 Pa, for 3 h, together with its X-ray diffraction pattern, also showing the austenite substrate.

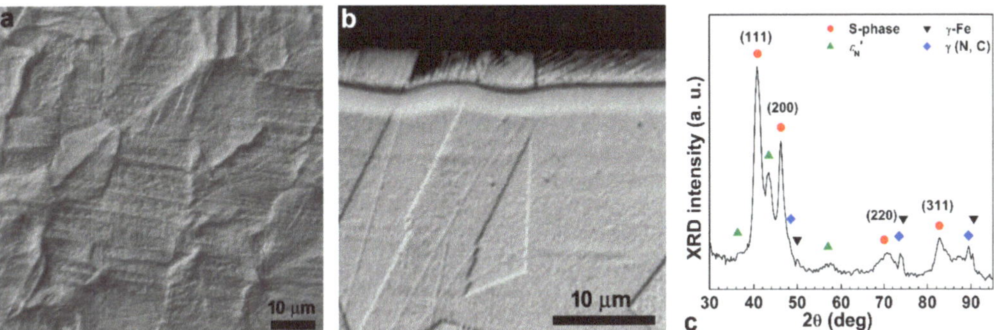

Figure 8. Surface morphology (**a**), cross-section micrograph (**b**) (etchant: acetic glyceregia) and X-ray diffraction pattern (**c**) of an AISI 202 austenitic stainless steel sample plasma nitrided at 380 °C, 340 Pa, for 5 h. (For further experimental details, see ref. [174]).

A characteristic feature of the surface morphology of low-temperature treated austenitic stainless steels is the etched appearance, with distinctly delineated grain boundaries, slip bands inside the grains and a swelling of the grains [11], as shown in Figure 8a. Even if plasma processes may enhance these features [174–176], they are also observed on gas nitrided samples [177], and they are caused by the local plastic deformations due to the expansion of the austenite lattice when interstitial atoms are solubilized (11 and references therein).

When the cross-section microstructure is examined, a featureless "white" layer, separated by the etched matrix, can be observed when an etchant as Marble's reagent is used [15]. The difficulty of etching expanded austenite supported the hypothesis of the formation an unknown nitride [15]. The use of etchants as glyceregia provides evidence that the nitrided layers are instead a modification of the austenite matrix, and that the grain boundaries of the expanded austenite are the continuation of the austenite matrix grain boundaries [11,178]. When nitriding is performed, a two-layer microstructure is delineated, with an outer thicker layer, consisting mainly of expanded austenite, and an inner layer without evident features (Figure 8b). The interfaces between these layers and between the inner layer and the matrix are fairly sharp. Also taking into account the suggestions of Christiansen et al. [169], it may be hypothesized that these interfaces are due to the sudden change of the N concentration (and as a response to the chemical etching as well), and they are not thermodynamic or crystallographic interfaces. The modified layers are fairly homogenous, even if grain-to-grain variations in thickness may be observed due to the different N diffusion in differently oriented grains [179]. Moreover, rotation of the grains with respect to the substrate is observable [180].

In the outer modified layer many lines may be observable, as shown in Figure 8b, and their number and extension in the substrate depend on nitriding conditions and alloy composition [174,181]. Tapered sections allow to appreciate the complex structure of these slip lines, due to localized plastic deformation phenomena occurring in the outer part of the modified layer [161,181]. A consequence of the plastic deformations is the formation of the so-called N-induced h.c.p. martensite, ε_N', which is a solid solution of nitrogen in the h.c.p. ε' martensite [182]. In fact, when plastic deformations occur in austenitic stainless steels, the fairly low stacking fault energy of these alloys, lowered further by N solubilization [183], may allow the formation of wide stacking faults, with the characteristic ABAB stacking of a h.c.p. structure [184]. According to Tao et al. [185], ε_N' forms from expanded austenite, and not from an unexpanded ε', through a martensitic shear transformation above a critical

N concentration. Tong et al. [186] objected that these h.c.p. regions should be regarded as "clustered" stacking faults, instead of plates of N-rich h.c.p. ε' martensite. The presence of ε_N' was observed by means of X-ray diffraction analysis [174,181,185], as depicted in Figure 8c, and in SAED patterns [182]. TEM analysis showed that the h.c.p. structure is coherent with the matrix, and it has the Shoji–Nishiyama orientation relationship typical of austenite to h.c.p. martensite transformation [176,182,185]. The formation of N-induced martensite, ε_N', is dependent on both solubilized N content and steel composition, and it is favored when the stacking fault energy of the steel is low, such as for the high-Mn austenitic stainless steels [174,176,181]. The formation of h.c.p. ε-Fe$_{2\text{-}3}$N nitride from ε_N', due to an ordering of N atoms and a distortion of ε_N' lattice, may also occur [187].

The inner modified layer consists of a solid solution of interstitial atoms (N, C) in slightly expanded austenite, $\gamma(N, C)$ [11,174,181]. It has been hypothesized that this layer is related to a local increase of C atoms, which were pushed by diffusing N atoms [188], or to the N concentration profile [134], or to the high residual stress, induced between the expanded austenite layer and the substrate [189].

A similar two-layer microstructure is obtained with treatments which use N- and C-containing gas mixtures in a single step (nitrocarburizing) [45,190], or sequentially [191]. The outer modified layer consists of N-rich expanded austenite, while in the inner layer C-rich expanded austenite is present. Low-temperature carburizing produces a single modified layer [45,134,192,193].

As the treatment temperature, duration, or the interstitial atoms feeding increase, Cr compounds precipitates tend to form, especially at grain boundaries [103].

The maximum observed N content in expanded austenite is ~38 at.% [58], while the maximum C content is ~19 at.% [59].

5.3. Expanded Austenite: Is It Really a Solid Solution?

The nature of the modified surface layer formed by low-temperature treatments in austenitic stainless steels has been debated from the beginning, thanks to the high corrosion resistance of the layer to chemical etching. When it became clear that the modified layer was a modification of the austenite matrix, the hypothesis that expanded austenite was a supersaturated solid solution of interstitial atoms in the f.c.c. austenite lattice was the natural consequence.

Most of the studies on expanded austenite are biased by the fact that this phase forms on the top of massive specimens. The high content of solubilized interstitial atoms, in particular N atoms, causes a huge expansion of the lattice, which cannot be accommodated only elastically, but causes local plastic deformations [11]; moreover, high compressive stresses develop [42]. TEM analysis showed the presence of dislocations, twins, stacking faults, slip bands, and a lamellar structure, ascribable to N-induced h.c.p. martensite, ε_N' [176,182,186]. Mössbauer spectroscopy observations suggested the presence, together with expanded austenite, of b.c.c. α' martensite with low N content [194]. The SAED patterns of expanded austenite have broad diffuse electron spots, therefore an almost isotropically expanded f.c.c. structure might be hypothesized [46,186]. Additional weak reflections, corresponding to forbidden reflections of a f.c.c. structure, were observed by some authors [186,195,196], thus it was hypothesized the formation of a simple cubic structure, in which N atoms occupy only one of the four interstitial octahedral sites, analogous to the γ'-Fe$_4$N structure. This phase was observed in the whole explored area in a domain structure [196] or in very small regions, suggesting a short-range ordering process [195]. When X-ray diffraction analysis of massive specimens was taken into account, neither isotropic expansion nor the presence of γ'-Fe$_4$N were observed. The X-ray diffraction pattern of expanded austenite in massive specimens has a peculiar feature, with peaks shifted towards lower angles as compared to those of austenite, in accordance with an expansion of the parent lattice, but with an apparently anisotropic expansion [45,103,197,198]. In fact, assuming a f.c.c. lattice as an indexing base, the lattice parameter value calculated from the d-spacing of the (200) plane is always greater than that calculated from the d-spacing of the (111), (220),

(311) or (222) planes [45,103,197,198]. This feature was observed both in N- and C-rich expanded austenite, even if the difference is greater for the N-rich expanded phase than for the C-rich one [197]. Different lattice structures (mixed f.c.c. lattice, tetragonal, monoclinic, rhombohedral, and equal-sided triclinic lattice) were tested without obtaining satisfying results [198]. Taking into account microstructure observations, it may be hypothesized that lattice expansion, residual stresses, and stacking faults, which all have an effect on the peak shifts, contribute to the anomalous shift of expanded austenite peaks (11 and references therein, 198).

Other observations add interesting details to the picture. X-ray photoelectron spectroscopy (XPS) analysis results suggest that a preferential bonding of N to Cr atoms establishes, but the binding energy of N is higher than that in Cr nitride [199,200]. Analyzing the N-rich expanded austenite present at the surface of AISI 304L samples, plasma nitrided at 400 °C for 3 h, Martinavičius et al. [201] observed, by means of conversion electron Mössbauer spectroscopy (CEMS) and X-ray absorption near edge structure spectroscopy (XANES) techniques, that for Cr there is a tendency toward a CrN-like environment with a Cr-N bonding, while the Fe environment in expanded austenite is very similar to that of γ'-Fe$_4$N, but with a larger disorder. Thus, it was hypothesized that the expanded austenite phase is composed by nanometric CrN-like precipitates, coherent with the matrix, incorporated in a γ'-Fe$_4$N matrix. Atom probe tomography (APT) observations seem to support this hypothesis [202]. On the basis of these studies it was suggested that the "colossal supersaturation" implies a spinodal decomposition, which produces the nano-sized precipitates [203]. In Fe–Cr–Ni alloys with a Cr content higher than 12 wt.%, nitrided at 380 °C for 3 h, the outer N-rich zone consisted of γ'-Fe$_4$N-like long-range order (LRO) regions and Cr-N short-range order (SRO) regions, which were indicated as γ_N', while the inner zone was an interstitially disordered expanded austenite with SRO regions of Cr and N atoms [204]. The two zones were indistinguishable from each other and formed a single "white" modified layer. It should be noted that γ_N' was regarded as an ordered phase having a N content far beyond that of γ'-Fe$_4$N (19.3–20 at. % [50]), i.e., a supersaturated phase [205].

The use of thin foils for producing expanded austenite allowed for the removal of the constraints of the matrix, and thus the effects of plastic deformations. X-ray diffraction analysis of thin foil specimens, consisting of expanded austenite of uniform composition, showed that both N-rich [58] and C-rich [206] expanded austenite have a f.c.c. lattice. Extended X-ray absorption fine structure (EXAFS) spectroscopy studies confirmed the presence of a preferential bonding between Cr and N atoms, suggesting the formation of SRO structures [207]. A similar result was found for carburized samples, with C atoms having a strong affinity with Cr atoms, thus probably promoting SRO [208]. X-ray diffraction analysis and CEMS observations of AISI 316 foils, gas nitrided in order to retain a homogenous content of 35.5 at.%, suggested the formation of a LRO structure analogous to γ'-Fe$_4$N [209]. A study that supports the hypothesis that at high N-content expanded austenite tends to become an ordered phase has been recently reported by Che and Lei [210]. By using a plasma-based low-energy ion implantation process, these authors nitrided at 380 °C for 10 min TEM foils of two Fe–Cr–Ni steels, one Ni-rich (Cr 17.66 wt.%, Ni 27 wt.%), and the other having a Ni content comparable to that of AISI 304 and AISI 316 (Cr 17.66 wt.%, Ni 12.6 wt.%), and compared the obtained results with massive specimens. For all of the samples the N content was estimated to be fairly high, in the range 30.3–33.4 at.%, far beyond that of γ'-Fe$_4$N. While in massive specimen twins (Ni-rich steel) or multiple stacking faults forming h.c.p. structures (AISI 304-like steel) were observed, the TEM foils did not show evidences of plastic deformations, thus they could be considered stress-free. The SAED patterns of expanded austenite in TEM foils of both steel types showed the strong diffraction spots of a f.c.c. structure together with the weak spots of a γ'-Fe$_4$N-like superstructure. Since the formation of this ordered structure was independent of stress and plastic strain, it was assumed that it was a characteristic feature of expanded austenite with a fairly high N content.

It is worth noting that the tendency to form a (short) range ordered structure, similar to that of γ'-Fe_4N, is present in the solid solution of N in f.c.c. γ-Fe [52], as reported in Section 2. The presence of Cr strengthens this tendency further on, since Cr has higher affinity for N than Fe. On the basis of the literature data, it may be hypothesized that, as N content increases in expanded austenite and the octahedral interstitial sites are occupied, there is a transition from a more disordered structure with Cr-N SRO regions to a structure in which γ'-Fe_4N-like LRO domains tend to form, while the f.c.c. lattice is maintained. For thin foils of gas nitrided AISI 304 and AISI 316, a maximum N content of ~38 at.% was detected [58]. A N content up to ~45 at.% was obtained in the expanded austenite that was produced as thin films by means of magnetron sputtering of AISI 316L in a N_2-Ar gas mixture [211]. As the N_2 volume fraction in the gas mixture was increased, a N content of 50 at.% was reached in the deposited phase. This MN-type phase maintained the f.c.c. lattice and had a constant lattice parameter of about 0.433 nm. It is known that the metastable FeN nitride may have two different structures, both having the Fe atoms arranged in a f.c.c. lattice, a ZnS-type, γ''-FeN, with each N atom coordinated by four Fe atoms at the corners of a regular tetrahedron, and a NaCl-type, γ'''-FeN, with each N atom coordinated by six Fe atoms at the corners of a regular octahedron, while CrN has a NaCl-type structure [211]. Unfortunately, experimental data on the arrangement of N atoms in this MN-type phase were not reported. Thus, it remains an open question whether, in austenitic stainless steels, N atoms are able to progressively occupy all the octahedral sites of f.c.c. lattice, thus there is a transition from a N-atoms disordered structure to a more ordered γ'-Fe_4N-like structure and then to a FeN/CrN NaCl-type structure, or whether also tetrahedral sites are occupied when a high N content is present, leading to a FeN ZnS-type structure.

5.4. Hardness of the Modified Surface Layers

The superior hardness of expanded austenite allows for a significant improvement in surface hardness of austenitic stainless steels. For N-rich expanded austenite, values up to ~1450 kg_f mm^{-2} were observed [212]. Values as high as 1600 kg_f mm^{-2} were reported when Cr nitride precipitates were also able to form and a small indenter load was used [38,103]. The microhardness profiles tend to reflect the average concentration profiles, as depicted in Figure 9. In nitrided samples, microhardness profiles usually show a plateau with high hardness values near the outer surface, and then a steep decrease to matrix values, related to the peculiar N-content profile [84,134,149,190,213]. C-rich expended austenite reaches values of ~1000 kg_f mm^{-2} [214], and the microhardness profile is smooth in carburized samples [84,134,149,190,213]. When treatments which use N- and C-containing gas mixtures are performed as a single step (nitrocarburizing) or in succession, the presence of both N-rich and C-rich expanded austenite allows to obtain high values at the surface with a gradual change towards matrix values [84,147,149,190,215].

5.5. Tribological Properties of the Modified Surface Layers

The hardened surface layers, obtained by means of low-temperature treatments, allow for an increase in the poor tribological properties of austenitic stainless steels. The efficiency in improving wear resistance depends on different factors: the microhardness profile, due to the treatment type, the test conditions, and the used counterface.

While untreated austenitic stainless steels usually suffer severe adhesive wear, an oxidative mild wear is observed for treated steels tested in pin-on-disk configuration, therefore wear loss significantly decreases in both dry [84,158,216] and wet [216,217] conditions. The importance of the materials constituting the tribosystem in influencing wear resistance was put in light for nitrided [216] and carburized [218] samples. For example, when AISI 316 samples, nitrided at 450 °C for 5 h, were tested in dry sliding conditions using a bearing steel as counterpart, the wear volume was approximately two orders of magnitude lower than that of the untreated steel, but using an alumina ball as counterpart the wear volume was only reduced by about a factor of eight [216]. The hard modified layer obtained with nitriding treatments improves efficiently abrasive resistance [213], and it allows for a low wear rate as long as the modified layer is not broken [197,215,219]. Carburized samples have a lower resistance to abrasion, as compared to nitrided samples [213], due to their lower hardness, but they show a higher load-bearing capacity [197,215], owing to the lower hardness gradient. The combination of high microhardness values at the surface with a fairly smooth hardness profile, obtained with nitrocarburizing [84] or by combining nitriding and carburizing [215], allows for a better load-bearing capacity than that obtained by nitriding or carburizing, hence a significant improvement of wear resistance was observed.

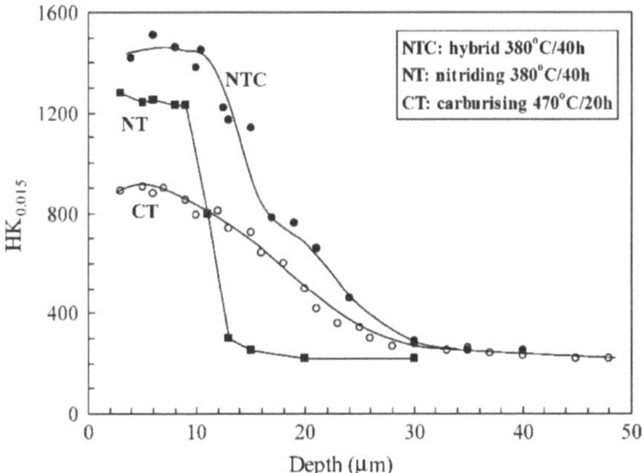

Figure 9. Microhardness profiles of AISI 321 samples plasma nitrided (380 °C, 40 h), carburized (470 °C, 20 h,) and nitrocarburized (NTC hybrid process) (380 °C, 40 h). For further details, see ref. [190]. (Reprinted from *Materials Science and Engineering A*, 404, Y. Sun, Hybrid plasma surface alloying of austenitic stainless steels with nitrogen and carbon, 124–129, Copyright (2005), with permission from Elsevier).

5.6. Corrosion Properties of the Modified Surface Layers

The presence of expanded austenite as the main phase in the outer modified surface layer, without the formation of significant amounts of compound precipitates, allows for the significant improvement in corrosion resistance of austenitic stainless steels in Cl$^-$-containing solutions. N-rich expanded austenite showed an excellent corrosion resistance in NaCl solutions [11,40,46,104,174,181,220,221] (Figure 10). It is known that N alloying of austenitic stainless steels promotes passivity, widens the passive range, and improves the resistance to intergranular corrosion and stress corrosion cracking [8]. It may be assumed that the enhancement of corrosion resistance for N-rich expanded austenite is due to mechanisms analogous to those suggested for N-containing austenitic stainless steels, in particular the reaction of released N atoms with H$^+$ to form ammonium ions, NH$_4^+$, that locally increase the pH where pits or crevices are forming, hence repassivation is promoted [46,220,222]. It was observed that the neutralizing effect of NH$_4^+$ ions is significant only above a certain range of pH [220]. In fact, Zhu and Lei [220] reported that the formation of an expanded austenite layer, produced by nitriding AISI 1Cr18Ni9Ti at 380 °C for 4 h, allowed to avoid pitting corrosion when the samples were put in contact with a 3% NaCl solution at pH in the range 4–13, while for the untreated steel active-passive transition occurred. At lower pH values, even if pitting phenomena occurred, higher pitting potential was registered for nitrided samples than for the untreated alloy.

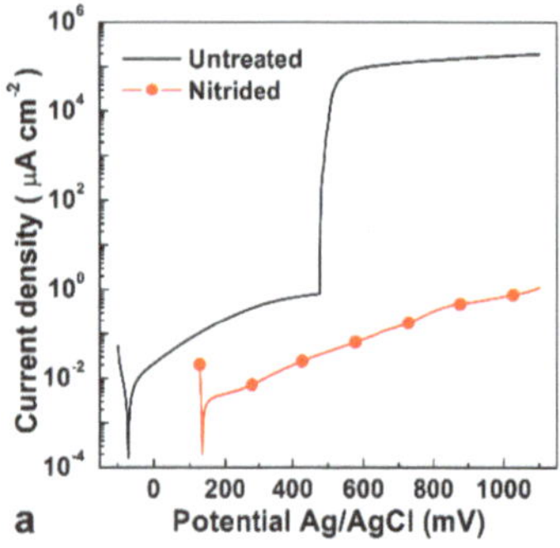

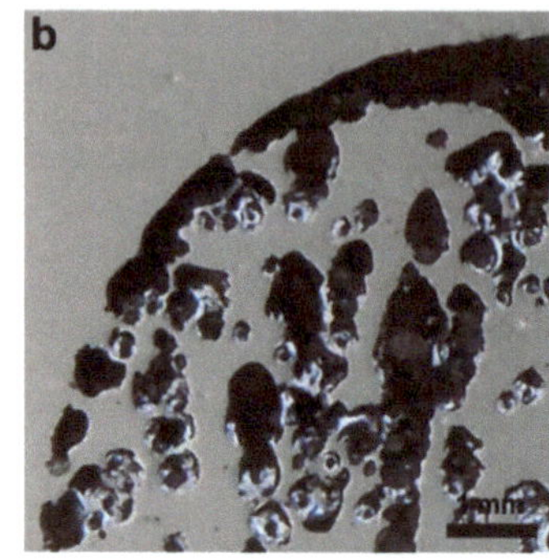

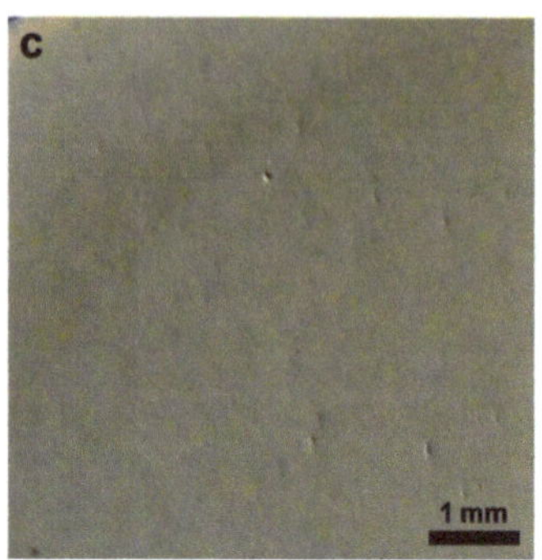

Figure 10. Corrosion behavior of AISI 202 samples untreated and plasma nitrided at 380 °C, 340 Pa, for 3 h: polarization curve (**a**), details of the surface morphology after corrosion test of untreated (**b**) and nitrided (**c**) samples (solution: 5% NaCl, aerated). (For further experimental details, see ref. [174]).

The protectiveness of the expanded austenite layer and its ability in hindering and delaying the occurrence of corrosion phenomena increase as the N content and the thickness of the modified layer become higher, and the precipitation of significant amounts of nitrides is avoided, therefore treatment conditions should be carefully chosen in order to obtain a N-rich, nitride-free, and thick modified layer [40,103,161,181,212,221]. The stability of the passive film was improved for up to 30 days of immersion in 3.5% NaCl for UNS S31254 super-austenitic stainless steel nitrided at 400 °C [104]. Electrochemical Impedance Spectroscopy (EIS) analysis showed that nitrided samples usually have higher impedance values than those of the untreated alloy, suggesting a better resistance to general corrosion [174,223,224]. Potentiodynamic tests demonstrated that nitrided samples can achieve a higher corrosion and pitting potential, and lower current density in the passive branch, in comparison with the untreated steel [158,174,177,221,225] (Figure 10a), thus the surface damage is reduced, as depicted in Figure 10b,c. By comparing the corrosion behavior of different austenitic stainless steels (AISI 316L, AISI 202 and Ni-free P558), nitrided at 360 and 380 °C for 3 h, it was hypothesized that the higher corrosion potential and lower anodic passive currents were influenced mainly by the N content in expanded austenite, while for pitting potential the role of alloy elements was also important [174]. Galvanostatic tests suggested that, after the occurrence of localized corrosion phenomena, repassivation may occur when the depth of pits or crevices is not too high [174]. N-rich expanded austenite layer also allows for an increase in the resistance to crevice corrosion [221].

Nitrided samples showed an enhanced corrosion resistance in Cl⁻ saline solutions, such as those used to simulate body fluids [40,226,227], and in HCl solution [225]. Moreover, nitrocarburizing [190,226] and carburizing [192,193,226] improve corrosion resistance in Cl⁻-containing solutions. For C-rich expanded austenite, it was suggested that the carburized layer decreases the mobility of charge carriers in the passive film, decreasing oxygen vacancy concentration as the C content is higher, due to the strong Cr–C bonds forming at the metal-passive-film interface [192].

The significant improvement of corrosion resistance in Cl⁻-containing solutions is reduced in NaCl + H_2SO_4 solutions [222], or in NaCl-free H_2SO_4 solutions, for which enhancement [228] and decrease [229] in corrosion resistance was observed. It was suggested that, in these conditions, the corrosion behavior is related to the presence of surface defects, as slip lines, microtwins, and dislocations, corresponding to high energy zones, which dissolve quickly and hinder repassivation [46].

6. Conclusions

Low-temperature thermochemical treatments of stainless steels allow to produce supersaturated solid solutions which are able to improve surface hardness, and hence wear resistance, without impairing corrosion resistance. The main characteristics of these "expanded" phases are summarized in Table 5.

Table 5. Main characteristics of expanded phases (N and C refer to nitriding and carburizing, respectively).

Properties	Expanded Martensite	Expanded Ferrite [1]	Expanded Austenite
Formation temperature (°C)	200–350 (N) 200–450 (C)	<380 (N) <470 (C)	300–450 (N) 300–550 (C)
Max. interstitial content (at.%)	22 (N) 10 (C)	24 (N) 10 (C)	38 (N) 19 (C)
Max. surface hardness [2] (kg_f mm^{-2})	1120 (N) 970 (C)	1090 (N) 1000 (C)	1450 (N) 1000 (C)
Hardness profile (without nitrides/carbides)	gradual change (N,C)	gradual change (N,C)	abrupt change (N) gradual change (C)
Localized corrosion resistance (without nitrides/carbides)	good (N,C)	good (N,C)	very good (N) good (C)
Tendency to form Fe-based compounds	high	high	low

[1] Data refer to the phase formed in ferritic stainless steels. [2] Values referred to the expanded phases only.

As highlighted by many studies, the characteristics and properties of the modified surface layers formed on stainless steels depend on both the treatment conditions and alloy composition. For hindering the precipitation of Cr nitrides/carbides, which cause a decrease of corrosion resistance, the treatment temperature should be sufficiently low that the diffusion of substitutional atoms has significantly slowed down. Due to the structure and alloy composition, for martensitic and ferritic stainless steels, the maximum treatment temperature is lower than that used for austenitic stainless steels. However, even if the precipitation of Cr compounds can be easily avoided, the possibility of obtaining a modified surface layer consisting of expanded martensite or expanded ferrite only may be a challenging task, since Fe-based nitrides/carbides tend to form together with these supersaturated solid solutions. Fe-based nitrides and carbides improve hardness and wear resistance in dry sliding conditions further on, but the presence of a multiphase structure at the surface influences the corrosion behavior. In austenitic stainless steels, the formation of Fe-based compounds can be usually avoided, and the expanded austenite–S-phase that forms in these steels has been recognized as a valuable mean to improve hardness, tribological properties, and corrosion resistance. It is worth noting that all of the "expanded" phases, as metastable phases, tend to transform when they are exposed to fairly high temperatures. Few data about the maximum service temperature are available for martensitic stainless steels, and they suggest that the decomposition of the modified surface layers may occur at 200 °C after few hours. On the contrary, expanded austenite seems to be more stable, and extrapolation of experimental data suggests that the modified layers consisting of this phase have a long-term stability (>10 years) when a maximum service temperature of about 200 °C is employed.

As a matter of fact, from a scientific point of view, many questions on expanded phases are still open, and they can stimulate further research activity on the low-temperature treatment of stainless steels. The thermodynamic and kinetic of formation of expanded phases, the influence of treatment conditions, as well as of alloy composition, and the structure of these phases are important topics, which need to be fully clarified for optimizing the thermochemical processes and for obtaining modified surface layers with characteristics tailored for the applications.

From a technological point of view, the numerous studies carried out on the low-temperature treatments of austenitic stainless steels have led to the scale-up of these treatments from research laboratories to production processes, and now the thermochemical treatments for obtaining modified surface layers consisting of expanded austenite are an industrial practice. The possibility of applying these treatments also to the other stainless steel types can allow to broaden the field of stainless steel components, which can benefit from improved surface hardness and good corrosion resistance for prolonging their service life.

Funding: This research received no external funding.

Institutional Review Board Statement: Not applicable.

Informed Consent Statement: Not applicable.

Data Availability Statement: Data sharing is not applicable.

Acknowledgments: IOP Publishing and Elsevier are acknowledged for the permission to use the figures.

Conflicts of Interest: The author declares no conflict of interest.

References

1. McGuire, M.F. *Stainless Steels for Design Engineers*; ASM International: Materials Park, OH, USA, 2008; ISBN 9780871707178.
2. Washko, S.D.; Aggen, G. Wrought Stainless Steels. In *ASM Handbook*; ASM International: Materials Park, OH, USA, 1997; Volume 1, pp. 841–907.
3. Jacob, A.; Povoden-Karadeniz, E.; Kozeschnik, E. Revised thermodynamic description of the Fe-Cr system based on an improved sublattice model of the σ phase. *Calphad* **2018**, *60*, 16–28. [CrossRef]
4. Lacombe, P.; Béranger, G. Structure and Equilibrium Diagrams of Various Stainless Steel Grades. In *Stainless Steels*; Lacombe, P., Baroux, B., Béranger, G., Eds.; Les Editions de Physique: Les Ulis, France, 1993; pp. 13–58. ISBN 9782868831897.
5. Talha, M.; Behera, C.K.; Sinha, O.P. A review on nickel-free nitrogen containing austenitic stainless steels for biomedical applications. *Mater. Sci. Eng. C* **2013**, *33*, 3563–3575. [CrossRef] [PubMed]
6. Lampman, S. Introduction to Surface Hardening of Steels. In *ASM Handbook*; ASM International: Materials Park, OH, USA, 1997; Volume 4, pp. 259–267.
7. O'Brien, J.M.; Goodman, D. Plasma (ion) nitriding. In *ASM Handbook*; ASM International: Materials Park, OH, USA, 1997; Volume 4, pp. 420–424.
8. Lo, K.H.; Shek, C.H.; Lai, J.K.L. Recent developments in stainless steels. *Mater. Sci. Eng. R Rep.* **2009**, *65*, 39–104. [CrossRef]
9. Davis, J.R. Surface Engineering of Stainless Steels. In *ASM Metal Handbook*; ASM International: Materials Park, OH, USA, 1994; Volume 5, pp. 741–761. [CrossRef]
10. Cardoso, R.P.; Mafra, M.; Brunatto, S.F. Low-temperature Thermochemical Treatments of Stainless Steels—An Introduction. In *Plasma Science and Technology—Progress in Physical States and Chemical Reactions*; Mieso, T., Ed.; InTech: Rijeka, Croatia, 2016; pp. 107–130, ISBN 9789535122807. [CrossRef]
11. Borgioli, F. From Austenitic Stainless Steel to Expanded Austenite-S Phase: Formation, Characteristics and Properties of an Elusive Metastable Phase. *Metals* **2020**, *10*, 187. [CrossRef]
12. Casteletti, L.C.; Neto, A.L.; Totten, G.E. Nitriding of Stainless Steels. *Metallogr. Microstruct. Anal.* **2014**, *3*, 477–508. [CrossRef]
13. Michal, G.M.; Ernst, F.; Kahn, H.; Cao, Y.; Oba, F.; Agarwal, N.; Heuer, A.H. Carbon supersaturation due to paraequilibrium carburization: Stainless steels with greatly improved mechanical properties. *Acta Mater.* **2006**, *54*, 1597–1606. [CrossRef]
14. Zhang, Z.L.; Bell, T. Structure and corrosion resistance of plasma-nitrided AISI 316 stainless steel. In *Heat Treatment Shanghai '83, Proceedings of the Third International Congress on Heat Treatment of Materials, Shanghai, China, 7–11 November 1983*; Bell, T., Ed.; The Metals Society: London, UK, 1984; p. 6.25.
15. Zhang, Z.L.; Bell, T. Structure and Corrosion Resistance of Plasma Nitrided Stainless Steel. *Surf. Eng.* **1985**, *1*, 131–136. [CrossRef]
16. Ichii, K.; Fujimura, K.; Takase, T. Microstructure, Corrosion Resistance, and Hardness of the Surface layer in Ion Nitrided 18-8 Stainless Steel. *Netsu Shori* **1985**, *25*, 191–195.
17. Ichii, K.; Fujimura, K.; Takase, T. Structure of the ion-nitrided layer of 18-8 stainless steel. *Technol. Rep. Kansai Univ.* **1986**, *27*, 135–144.
18. De Benedetti, B.; Angelini, E.; Zucchi, F. Nitrurazione ionica di acciai inossidabili. In Proceedings of the 10th National Congress on Thermal Treatments, Salsomaggiore, Italy, 16–18 October 1984; Associazione Italiana di Metallurgia: Milan, Italy, 1984; pp. 81–88.
19. De Benedetti, B.; Angelini, E.; Zucchi, F. Nitrurazione ionica di acciai inossidabili. *Metall. Ital.* **1985**, *77*, 615–621.
20. Angelini, E.; Burdese, A.; De Benedetti, B. Ion-nitriding of austenitic stainless steels. *Metall. Sci. Technol.* **1988**, *6*, 33–39.
21. Yasumaru, N.; Kamachi, K. Nitrogen-induced Phase Transformation in Type 304 Austenitic Stainless Steel. *J. Jpn. Inst. Met.* **1986**, *50*, 362–368. [CrossRef]
22. Lerner, R.M. Glow-discharge nitriding of Nitralloy 135M and AISI304 stainless steel. *J. Iron Steel Inst.* **1972**, *210*, 631–632.
23. Spalvins, T. Tribological and microstructural characteristics of ion-nitrided steels. *Thin Solid Film.* **1983**, *108*, 157–163. [CrossRef]
24. Kolster, B.H. Verschleiß- und korrosionsfeste Schichten auf austenitischen Stälen. *VDI-Berichte* **1983**, *506*, 107–113.
25. Ozbaysal, K.; Inal, O.T. Structure and properties of ion-nitrided stainless steels. *J. Mater. Sci.* **1986**, *21*, 4318–4326. [CrossRef]
26. Dingremont, N.; Bergmann, E.; Hans, M.; Collignon, P. Comparison of the corrosion resistance of different steel grades nitrided, coated and duplex treated. *Surf. Coat. Technol.* **1995**, *76–77*, 218–224. [CrossRef]
27. Sun, Y.; Bell, T.; Wood, G. Wear behaviour of plasma-nitrided martensitic stainless steel. *Wear* **1994**, *178*, 131–138. [CrossRef]

28. Tesi, B.; Bacci, T.; Poli, G. Analysis of surface structures and of size and shape variations in ionitrided precipitation hardening stainless steel samples. *Vacuum* **1985**, *35*, 307–314. [CrossRef]
29. Marchev, K.; Cooper, C.V.; Giessen, B.C. Observation of a compound layer with very low friction coefficient in ion-nitrided martensitic 410 stainless steel. *Surf. Coat. Technol.* **1998**, *99*, 229–233. [CrossRef]
30. Alphonsa, I.; Chainani, A.; Raole, P.M.; Ganguli, B.; John, P.I. A study of martensitic stainless steel AISI 420 modified using plasma nitriding. *Surf. Coat. Technol.* **2002**, *150*, 263–268. [CrossRef]
31. Bacci, T.; Borgioli, F.; Galvanetto, E.; Pradelli, G. Glow-discharge nitriding of sintered stainless steels. *Surf. Coat. Technol.* **2001**, *139*, 251–256. [CrossRef]
32. Borgioli, F.; Galvanetto, E.; Bacci, T.; Pradelli, G. Influence of the treatment atmosphere on the characteristics of glow-discharge treated sintered stainless steels. *Surf. Coat. Technol.* **2002**, *149*, 192–197. [CrossRef]
33. Byeli, A.V.; Kukareko, V.A.; Lobodaeva, O.V.; Shykh, S.K.; Davis, J.A.; Wilbur, P.J. Structure and wear resistance of 20X13 steel ion-implanted with nitrogen at a high beam-current density. *Surf. Coat. Technol.* **1997**, *96*, 255–261. [CrossRef]
34. Kim, S.K.; Yoo, J.S.; Priest, J.M.; Fewell, M.P. Martensitic Stainless Steel Nitrided in a Low-Pressure rf Plasma. In *Stainless Steel 2000: Proceedings of an International Current Status Seminar on Thermochemical Surface Engineering of Stainless Steel*; Bell, T., Akamatsu, K., Eds.; CRC Press: New York, NY, USA, 2001; pp. 149–158. [CrossRef]
35. Corengia, P.; Ybarra, G.; Moina, C.; Cabo, A.; Broitman, E. Microstructure and corrosion behaviour of DC-pulsed plasma nitrided AISI 410 martensitic stainless steel. *Surf. Coat. Technol.* **2004**, *187*, 63–69. [CrossRef]
36. de Oliveira, A.M.; Lombardi Neto, A.; Tottem, G.E.; Casteletti, L.C. Ion nitriding and nitrocarburizing of stainless steels and its influence in the corrosion resistance. In Proceedings of the 20th International Conference on Surface Modification Technologies, Vienna, Austria, 25–29 September 2006; Sudarshan, T.S., Stiglich, J.J., Eds.; ASM International: Materials Park, OH, USA, 2007; pp. 118–123.
37. Somers, M.A.J.; Christiansen, T. Kinetics of microstructure evolution during gaseous thermochemical surface treatment. *J. Phase Equilibria Diffus.* **2005**, *26*, 520–528. [CrossRef]
38. Bell, T. Surface engineering of austenitic stainless steel. *Surf. Eng.* **2002**, *18*, 415–422. [CrossRef]
39. Bell, T. Current Status of Supersaturated Surface Engineered S-Phase Materials. *Key Eng. Mater.* **2008**, *373–374*, 289–295. [CrossRef]
40. Fossati, A.; Galvanetto, E.; Bacci, T.; Borgioli, F. Improvement of corrosion resistance of austenitic stainless steels by means of glow-discharge nitriding. *Corros. Rev.* **2011**, *29*, 209–221. [CrossRef]
41. Luo, Q.; Yang, S. From Micro to Nano Scales -Recent Progress in the Characterization of Nitrided Austenitic Stainless Steels. *Int. J. Nanomed. Nanosurgery* **2015**, *1*, 1. [CrossRef]
42. Somers, M.; Kücükyildiz, Ö.; Ormstrup, C.; Alimadadi, H.; Hattel, J.; Christiansen, T.; Winther, G. Residual Stress in Expanded Austenite on Stainless Steel; Origin, Measurement, and Prediction. *Mater. Perform. Charact.* **2018**, *7*, 693–716. [CrossRef]
43. Somers, M.A.J.; Christiansen, T.L. Low Temperature Surface Hardening of Stainless Steel. In *Thermochemical Surface Engineering of Steels*; Mittemeijer, E.J., Somers, M.A.J., Eds.; Woodhead Publishing: Oxford, UK, 2015; pp. 557–579, ISBN 9780857095923. [CrossRef]
44. Czerwiec, T.; Renevier, N.; Michel, H. Low-temperature plasma-assisted nitriding. *Surf. Coat. Technol.* **2000**, *131*, 267–277. [CrossRef]
45. Czerwiec, T.; He, H.; Marcos, G.; Thiriet, T.; Weber, S.; Michel, H. Fundamental and Innovations in Plasma Assisted Diffusion of Nitrogen and Carbon in Austenitic Stainless Steels and Related Alloys. *Plasma Process. Polym.* **2009**, *6*, 401–409. [CrossRef]
46. Dong, H. S-phase surface engineering of Fe-Cr, Co-Cr and Ni-Cr alloys. *Int. Mater. Rev.* **2010**, *55*, 65–98. [CrossRef]
47. Christiansen, T.L.; Somers, M.A.J. Low-temperature gaseous surface hardening of stainless steel: The current status. *Int. J. Mater. Res.* **2009**, *100*, 1361–1377. [CrossRef]
48. Collins, S.R.; Williams, P.C.; Marx, S.V.; Heuer, A.; Ernst, F.; Kahn, H. Low-Temperature Carburization of Austenitic Stainless Steels. In *ASM Handbook*; Dosset, J., Totten, G.E., Eds.; ASM International: Materials Park, OH, USA, 2014; Volume 4D, pp. 451–460.
49. Scheuer, C.J.; Zanetti, F.I.; Cardoso, R.P.; Brunatto, S.F. Ultra-low—to high-temperature plasma-assisted nitriding: Revisiting and going further on the martensitic stainless steel treatment. *Mater. Res. Express* **2019**, *6*, 026529. [CrossRef]
50. Wriedt, H.A.; Gokcen, N.A.; Nafziger, R.H. The Fe-N (Iron-Nitrogen) system. *Bull. Alloy Phase Diagr.* **1987**, *8*, 355–377. [CrossRef]
51. Okamoto, H. The C-Fe (carbon-iron) system. *J. Phase Equilibria* **1992**, *13*, 543–565. [CrossRef]
52. Gavriljuk, V.G. Carbon and nitrogen in iron-based austenite and martensite: An attempt at comparative analysis. *J. Phys. IV Fr.* **2003**, *112*, 51–59. [CrossRef]
53. Shanina, B.D.; Gavriljuk, V.G.; Konchits, A.A.; Kolesnik, S.P. The influence of substitutional atoms upon the electron structure of the iron-based transition metal alloys. *J. Phys. Condens. Matter* **1998**, *10*, 1825–1838. [CrossRef]
54. Cao, Y.; Ernst, F.; Michal, G.M. Colossal carbon supersaturation in austenitic stainless steels carburized at low temperature. *Acta Mater.* **2003**, *51*, 4171–4181. [CrossRef]
55. Lengauer, W. Nitrides: Transition Metal Solid-state Chemistry. In *Encyclopedia of Inorganic and Bioinorganic Chemistry*; Scott, R.A., Ed.; John Wiley & Sons, Ltd.: Hoboken, NJ, USA, 2015; pp. 197–213. [CrossRef]
56. Shatynski, S.R. The thermochemistry of transition metal carbides. *Oxid. Met.* **1979**, *13*, 105–118. [CrossRef]
57. Williamson, D.L.; Ozturk, O.; Wei, R.; Wilbur, P.J. Metastable phase formation and enhanced diffusion in f.c.c. alloys under high dose, high flux nitrogen implantation at high and low ion energies. *Surf. Coat. Technol.* **1994**, *65*, 15–23. [CrossRef]

58. Christiansen, T.; Somers, M.A.J. Controlled dissolution of colossal quantities of nitrogen in stainless steel. *Metall. Mater. Trans. A* **2006**, *37*, 675–682. [CrossRef]

59. Christiansen, T.L.; Ståhl, K.; Brink, B.K.; Somers, M.A.J. On the Carbon Solubility in Expanded Austenite and Formation of Hägg Carbide in AISI 316 Stainless Steel. *Steel Res. Int.* **2016**, *87*, 1395–1405. [CrossRef]

60. Manova, D.; Thorwarth, G.; Mändl, S.; Neumann, H.; Stritzker, B.; Rauschenbach, B. Variable lattice expansion in martensitic stainless steel after nitrogen ion implantation. *Nucl. Instrum. Methods Phys. Res. Sect. B Beam Interact. Mater. At.* **2006**, *242*, 285–288. [CrossRef]

61. Michal, G.M.; Gu, X.; Jennings, W.D.; Kahn, H.; Ernst, F.; Heuer, A.H. Paraequilibrium Carburization of Duplex and Ferritic Stainless Steels. *Metall. Mater. Trans. A* **2009**, *40*, 1781–1790. [CrossRef]

62. Frandsen, R.B.; Christiansen, T.; Somers, M.A.J. Simultaneous surface engineering and bulk hardening of precipitation hardening stainless steel. *Surf. Coat. Technol.* **2006**, *200*, 5160–5169. [CrossRef]

63. Baniasadi, F.; Bahmannezhad, B.; Nikpoor, N.; Asgari, S. Thermal stability investigation of expanded martensite. *Surf. Coat. Technol.* **2016**, *300*, 87–94. [CrossRef]

64. Kim, S.K.; Yoo, J.S.; Priest, J.M.; Fewell, M.P. Characteristics of martensitic stainless steel nitrided in a low-pressure RF plasma. *Surf. Coat. Technol.* **2003**, *163–164*, 380–385. [CrossRef]

65. Brühl, S.P.; Charadia, R.; Simison, S.; Lamas, D.G.; Cabo, A. Corrosion behavior of martensitic and precipitation hardening stainless steels treated by plasma nitriding. *Surf. Coat. Technol.* **2010**, *204*, 3280–3286. [CrossRef]

66. Espitia, L.A.; Varela, L.; Pinedo, C.E.; Tschiptschin, A.P. Cavitation erosion resistance of low temperature plasma nitrided martensitic stainless steel. *Wear* **2013**, *301*, 449–456. [CrossRef]

67. Espitia, L.A.; Dong, H.; Li, X.-Y.; Pinedo, C.E.; Tschiptschin, A.P. Cavitation erosion resistance and wear mechanisms of active screen low temperature plasma nitrided AISI 410 martensitic stainless steel. *Wear* **2015**, *332–333*, 1070–1079. [CrossRef]

68. Wang, J.; Lin, Y.; Zeng, D.; Yan, J.; Fan, H. Effects of the Process Parameters on the Microstructure and Properties of Nitrided 17–4PH Stainless Steel. *Metall. Mater. Trans. B* **2013**, *44*, 414–422. [CrossRef]

69. Bottoli, F.; Winther, G.; Christiansen, T.L.; Somers, M.A.J. Influence of Microstructure and Process Conditions on Simultaneous Low-Temperature Surface Hardening and Bulk Precipitation Hardening of Nanoflex®. *Metall. Mater. Trans. A* **2015**, *46*, 5201–5216. [CrossRef]

70. Pinedo, C.E.; Larrotta, S.I.V.; Nishikawa, A.S.; Dong, H.; Li, X.-Y.; Magnabosco, R.; Tschiptschin, A.P. Low temperature active screen plasma nitriding of 17–4 PH stainless steel. *Surf. Coat. Technol.* **2016**, *308*, 189–194. [CrossRef]

71. Dong, H.; Esfandiari, M.; Li, X.Y. On the microstructure and phase identification of plasma nitrided 17–4PH precipitation hardening stainless steel. *Surf. Coat. Technol.* **2008**, *202*, 2969–2975. [CrossRef]

72. Somers, M.A.J.; Lankreijer, R.M.; Mittemeijer, E.J. Excess nitrogen in the ferrite matrix of nitrided binary iron-based alloys. *Philos. Mag. A* **1989**, *59*, 353–378. [CrossRef]

73. Pinedo, C.E.; Monteiro, W.A. On the kinetics of plasma nitriding a martensitic stainless steel type AISI 420. *Surf. Coat. Technol.* **2004**, *179*, 119–123. [CrossRef]

74. Li, C.X.; Bell, T. A comparative study of low temperature plasma nitriding, carburising and nitrocarburising of AISI 410 martensitic stainless steel. *Mater. Sci. Technol.* **2007**, *23*, 355–361. [CrossRef]

75. Barua, A.; Kim, H.; Lee, I. Effect of the amount of CH_4 gas on the characteristics of surface layers of low temperature plasma nitrided martensitic precipitation-hardening stainless steel. *Surf. Coat. Technol.* **2016**, *307*, 1034–1040. [CrossRef]

76. Mändl, S.; Fritzsche, B.; Manova, D.; Hirsch, D.; Neumann, H.; Richter, E.; Rauschenbach, B. Wear reduction in AISI 630 martensitic stainless steel after energetic nitrogen ion implantation. *Surf. Coat. Technol.* **2005**, *195*, 258–263. [CrossRef]

77. Scheuer, C.J.; Cardoso, R.P.; Zanetti, F.I.; Amaral, T.; Brunatto, S.F. Low-temperature plasma carburizing of AISI 420 martensitic stainless steel: Influence of gas mixture and gas flow rate. *Surf. Coat. Technol.* **2012**, *206*, 5085–5090. [CrossRef]

78. Scheuer, C.J.; Cardoso, R.P.; Brunatto, S.F. Sequential low-temperature plasma-assisted thermochemical treatments of the AISI 420 martensitic stainless steel. *Surf. Coat. Technol.* **2021**, *421*, 127459. [CrossRef]

79. Scheuer, C.J.; Cardoso, R.P.; Brunatto, S.F. Low-temperature Plasma Assisted Thermochemical Treatments of AISI 420 Steel: Comparative Study of Obtained Layers. *Mater. Res.* **2015**, *18*, 1392–1399. [CrossRef]

80. Anjos, A.D.; Scheuer, C.J.; Brunatto, S.F.; Cardoso, R.P. Low-temperature plasma nitrocarburizing of the AISI 420 martensitic stainless steel: Microstructure and process kinetics. *Surf. Coat. Technol.* **2015**, *275*, 51–57. [CrossRef]

81. Scheuer, C.J.; Cardoso, R.P.; Mafra, M.; Brunatto, S.F. AISI 420 martensitic stainless steel low-temperature plasma assisted carburizing kinetics. *Surf. Coat. Technol.* **2013**, *214*, 30–37. [CrossRef]

82. Scheuer, C.J.; Possoli, F.A.A.; Borges, P.C.; Cardoso, R.P.; Brunatto, S.F. AISI 420 martensitic stainless steel corrosion resistance enhancement by low-temperature plasma carburizing. *Electrochim. Acta* **2019**, *317*, 70–82. [CrossRef]

83. Fernandes, F.A.P.; Picone, C.A.; Totten, G.E.; Casteletti, L.C. Corrosion Behavior of Plasma Nitrided and Nitrocarburised Supermartensitic Stainless Steel. *Mater. Res.* **2018**, *21*, e20160793. [CrossRef]

84. Cheng, Z.; Li, C.X.; Dong, H.; Bell, T. Low temperature plasma nitrocarburising of AISI 316 austenitic stainless steel. *Surf. Coat. Technol.* **2005**, *191*, 195–200. [CrossRef]

85. Ferreira, L.M.; Brunatto, S.F.; Cardoso, R.P. Martensitic Stainless Steels Low-temperature Nitriding: Dependence of Substrate Composition. *Mater. Res.* **2015**, *18*, 622–627. [CrossRef]

86. Figueroa, C.A.; Alvarez, F.; Mitchell, D.R.G.; Collins, G.A.; Short, K.T. Previous heat treatment inducing different plasma nitriding behaviors in martensitic stainless steels. *J. Vac. Sci. Technol. A* **2006**, *24*, 1795–1801. [CrossRef]
87. Riazi, H.; Ashrafizadeh, F.; Hosseini, S.R.; Ghomashchi, R.; Liu, R. Characterization of simultaneous aged and plasma nitrided 17–4 PH stainless steel. *Mater. Charact.* **2017**, *133*, 33–43. [CrossRef]
88. Manova, D.; Mändl, S.; Neumann, H.; Rauschenbach, B. Influence of annealing conditions on ion nitriding of martensitic stainless steel. *Surf. Coat. Technol.* **2006**, *200*, 6563–6567. [CrossRef]
89. Sun, Y.; Bell, T. Low Temperature Plasma Nitriding Characteristics of Precipitation Hardening Stainless Steel. *Surf. Eng.* **2003**, *19*, 331–336. [CrossRef]
90. Manova, D.; Mändl, S.; Neumann, H.; Rauschenbach, B. Wear behaviour of martensitic stainless steel after PIII surface treatment. *Surf. Coat. Technol.* **2005**, *200*, 137–140. [CrossRef]
91. Borgioli, F.; Fossati, A.; Raugei, L.; Galvanetto, E.; Bacci, T. Low temperature glow-discharge nitriding of stainless steels. In Proceedings of the 7th European Stainless Steel Conference: Science and Market, Como, Italy, 21–23 September 2011; Associazione Italiana di Metallurgia: Milan, Italy, 2011.
92. Brunatto, S.F.; Scheuer, C.J.; Boromei, I.; Martini, C.; Ceschini, L.; Cardoso, R.P. Martensite coarsening in low-temperature plasma carburizing. *Surf. Coat. Technol.* **2018**, *350*, 161–171. [CrossRef]
93. Figueroa, C.A.; Alvarez, F.; Zhang, Z.; Collins, G.A.; Short, K.T. Structural modifications and corrosion behavior of martensitic stainless steel nitrided by plasma immersion ion implantation. *J. Vac. Sci. Technol. A* **2005**, *23*, 693–698. [CrossRef]
94. Kurelo, B.C.E.S.; de Souza, G.B.; Serbena, F.C.; de Oliveira, W.R.; Marino, C.E.B.; Taminato, L.A. Performance of nitrogen ion-implanted supermartensitic stainless steel in chlorine- and hydrogen-rich environments. *Surf. Coat. Technol.* **2018**, *351*, 29–41. [CrossRef]
95. Allenstein, A.N.; Cardoso, R.P.; Machado, K.D.; Weber, S.; Pereira, K.M.P.; dos Santos, C.A.L.; Panossian, Z.; Buschinelli, A.J.A.; Brunatto, S.F. Strong evidences of tempered martensite-to-nitrogen-expanded austenite transformation in CA-6NM steel. *Mater. Sci. Eng. A* **2012**, *552*, 569–572. [CrossRef]
96. Zangiabadi, A. Low-Temperature Interstitial Hardening of 15-5 Precipitation Hardening Martensitic Stainless Steel. Ph.D. Thesis, Department of Materials Science and Engineering, Case Western Reserve University, Cleveland, OH, USA, 2016.
97. Zangiabadi, A.; Dalton, J.C.; Wang, D.; Ernst, F.; Heuer, A.H. The Formation of Martensitic Austenite During Nitridation of Martensitic and Duplex Stainless Steels. *Metall. Mater. Trans. A* **2017**, *48*, 8–13. [CrossRef]
98. Maugis, P.; Connétable, D.; Eyméoud, P. Stability of Zener order in martensite: An atomistic evidence. *Scr. Mater.* **2021**, *194*, 113632. [CrossRef]
99. Tanaka, T.; Maruyama, N.; Nakamura, N.; Wilkinson, A.J. Tetragonality of Fe-C martensite—A pattern matching electron backscatter diffraction analysis compared to X-ray diffraction. *Acta Mater.* **2020**, *195*, 728–738. [CrossRef]
100. Tsuchiyama, T.; Inoue, K.; Hyodo, K.; Akama, D.; Nakada, N.; Takaki, S.; Koyano, T. Comparison of Microstructure and Hardness between High-carbon and High-nitrogen Martensites. *ISIJ Int.* **2019**, *59*, 161–168. [CrossRef]
101. Siqueira, J.S.; de Abreu Alves, M.R.; Marcial Luiz, T.; Renzetti, R.A. Effect of heat treatment on the chromium-depleted zones of a high carbon martensitic stainless steel. *Mater. Corros.* **2021**, *72*, 1752–1761. [CrossRef]
102. Mändl, S.; Rauschenbach, B. Comparison of expanded austenite and expanded martensite formed after nitrogen PIII. *Surf. Coat. Technol.* **2004**, *186*, 277–281. [CrossRef]
103. Borgioli, F.; Fossati, A.; Galvanetto, E.; Bacci, T.; Pradelli, G. Glow discharge nitriding of AISI 316L austenitic stainless steel: Influence of treatment pressure. *Surf. Coat. Technol.* **2006**, *200*, 5505–5513. [CrossRef]
104. Luiz, L.A.; Kurelo, B.C.E.S.; de Souza, G.B.; de Andrade, J.; Marino, C.E.B. Effect of nitrogen plasma immersion ion implantation on the corrosion protection mechanisms of different stainless steels. *Mater. Today Commun.* **2021**, *28*, 102655. [CrossRef]
105. Liu, R.L.; Yan, F.; Wei, C.Y.; Yan, M.F. Characteristics of carbon-expanded α phase layer on AISI 431 stainless steel using experimental and fist-principles calculation methods. *Surf. Coat. Technol.* **2019**, *375*, 66–73. [CrossRef]
106. Li, C.X.; Bell, T. Corrosion properties of plasma nitrided AISI 410 martensitic stainless steel in 3.5% NaCl and 1% HCl aqueous solutions. *Corros. Sci.* **2006**, *48*, 2036–2049. [CrossRef]
107. Rovani, A.C.; Breganon, R.; de Souza, G.S.; Brunatto, S.F.; Pintaúde, G. Scratch resistance of low-temperature plasma nitrided and carburized martensitic stainless steel. *Wear* **2017**, *376–377*, 70–76. [CrossRef]
108. Nakasa, K.; Yamamoto, A.; Wang, R.; Sumomogi, T. Effect of plasma nitriding on the strength of fine protrusions formed by sputter etching of AISI type 420 stainless steel. *Surf. Coat. Technol.* **2015**, *272*, 298–308. [CrossRef]
109. Xi, Y.; Liu, D.; Han, D. Improvement of erosion and erosion–corrosion resistance of AISI420 stainless steel by low temperature plasma nitriding. *Appl. Surf. Sci.* **2008**, *254*, 5953–5958. [CrossRef]
110. Corengia, P.; Walther, F.; Ybarra, G.; Sommadossi, S.; Corbari, R.; Broitman, E. Friction and rolling–sliding wear of DC-pulsed plasma nitrided AISI 410 martensitic stainless steel. *Wear* **2006**, *260*, 479–485. [CrossRef]
111. Bernardelli, E.A.; Borges, P.C.; Fontana, L.C.; Floriano, J.B. Role of plasma nitriding temperature and time in the corrosion behaviour and microstructure evolution of 15–5 PH stainless steel. *Kov. Mater.* **2010**, *48*, 110–115. [CrossRef]
112. Blawert, C.; Mordike, B.L.; Jirásková, Y.; Schneeweiss, O. Structure and composition of expanded austenite produced by nitrogen plasma immersion ion implantation of stainless steels X6CrNiTi1810 and X2CrNiMoN2253. *Surf. Coat. Technol.* **1999**, *116–119*, 189–198. [CrossRef]

113. Kurelo, B.C.E.S.; de Souza, G.B.; Serbena, F.C.; Lepienski, C.M.; Borges, P.C. Mechanical properties and corrosion resistance of α_N-rich layers produced by PIII on a super ferritic stainless steel. *Surf. Coat. Technol.* **2020**, *403*, 126388. [CrossRef]

114. Alphonsa, J.; Mukherjee, S.; Raja, V.S. Study of plasma nitriding and nitrocarburising of AISI 430F stainless steel for high hardness and corrosion resistance. *Corros. Eng. Sci. Technol.* **2018**, *53*, 51–58. [CrossRef]

115. de Sousa, R.R.M.; de Araújo, F.O.; da Costa, J.A.P.; de Oliverira, A.M.; Melo, M.S.; Alves, C., Jr. Cathodic Cage Nitriding of AISI 409 Ferritic Stainless Steel with the Addition of CH_4. *Mater. Res.* **2012**, *15*, 260–265. [CrossRef]

116. Larisch, B.; Brusky, U.; Spies, H.-J. Plasma nitriding of stainless steels at low temperatures. *Surf. Coat. Technol.* **1999**, *116–119*, 205–211. [CrossRef]

117. Feugeas, J.; Gómez, B.; Craievich, A. Ion nitriding of stainless steels. Real time surface characterization by synchrotron X-ray diffraction. *Surf. Coat. Technol.* **2002**, *154*, 167–175. [CrossRef]

118. Blawert, C.; Mordike, B.L.; Rensch, U.; Wünsch, R.; Wiedemann, R.; Oettel, H. Influence of the material composition on the nitriding result of steels by plasma immersion ion implantation. *Surf. Coat. Technol.* **2000**, *131*, 334–339. [CrossRef]

119. Schreiber, G.; Rensch, U.; Oettel, H.; Blawert, C.; Mordike, B.L. Thermal stability of PI^3 nitrided surface layers on ferritic steels. *Surf. Coat. Technol.* **2003**, *169–170*, 447–451. [CrossRef]

120. Spies, H.-J. Corrosion Behaviour of Nitrided, nitrocarburised and Carburised Steels. In *Thermochemical Surface Engineering of Steel*; Mittemeijer, E.J., Somers, M.A.J., Eds.; Woodhead Publishing: Oxford, UK, 2015; pp. 267–309, ISBN 9780857095923. [CrossRef]

121. Kliauga, A.M.; Pohl, M.; Cordier-Robert, C.; Foct, J. Phase transformations in a super ferritic stainless steel containing 28% Cr after nitrogen ion implantation. *J. Mater. Sci.* **1999**, *34*, 4065–4073. [CrossRef]

122. Jirásková, Y.; Blawert, C.; Schneeweiss, O. Thermal Stability of Stainless Steel Surfaces Nitrided by Plasma Immersion Ion Implantation. *Phys. Status Solidi* **1999**, *175*, 537–548. [CrossRef]

123. de Araújo Junior, E.; Bandeira, R.M.; Manfrinato, M.D.; Moreto, J.A.; Borges, R.; dos Santos Vales, S.; Suzuki, P.A.; Rossino, L.S. Effect of ionic plasma nitriding process on the corrosion and micro-abrasive wear behavior of AISI 316L austenitic and AISI 470 super-ferritic stainless steels. *J. Mater. Res. Technol.* **2019**, *8*, 2180–2191. [CrossRef]

124. Tuckart, W.; Gregorio, M.; Iurman, L. Sliding wear of plasma nitrided AISI 405 ferritic stainless steel. *Surf. Eng.* **2010**, *26*, 185–190. [CrossRef]

125. Bielawski, J.; Baranowska, J.; Szczecinski, K. Microstructure and properties of layers on chromium steel. *Surf. Coat. Technol.* **2006**, *200*, 6572–6577. [CrossRef]

126. Michler, T.; Grischke, M.; Bewilogua, K.; Dimigen, H. Properties of duplex coatings prepared by plasma nitriding and PVD Ti–C:H deposition on X20Cr13 ferritic stainless steel. *Thin Solid Film.* **1998**, *322*, 206–212. [CrossRef]

127. Spies, H.-J.; Eckstein, C.; Zimdars, H. Structure and corrosion behaviour of stainless steels after plasma and gas nitriding. *Surf. Eng.* **2002**, *18*, 459–460. [CrossRef]

128. Gontijo, L.C.; Machado, R.; Casteletti, L.C.; Kuri, S.E.; Nascente, P.A.P. X-ray diffraction characterisation of expanded austenite and ferrite in plasma nitrided stainless steels. *Surf. Eng.* **2010**, *26*, 265–270. [CrossRef]

129. Bielawski, J.; Baranowska, J. Formation of nitrided layers on duplex steel—Influence of multiphase substrate. *Surf. Eng.* **2010**, *26*, 299–304. [CrossRef]

130. Pinedo, C.E.; Varela, L.B.; Tschiptschin, A.P. Low-temperature plasma nitriding of AISI F51 duplex stainless steel. *Surf. Coat. Technol.* **2013**, *232*, 839–843. [CrossRef]

131. Chiu, L.H.; Su, Y.Y.; Chen, F.S.; Chang, H. Microstructure and Properties of Active Screen Plasma Nitrided Duplex Stainless Steel. *Mater. Manuf. Process.* **2010**, *25*, 316–323. [CrossRef]

132. Tschiptschin, A.P.; Varela, L.B.; Pinedo, C.E.; Li, X.Y.; Dong, H. Development and microstructure characterization of single and duplex nitriding of UNS S31803 duplex stainless steel. *Surf. Coat. Technol.* **2017**, *327*, 83–92. [CrossRef]

133. de Oliveira, W.R.; Kurelo, B.C.E.S.; Ditzel, D.G.; Serbena, F.C.; Foerster, C.E.; de Souza, G.B. On the S-phase formation and the balanced plasma nitriding of austenitic-ferritic super duplex stainless steel. *Appl. Surf. Sci.* **2018**, *434*, 1161–1174. [CrossRef]

134. Christiansen, T.; Somers, M.A.J. Low temperature gaseous nitriding and carburising of stainless steel. *Surf. Eng.* **2005**, *21*, 445–455. [CrossRef]

135. Blawert, C.; Weisheit, A.; Mordike, B.L.; Knoop, R.M. Plasma immersion ion implantation of stainless steel: Austenitic stainless steel in comparison to austenitic-ferritic stainless steel. *Surf. Coat. Technol.* **1996**, *85*, 15–27. [CrossRef]

136. Calabokis, O.P.; de la Rosa, Y.N.; Lepienski, C.M.; Cardoso, R.P.; Borges, P.C. Crevice and pitting corrosion of low temperature plasma nitrided UNS S32750 super duplex stainless steel. *Surf. Coat. Technol.* **2021**, *413*, 127095. [CrossRef]

137. Pinedo, C.E.; Tschiptschin, A.P. Low temperature plasma carburizing of AISI 316L austenitic stainless steel and AISI F51 duplex stainless steel. *Rev. Esc. Minas* **2013**, *66*, 209–214. [CrossRef]

138. Alphonsa, J.; Raja, V.S.; Mukherjee, S. Study of plasma nitriding and nitrocarburizing for higher corrosion resistance and hardness of 2205 duplex stainless steel. *Corros. Sci.* **2015**, *100*, 121–132. [CrossRef]

139. Bobadilla, M.; Tschiptschin, A.P. On the Nitrogen Diffusion in a Duplex Stainless Steel. *Mater. Res.* **2015**, *18*, 390–394. [CrossRef]

140. Dalton, J.C.; Ernst, F.; Heuer, A.H. Low-Temperature Nitridation of 2205 Duplex Stainless Steel. *Metall. Mater. Trans. A* **2020**, *51*, 608–617. [CrossRef]

141. Wang, D.; Chen, C.-W.; Dalton, J.C.; Yang, F.; Sharghi-Moshtaghin, R.; Kahn, H.; Ernst, F.; Williams, R.E.A.; McComb, D.W.; Heuer, A.H. "Colossal" interstitial supersaturation in delta ferrite in stainless steels—I. Low-temperature carburization. *Acta Mater.* **2015**, *86*, 193–207. [CrossRef]

142. Sasidhar, K.N.; Meka, S.R. What causes the colossal C supersaturation of δ-ferrite in stainless steel during low-temperature carburization? *Scr. Mater.* **2019**, *162*, 118–120. [CrossRef]
143. Pintaude, G.; Rovani, A.C.; das Neves, J.C.K.; Lagoeiro, L.E.; Li, X.; Dong, H.S. Wear and Corrosion Resistances of Active Screen Plasma-Nitrided Duplex Stainless Steels. *J. Mater. Eng. Perform.* **2019**, *28*, 3673–3682. [CrossRef]
144. Kahn, H.; Heuer, A.H.; Michal, G.M.; Ernst, F.; Sharghi-Moshtaghin, R.; Ge, Y.; Natishan, P.M.; Rayne, R.J.; Martin, F.J. Interstitial hardening of duplex 2205 stainless steel by low temperature carburisation: Enhanced mechanical and electrochemical performance. *Surf. Eng.* **2012**, *28*, 213–219. [CrossRef]
145. Hannula, S.-P.; Nenonen, P.; Hirvonen, J.-P. Surface structure and properties of ion-nitrided austenitic stainless steels. *Thin Solid Film.* **1989**, *181*, 343–350. [CrossRef]
146. Adachi, S.; Yamaguchi, T.; Ueda, N. Formation and Properties of Nitrocarburizing S-Phase on AISI 316L Stainless Steel-Based WC Composite Layers by Low-Temperature Plasma Nitriding. *Metals* **2021**, *11*, 1538. [CrossRef]
147. Adachi, S.; Ueda, N. Combined plasma carburizing and nitriding of sprayed AISI 316L steel coating for improved wear resistance. *Surf. Coat. Technol.* **2014**, *259*, 44–49. [CrossRef]
148. Adachi, S.; Egawa, M.; Yamaguchi, T.; Ueda, N. Low-Temperature Plasma Nitriding for Austenitic Stainless Steel Layers with Various Nickel Contents Fabricated via Direct Laser Metal Deposition. *Coatings* **2020**, *10*, 365. [CrossRef]
149. Adachi, S.; Ueda, N. Wear and Corrosion Properties of Cold-Sprayed AISI 316L Coatings Treated by Combined Plasma Carburizing and Nitriding at Low Temperature. *Coatings* **2018**, *8*, 456. [CrossRef]
150. Adachi, S.; Ueda, N. Formation of Expanded Austenite on a Cold-Sprayed AISI 316L Coating by Low-Temperature Plasma Nitriding. *J. Therm. Spray Technol.* **2015**, *24*, 1399–1407. [CrossRef]
151. Adachi, S.; Ueda, N. Formation of S-phase layer on plasma sprayed AISI 316L stainless steel coating by plasma nitriding at low temperature. *Thin Solid Film.* **2012**, *523*, 11–14. [CrossRef]
152. Lindner, T.; Mehner, T.; Lampke, T. Surface modification of austenitic thermal-spray coatings by low-temperature nitrocarburizing. *IOP Conf. Ser. Mater. Sci. Eng.* **2016**, *118*, 12008. [CrossRef]
153. Lindner, T.; Kutschmann, P.; Löbel, M.; Lampke, T. Hardening of HVOF-Sprayed Austenitic Stainless-Steel Coatings by Gas Nitriding. *Coatings* **2018**, *8*, 348. [CrossRef]
154. Kutschmann, P.; Lindner, T.; Börner, K.; Reese, U.; Lampke, T. Effect of Adjusted Gas Nitriding Parameters on Microstructure and Wear Resistance of HVOF-Sprayed AISI 316L Coatings. *Materials* **2019**, *12*, 1760. [CrossRef]
155. Esfandiari, M.; Dong, H. Plasma surface engineering of precipitation hardening stainless steels. *Surf. Eng.* **2006**, *22*, 86–92. [CrossRef]
156. Esfandiari, M.; Dong, H. Improving the surface properties of A286 precipitation-hardening stainless steel by low-temperature plasma nitriding. *Surf. Coat. Technol.* **2007**, *201*, 6189–6196. [CrossRef]
157. Wang, S.; Cai, W.; Li, J.; Wei, W.; Hu, J. A novel rapid D.C. plasma nitriding at low gas pressure for 304 austenitic stainless steel. *Mater. Lett.* **2013**, *105*, 47–49. [CrossRef]
158. Lu, S.; Zhao, X.; Wang, S.; Li, J.; Wei, W.; Hu, J. Performance enhancement by plasma nitriding at low gas pressure for 304 austenitic stainless steel. *Vacuum* **2017**, *145*, 334–339. [CrossRef]
159. Buhagiar, J.; Li, X.; Dong, H. Formation and microstructural characterisation of S-phase layers in Ni-free austenitic stainless steels by low-temperature plasma surface alloying. *Surf. Coat. Technol.* **2009**, *204*, 330–335. [CrossRef]
160. Borgioli, F.; Fossati, A.; Matassini, G.; Galvanetto, E.; Bacci, T. Low temperature glow-discharge nitriding of a low nickel austenitic stainless steel. *Surf. Coat. Technol.* **2010**, *204*, 3410–3417. [CrossRef]
161. Borgioli, F.; Galvanetto, E.; Bacci, T. Surface Modification of a Nickel-Free Austenitic Stainless Steel by Low-Temperature Nitriding. *Metals* **2021**, *11*, 1845. [CrossRef]
162. Tao, X.; Li, X.; Dong, H.; Matthews, A.; Leyland, A. Evaluation of the sliding wear and corrosion performance of triode-plasma nitrided Fe-17Cr-20Mn-0.5N high-manganese and Fe-19Cr-35Ni-1.2Si high-nickel austenitic stainless steels. *Surf. Coat. Technol.* **2021**, *409*, 126890. [CrossRef]
163. Ernst, F.; Avishai, A.; Kahn, H.; Gu, X.; Michal, G.M.; Heuer, A.H. Enhanced Carbon Diffusion in Austenitic Stainless Steel Carburized at Low Temperature. *Metall. Mater. Trans. A* **2009**, *40*, 1768–1780. [CrossRef]
164. Wu, D.; Ge, Y.; Kahn, H.; Ernst, F.; Heuer, A.H. Diffusion profiles after nitrocarburizing austenitic stainless steel. *Surf. Coat. Technol.* **2015**, *279*, 180–185. [CrossRef]
165. Mändl, S.; Scholze, F.; Neumann, H.; Rauschenbach, B. Nitrogen diffusivity in expanded austenite. *Surf. Coat. Technol.* **2003**, *174–175*, 1191–1195. [CrossRef]
166. Williamson, D.L.; Ivanov, I.; Wei, R.; Wilbur, P.J. Role of Chromium in High-Dose, High-Rate, Elevated Temperature Nitrogen Implantation of Austenitic Stainless Steels. *MRS Proc.* **1991**, *235*, 473. [CrossRef]
167. Parascandola, S.; Günzel, R.; Grötzschel, R.; Richter, E.; Möller, W. Analysis of deuterium induced nuclear reactions giving criteria for the formation process of expanded austenite. *Nucl. Instrum. Methods Phys. Res. Sect. B Beam Interact. Mater. At.* **1998**, *136–138*, 1281–1285. [CrossRef]
168. Christiansen, T.L.; Drouet, M.; Martinavičius, A.; Somers, M.A.J. Isotope exchange investigation of nitrogen redistribution in expanded austenite. *Scr. Mater.* **2013**, *69*, 582–585. [CrossRef]
169. Christiansen, T.; Dahl, K.V.; Somers, M.A.J. Nitrogen diffusion and nitrogen depth profiles in expanded austenite: Experimental assessment, numerical simulation and role of stress. *Mater. Sci. Technol.* **2008**, *24*, 159–167. [CrossRef]

170. Jespersen, F.N.; Hattel, J.H.; Somers, M.A.J. Modelling the evolution of composition-and stress-depth profiles in austenitic stainless steels during low-temperature nitriding. *Model. Simul. Mater. Sci. Eng.* **2016**, *24*, 25003. [CrossRef]
171. Galdikas, A.; Moskalioviene, T. The Anisotropic Stress-Induced Diffusion and Trapping of Nitrogen in Austenitic Stainless Steel during Nitriding. *Metals* **2020**, *10*, 1319. [CrossRef]
172. Li, X.Y. Low Temperature Plasma Nitriding of 316 Stainless Steel—Nature of S Phase and Its Thermal Stability. *Surf. Eng.* **2001**, *17*, 147–152. [CrossRef]
173. Li, X.Y.; Thaiwatthana, S.; Dong, H.; Bell, T. Thermal stability of carbon S phase in 316 stainless steel. *Surf. Eng.* **2002**, *18*, 448–451. [CrossRef]
174. Borgioli, F.; Galvanetto, E.; Bacci, T. Corrosion behaviour of low temperature nitrided nickel-free, AISI 200 and AISI 300 series austenitic stainless steels in NaCl solution. *Corros. Sci.* **2018**, *136*, 352–365. [CrossRef]
175. Borgioli, F.; Galvanetto, E.; Bacci, T. Influence of surface morphology and roughness on water wetting properties of low temperature nitrided austenitic stainless steels. *Mater. Charact.* **2014**, *95*, 278–284. [CrossRef]
176. Tao, X.; Liu, X.; Matthews, A.; Leyland, A. The influence of stacking fault energy on plasticity mechanisms in triode-plasma nitrided austenitic stainless steels: Implications for the structure and stability of nitrogen-expanded austenite. *Acta Mater.* **2019**, *164*, 60–75. [CrossRef]
177. Baranowska, J.; Arnold, B. Corrosion resistance of nitrided layers on austenitic steel. *Surf. Coat. Technol.* **2006**, *200*, 6623–6628. [CrossRef]
178. Borgioli, F.; Fossati, A.; Galvanetto, E.; Bacci, T. Glow-discharge nitriding of AISI 316L austenitic stainless steel: Influence of treatment temperature. *Surf. Coat. Technol.* **2005**, *200*, 2474–2480. [CrossRef]
179. Wu, D.; Kahn, H.; Dalton, J.C.; Michal, G.M.; Ernst, F.; Heuer, A.H. Orientation dependence of nitrogen supersaturation in austenitic stainless steel during low-temperature gas-phase nitriding. *Acta Mater.* **2014**, *79*, 339–350. [CrossRef]
180. Templier, C.; Stinville, J.C.; Villechaise, P.; Renault, P.O.; Abrasonis, G.; Rivière, J.P.; Martinavičius, A.; Drouet, M. On lattice plane rotation and crystallographic structure of the expanded austenite in plasma nitrided AISI 316L steel. *Surf. Coat. Technol.* **2010**, *204*, 2551–2558. [CrossRef]
181. Borgioli, F.; Galvanetto, E.; Bacci, T. Low temperature nitriding of AISI 300 and 200 series austenitic stainless steels. *Vacuum* **2016**, *127*, 51–60. [CrossRef]
182. Lei, M.K. Phase transformations in plasma source ion nitrided austenitic stainless steel at low temperature. *J. Mater. Sci.* **1999**, *34*, 5975–5982. [CrossRef]
183. Yakubtsov, I.A.; Ariapour, A.; Perovic, D.D. Effect of nitrogen on stacking fault energy of f.c.c. iron-based alloys. *Acta Mater.* **1999**, *47*, 1271–1279. [CrossRef]
184. Talonen, J.; Hänninen, H. Formation of shear bands and strain-induced martensite during plastic deformation of metastable austenitic stainless steels. *Acta Mater.* **2007**, *55*, 6108–6118. [CrossRef]
185. Tao, X.; Qi, J.; Rainforth, M.; Matthews, A.; Leyland, A. On the interstitial induced lattice inhomogeneities in nitrogen-expanded austenite. *Scr. Mater.* **2020**, *185*, 146–151. [CrossRef]
186. Tong, K.; Ye, F.; Che, H.; Lei, M.K.; Miao, S.; Zhang, C. High-density stacking faults in a supersaturated nitrided layer on austenitic stainless steel. *J. Appl. Crystallogr.* **2016**, *49*, 1967–1971. [CrossRef]
187. Lei, M.K.; Huang, Y.; Zhang, Z.L. In situ Transformation of Nitrogen-induced h.c.p. Martensite in Plasma Source Ion–nitrided Austenitic Stainless Steel. *J. Mater. Sci. Lett.* **1998**, *17*, 1165–1167. [CrossRef]
188. Tsujikawa, M.; Yamauchi, N.; Ueda, N.; Sone, T.; Hirose, Y. Behavior of carbon in low temperature plasma nitriding layer of austenitic stainless steel. *Surf. Coat. Technol.* **2005**, *193*, 309–313. [CrossRef]
189. Williamson, D.L.; Davis, J.A.; Wilbur, P.J. Effect of austenitic stainless steel composition on low-energy, high-flux, nitrogen ion beam processing. *Surf. Coat. Technol.* **1998**, *103–104*, 178–184. [CrossRef]
190. Sun, Y. Hybrid plasma surface alloying of austenitic stainless steels with nitrogen and carbon. *Mater. Sci. Eng. A* **2005**, *404*, 124–129. [CrossRef]
191. Tsujikawa, M.; Yoshida, D.; Yamauchi, N.; Ueda, N.; Sone, T.; Tanaka, S. Surface material design of 316 stainless steel by combination of low temperature carburizing and nitriding. *Surf. Coat. Technol.* **2005**, *200*, 507–511. [CrossRef]
192. Niu, W.; Lillard, R.S.; Li, Z.; Ernst, F. Properties of the Passive Film Formed on Interstitially Hardened AISI 316L Stainless Steel. *Electrochim. Acta* **2015**, *176*, 410–419. [CrossRef]
193. Sun, Y. Corrosion behaviour of low temperature plasma carburised 316L stainless steel in chloride containing solutions. *Corros. Sci.* **2010**, *52*, 2661–2670. [CrossRef]
194. Czerwiec, T.; Andrieux, A.; Marcos, G.; Michel, H.; Bauer, P. Is "expanded austenite" really a solid solution? Mössbauer observation of an annealed AISI 316L nitrided sample. *J. Alloy. Compd.* **2019**, *811*, 151972. [CrossRef]
195. Stróż, D.; Psoda, M. TEM studies of plasma nitrided austenitic stainless steel. *J. Microsc.* **2010**, *237*, 227–231. [CrossRef]
196. Jiang, J.C.; Meletis, E.I. Microstructure of the nitride layer of AISI 316 stainless steel produced by intensified plasma assisted processing. *J. Appl. Phys.* **2000**, *88*, 4026–4031. [CrossRef]
197. Blawert, C.; Kalvelage, H.; Mordike, B.L.; Collins, G.A.; Short, K.T.; Jirásková, Y.; Schneeweiss, O. Nitrogen and carbon expanded austenite produced by PI^3. *Surf. Coat. Technol.* **2001**, *136*, 181–187. [CrossRef]
198. Fewell, M.P.; Priest, J.M. High-order diffractometry of expanded austenite using synchrotron radiation. *Surf. Coat. Technol.* **2008**, *202*, 1802–1815. [CrossRef]

199. Liang, W.; Xiaolei, X.; Jiujun, X.; Yaqin, S. Characteristics of low pressure plasma arc source ion nitrided layer on austenitic stainless steel at low temperature. *Thin Solid Film.* **2001**, *391*, 11–16. [CrossRef]
200. Riviere, J.P.; Cahoreau, M.; Meheust, P. Chemical bonding of nitrogen in low energy high flux implanted austenitic stainless steel. *J. Appl. Phys.* **2002**, *91*, 6361–6366. [CrossRef]
201. Martinavičius, A.; Abrasonis, G.; Scheinost, A.C.; Danoix, R.; Danoix, F.; Stinville, J.C.; Talut, G.; Templier, C.; Liedke, O.; Gemming, S.; et al. Nitrogen interstitial diffusion induced decomposition in AISI 304L austenitic stainless steel. *Acta Mater.* **2012**, *60*, 4065–4076. [CrossRef]
202. Martinavičius, A.; Danoix, R.; Drouet, M.; Templier, C.; Hannoyer, B.; Danoix, F. Atom probe tomography characterization of nitrogen induced decomposition in low temperature plasma nitrided 304L austenitic stainless steel. *Mater. Lett.* **2015**, *139*, 153–156. [CrossRef]
203. Sasidhar, K.N.; Meka, S.R. Thermodynamic reasoning for colossal N supersaturation in austenitic and ferritic stainless steels during low-temperature nitridation. *Sci. Rep.* **2019**, *9*, 7996. [CrossRef]
204. Che, H.L.; Tong, S.; Wang, K.S.; Lei, M.K.; Somers, M.A.J. Co-existence of γ_N' phase and γ_N phase on nitrided austenitic Fe–Cr–Ni alloys- I. experiment. *Acta Mater.* **2019**, *177*, 35–45. [CrossRef]
205. Che, H.L.; Lei, M.K.; Somers, M.A.J. A simple model for nitrogen-induced lattice expansion of γ_N' and γ_N phases in Fe–Cr–Ni alloys with different chromium contents. *Philos. Mag. Lett.* **2020**, *100*, 435–441. [CrossRef]
206. Hummelshøj, T.S.; Christiansen, T.L.; Somers, M.A.J. Lattice expansion of carbon-stabilized expanded austenite. *Scr. Mater.* **2010**, *63*, 761–763. [CrossRef]
207. Oddershede, J.; Christiansen, T.L.; Ståhl, K.; Somers, M.A.J. Extended X-ray absorption fine structure investigation of nitrogen stabilized expanded austenite. *Scr. Mater.* **2010**, *62*, 290–293. [CrossRef]
208. Oddershede, J.; Christiansen, T.L.; Ståhl, K.; Somers, M.A.J. Extended X-ray absorption fine structure investigation of carbon stabilized expanded austenite and carbides in stainless steel AISI 316. *Steel Res. Int.* **2011**, *82*, 1248–1254. [CrossRef]
209. Brink, B.K.; Ståhl, K.; Christiansen, T.L.; Frandsen, C.; Hansen, M.F.; Somers, M.A.J. Composition-dependent variation of magnetic properties and interstitial ordering in homogeneous expanded austenite. *Acta Mater.* **2016**, *106*, 32–39. [CrossRef]
210. Che, H.L.; Lei, M.K. Microstructure of perfect nitrogen-expanded austenite formed by unconstrained nitriding. *Scr. Mater.* **2021**, *194*, 113705. [CrossRef]
211. Kappaganthu, S.R.; Sun, Y. Formation of an MN-type cubic nitride phase in reactively sputtered stainless steel-nitrogen films. *J. Cryst. Growth* **2004**, *267*, 385–393. [CrossRef]
212. Fossati, A.; Borgioli, F.; Galvanetto, E.; Bacci, T. Glow-discharge nitriding of AISI 316L austenitic stainless steel: Influence of treatment time. *Surf. Coat. Technol.* **2006**, *200*, 3511–3517. [CrossRef]
213. Thaiwatthana, S.; Li, X.Y.; Dong, H.; Bell, T. Comparison studies on properties of nitrogen and carbon S phase on low temperature plasma alloyed AISI 316 stainless steel. *Surf. Eng.* **2002**, *18*, 433–437. [CrossRef]
214. Sun, Y.; Li, X.; Bell, T. Low temperature plasma carburising of austenitic stainless steels for improved wear and corrosion resistance. *Surf. Eng.* **1999**, *15*, 49–54. [CrossRef]
215. Duarte, M.C.S.; Godoy, C.; Wilson, J.C.A.-B. Analysis of sliding wear tests of plasma processed AISI 316L steel. *Surf. Coat. Technol.* **2014**, *260*, 316–325. [CrossRef]
216. Sun, Y.; Bell, T. Sliding wear characteristics of low temperature plasma nitrided 316 austenitic stainless steel. *Wear* **1998**, *218*, 34–42. [CrossRef]
217. Baranowska, J.; Franklin, S.E.; Kochmańska, A. Wear behaviour of low-temperature gas nitrided austenitic stainless steel in a corrosive liquid environment. *Wear* **2007**, *263*, 669–673. [CrossRef]
218. Ceschini, L.; Chiavari, C.; Marconi, A.; Martini, C. Influence of the countermaterial on the dry sliding friction and wear behaviour of low temperature carburized AISI 316L steel. *Tribol. Int.* **2013**, *67*, 36–43. [CrossRef]
219. De Las Heras, E.; Santamaría, D.G.; García-Luis, A.; Cabo, A.; Brizuela, M.; Ybarra, G.; Mingolo, N.; Brühl, S.; Corengia, P. Microstructure and Wear Behavior of DC-Pulsed Plasma Nitrided AISI 316L Austenitic Stainless Steel. *Plasma Process. Polym.* **2007**, *4*, S741–S745. [CrossRef]
220. Zhu, X.M.; Lei, M.K. Pitting corrosion resistance of high nitrogen f.c.c. phase in plasma source ion nitrided austenitic stainless steel. *Surf. Coat. Technol.* **2000**, *131*, 400–403. [CrossRef]
221. Fossati, A.; Borgioli, F.; Galvanetto, E.; Bacci, T. Corrosion resistance properties of glow-discharge nitrided AISI 316L austenitic stainless steel in NaCl solutions. *Corros. Sci.* **2006**, *48*, 1513–1527. [CrossRef]
222. Flis-Kabulska, I.; Sun, Y.; Flis, J. Monitoring the near-surface pH to probe the role of nitrogen in corrosion behaviour of low-temperature plasma nitrided 316L stainless steel. *Electrochim. Acta* **2013**, *104*, 208–215. [CrossRef]
223. Abreu, C.M.; Cristóbal, M.J.; Merino, P.; Nóvoa, X.R.; Pena, G.; Pérez, M.C. Electrochemical behaviour of an AISI 304L stainless steel implanted with nitrogen. *Electrochim. Acta* **2008**, *53*, 6000–6007. [CrossRef]
224. Zhu, X.M.; Guo, Y.; Xing, Z.Q.; Lei, M.K. Effect of Nitrogen on Semiconducting Properties of Passive Films of a High Nitrogen Face-Centered-Cubic Phase Formed on Austenitic Stainless Steel. *J. Electrochem. Soc.* **2012**, *159*, C319–C325. [CrossRef]
225. Li, C.X.; Bell, T. Corrosion properties of active screen plasma nitrided 316 austenitic stainless steel. *Corros. Sci.* **2004**, *46*, 1527–1547. [CrossRef]
226. Buhagiar, J.; Dong, H. Corrosion properties of S-phase layers formed on medical grade austenitic stainless steel. *J. Mater. Sci. Mater. Med.* **2012**, *23*, 271–281. [CrossRef]

227. Stio, M.; Martinesi, M.; Treves, C.; Borgioli, F. Cultures and co-cultures of human blood mononuclear cells and endothelial cells for the biocompatibility assessment of surface modified AISI 316L austenitic stainless steel. *Mater. Sci. Eng. C* **2016**, *69*, 1081–1091. [CrossRef]

228. Lei, M.K.; Zhang, Z.L.; Zhu, X.M. Effects of nitrogen-induced hcp martensite formation on corrosion resistance of plasma source ion nitrided austenitic stainless steel. *J. Mater. Sci. Lett.* **1999**, *18*, 1537–1538. [CrossRef]

229. Gontijo, L.C.; Machado, R.; Kuri, S.E.; Casteletti, L.C.; Nascente, P.A.P. Corrosion resistance of the layers formed on the surface of plasma-nitrided AISI 304L steel. *Thin Solid Films* **2006**, *515*, 1093–1096. [CrossRef]

Article

Comparison of Nitriding Behavior for Austenitic Stainless Steel 316Ti and Super Austenitic Stainless Steel 904L

Stephan Mändl * and Darina Manova

Leibniz Institute of Surface Engineering (IOM), Permoserstr. 15, 04318 Leipzig, Germany; darina.manova@iom-leipzig.de
* Correspondence: stephan.maendl@iom-leipzig.de

Abstract: In situ X-ray diffraction (XRD) was used to compare nitrogen low-energy ion implantation (LEII) into austenitic stainless steel 316Ti and super austenitic stainless steel 904L. While the diffusion and layer growth were very similar, as derived from the decreasing intensity of the substrate reflection, strong variations in the observed lattice expansion—as a function of orientation, the steel alloy, and nitriding temperature—were observed. Nevertheless, a similar resulting nitrogen content was measured using time-of-flight secondary ion mass spectrometry (ToF-SIMS). Furthermore, for some conditions, the formation of a double layer with two distinct lattice expansions was observed, especially for steel 904L. Regarding the stability of expanded austenite, 316Ti had already decayed in CrN during nitriding at 500 °C, while no such effect was observed for 904L. Thus, the alloy composition has a strong influence only on the lattice expansion and the stability of expanded austenite—but not the diffusion and nitrogen content.

Keywords: expanded austenite; austenitic stainless steel; nitriding; in situ XRD

1. Introduction

Austenitic stainless steel is a rather soft material and is prone to high wear rates. Thus, the nitriding of austenitic stainless steel with the aim of increasing its tribological properties has been developed [1–3]. Here, nitrogen insertion in the temperature range between 300 and 450 °C leads to a super-saturated phase with nitrogen still in solid solution [4–9]. This so-called expanded austenite is a metastable phase which can accommodate nitrogen up to 38 at.%, resulting in anisotropic lattice expansion up to 12% [10]. However, it starts to decay into CrN/Cr_2N and a nitrogen-free Fe-Ni phase at around 450 °C [11].

Nevertheless, most of the existing work describes the behavior of the common alloys, such as 304 and 316L. Here, the fcc structure is stabilized at room temperature by adding just enough austenite stabilizing elements, such as Ni and Mn, while maintaining a high Cr content to allow passivation of the surface. For alloys with even higher Ni contents with improved and better corrosion resistance, which are commonly classified as Ni-base alloys, a similar formation of expanded austenite is observed [6]. In between the austenitic stainless steel and Ni-base alloys, there reside super austenitic stainless steels; they are considered to be steels with a Pitting Resistance Equivalent Number (PREN) greater than 40. Thus, much higher additions of Ni and Mo are necessary, combined with a lower Fe content, to obtain an even better corrosion resistance than that of austenitic stainless steel.

While there is a huge amount of literature on the nitriding process, with more than a dozen reviews summarizing the results for austenitic stainless steel [1,12–15], there is almost no information on the phase formation in super austenitic stainless steel. The nitriding of steel UNS S31254 between 400 and 500 °C resulted in expanded austenite with a lattice expansion of 5–10% [16]. At a treatment time of 5 h, the layer thickness after nitriding increased from 5 to 20 μm, with nitrocarburizing producing thicker layers [17]. At the same time, CrN precipitates have been identified with TEM even at 400 °C. Using the same steel alloy, nitriding at lower temperatures between 300 and 400 °C resulted in an expanded phase [18]. However, at 400 °C, dispersed nitrides in a superficial region of about 400 nm have been observed using glancing angle XRD. Thus, it appears that the higher Ni content in super austenitic stainless steel may have some influence on the stability of the expanded phase, while no direct comparison of the diffusion and the actual nitrogen near the surface is possible from the literature data.

Hence, this manuscript compares the nitriding behavior and the stability of expanded austenite for AISI 316Ti and 904L using in situ X-ray diffraction during nitriding as well as during depth profiling by sputter etching. Thus, direct access to the diffusion, phase formation, and phase stability was possible without manufacturing a large set of samples. Furthermore, this assures that the time series is actually produced under identical conditions.

2. Materials and Methods

Two different stainless steels, the austenitic AISI 316Ti (X6CrNiMoTi17-12-2, equivalent to DIN 1.4571, according to DIN/EN (in wt.%) C $\leq$ 0.08, Cr 16.5–18.5, Ni 10.5–13.5, Mo 2.0–2.5, Ti 5 × C $\leq$ 0.70) and the super austenitic AISI 904L (X1NiCrMoCu25.20.5, equivalent to DIN 1.4539, according to DIN/EN (in wt.%) C $\leq$ 0.02, Cr 19.0–21.0, Ni 24.0–26.0, Mo 4.0–5.0, Cu 1.2–2.0) were used. Rods with 15 mm and 20 mm diameters, for 316Ti and 904L, respectively, furnished coupons with 2.5 mm thickness, which were ground and polished up to a mirror finish.

The experiments are a combination of low-energy ion implantation (LEII) and in situ XRD measurements in a vacuum system consisting of two chambers, an ion source chamber and an XRD chamber, with separated pumping systems for independent control of the vacuum conditions. Both chambers are connected by a gate valve. The XRD chamber contains a diffractometer in Bragg–Brentano geometry with an external sample heating system. Thus, it is possible to make an adjustment of ion current density independently of the process temperature. The sample temperature is measured closely to the sample surface during the whole experiment and can be kept constant via a fast control loop. More details on the experimental equipment can be found in [19]. At the same time, the narrow XRD substrate peaks for stainless steel can be used for a temperature control independently of the thermocouple with a resolution of better than 5 K.

An ion beam with an energy of up to 2 keV was produced by a broad beam Kaufman-type source [20,21]. It is movable inside the ion source chamber, allowing an in situ adjustment of the beam with respect to the sample holder in the XRD chamber. Before and after every experiment, the ion current density is measured using a Faraday cup, with the identical size and shape as the sample holder, placed on the position of the sample holder. The measurements confirmed the high stability of the ion source [19]. The acquisition time for one diffractogram encompassing a 2θ range from 35 to 54 deg is slightly less than 2 min as a position-sensitive linear X-ray detector is used. The wavelength of 0.15418 nm corresponds to Cu K$_\alpha$ radiation.

Depending on the ion species, nitriding and sputter etching can be performed. Nitrogen ions are formed and accelerated for nitriding, whereas noble gas ions, e.g., Ar, are employed for the sputter etching process. In the former case, the time-resolved X-ray data are indicative of the phase changes occurring during nitriding, averaged across the whole information depth. For austenitic stainless steel, this corresponds to a layer thickness of around 2.5 μm for the current angular range [22]. In the latter case, depth-resolved information on the phase formation can be obtained during sputter etching where no changes within the samples are supposed to occur (neither diffusion nor phase transformations). Hence, this allows us to convert the time series of the acquired X-ray diffractograms into a depth series with a depth resolution of 50 nm or better [23].

The studied samples were nitrided with 0.8 keV nitrogen ions. The temperature during the nitriding experiments was between 350 and 500 °C, with the samples being heated to this temperature before the process was begun. The ion current density was fixed at 180 μA/cm^2. The nitriding time varied between 90 and 180 min. For the sputter etching experiments, 0.9 keV argon ions were used. Here, the process temperature was maintained at 180 °C; this elevated temperature was necessary to avoid the artificial peak shifts caused by a slow but noticeable heating of the sample during the whole experiment. However, no diffusion or phase transition processes were observed at this temperature, in agreement with the diffusion rates calculated from the activation energies reported in the literature [24]. Using the temporal evolution of the substrate X-ray reflections, the (orientation-dependent) sputter rate can be directly determined [23]. This calculation also serves as an independent check on the stability of the ion current.

ToF-SIMS measurements (IONTOF SIMS V, Münster, Germany) of the samples were carried out with 15 keV ^{69}Ga$^+$ ions for the analysis and 2 keV O$_2$$^+$ ions for sputter profiling to obtain the nitrogen depth distribution. The scan areas were chosen to be 100×100 µm^2 and 300×300 µm^2, respectively, to avoid crater edge effects of the analysis beam. The ion beam current for analysis was about 2 pA, while the ion beam current for sputtering was around 700 nA, resulting in a surface removal rate of about 0.7–0.9 nm/s for the selected scan size. The thickness of the nitrided layers was determined from the respective crater depth, assuming a constant sputter rate within the implanted layer.

For the calibration of the count rate in SIMS, selected samples were additionally measured using GDOES by an external company (TAZ GmbH, Aichach, Germany accreditation according to [25]—a Europewide quality standard for analytical testing laboratories) with the established procedures used for nitrided steels; the following elements were measured: C, N, O, Fe, Cr, Ni, Mo, and S. Here, a typical circular crater size with a 2.5 mm diameter was formed.

Additionally, ex situ XRD measurements were conducted at room temperature after the end of the experiments using another XRD set-up (Rigaku SmartLab multi-purpose diffractometer system, Tokyo, Japan). The system is equipped with a 9 kW rotating Cu anode X-ray source, a primary Johansson Ge(111) Kα_1 monochromator (wavelength Cu Kα_1: 0.15406 nm), a 5-axis-goniometer, and a solid state area detector (HyPix-3000, Rigaku, Tokyo, Japan) with 0D, 1D, and 2D detection modes. The samples were measured in parallel beam geometry performing a continuous/accumulating 1D scan over the 2θ range of interest. The XRD measurements were recorded and evaluated using the Rigaku SmartLab Studio II software package.

3. Results

As the in situ measurements yield a plethora of information, several different aspects of the nitriding process can be investigated (as a function of time or a function of depth), using very few samples; furthermore, the introduction of inevitable errors can be avoided by using ex situ measurements for a larger number of samples within a time series. In the following subsections, (i) diffusion, (ii) lattice expansion, and (iii) peak splitting, as well as (iv) the stability of the expanded phase, are investigated in detail.

3.1. Diffusion

Figure 1a,b present the nitrogen depth profiles for both steel grades—904L and 316Ti—as a function of temperature, together with a calibration curve from GDOES as an insert. While the treatment time varies for the 904L samples by up to a factor of two between 90 and 186 min, the time is always 90 min for 316Ti. Nevertheless, some general tendencies can be extracted from the data. At 350 °C, a slower diffusion is observed for 904L, while similar diffusivities were encountered for the higher temperatures. The surface nitrogen content appears to have a maximum near 400 °C for both alloys, where up to 40 at.% was measured. The rather high nitrogen flux aggregates near the surface at this temperature as the diffusion is not fast enough to transport the nitrogen into deeper regions. As shown in panel Figure 1c, there is a very good correlation between the strong decrease in the nitrogen content and the region near the surface, where a different etching contrast is observed. As the original grains are still visible, it can be assumed that no CrN precipitates are formed.

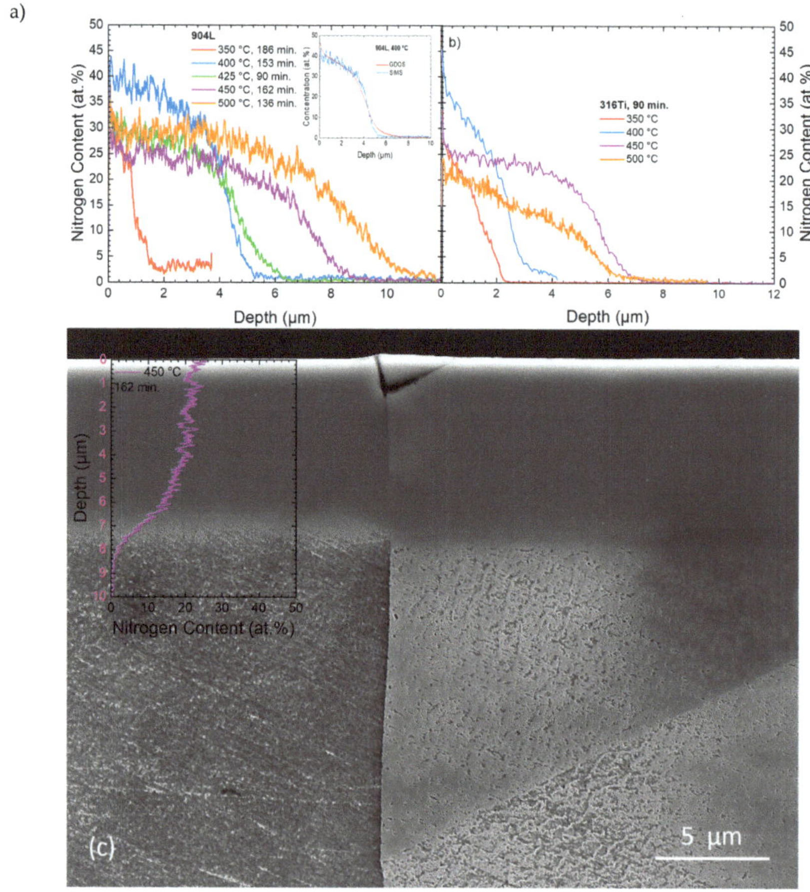

Figure 1. Nitrogen depth profiles after nitriding of steel grade (**a**) 904L and (**b**) 316Ti at different process temperatures and different process times, as obtained from SIMS measurements. In insert, nitrogen depth profile for 904L sample nitrided at 400 °C, measured by SIMS and by GDOES, is shown. (**c**) Micrograph of cross-section for 904L nitrided at 450 °C with the superimposed nitrogen depth profile.

At higher temperatures, a faster diffusion leads to a rather constant surface concentration around 25–30 at.%. The behavior at 350 °C is peculiar. Despite lower diffusivities, less nitrogen is incorporated. Hence, different surface processes—beyond sputtering, especially preferential sputtering (which should be temperature-independent)—must be active. Typically, a temperature-dependent desorption or sticking coefficient is observed in catalytic surface processes, which could also be the underlying effect here.

However, the nitrogen is implanted below the surface, circumventing any absorption and dissociation pathways during nitriding, while the oxide layer is sputtered away within the first minute of the treatment. This should not lead to a delayed nitriding for 904L only at a temperature of 350 °C—even for a hypothetical thicker oxide layer. The reason is that this effect is (again) not temperature-dependent.

Figure 2 depicts the typical time evolution of the XRD diffractograms obtained during nitriding; these are examples of two samples only, both of which were nitrided at 400 °C for different times. As 81 diffractograms were collected during the nitriding time of 160 min, any presentation of all the data in a conventional plot, i.e., "intensity vs. angle" (even with vertical offsets, cf. the graphical abstract), will lead to a skewed or incomplete presentation. Thus, 2D contour plots were chosen to present the results. Here, the intensity is color-coded (on a logarithmic scale) as a function of time and angle. Thus, the intensity evolution (increase/decrease), as well as the gradual changes in peak positions, is readily identified. However, for a more detailed picture, all the diffractograms must be analyzed in detail.

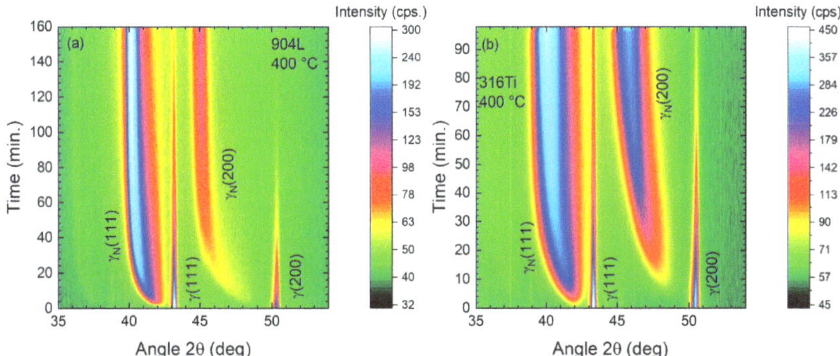

Figure 2. Contour plot of X-ray diffractograms measured during nitriding at 400 °C of stainless steel grade (**a**) 904L and (**b**) 316Ti. Please note the different treatment times of 160 min for 904L and 100 min for 316Ti.

As a starting point, the behavior of both alloys at 400 °C is presented, where no qualitative differences exist between them. For both alloys, a combination of decreasing substrate intensity with the onset of a continuous lattice expansion is observed. Other effects are seen for higher temperatures and are shown below; they include a jump in the lattice expansion at a certain time or the decreasing intensity of the expanded phase after longer times (indicative of the decay and the formation of nanocrystalline, X-ray amorphous CrN precipitates in an FeNi matrix). However, we start now with the growth of the expanded phase as a function of time.

As the expanded phase is always found to have a sharp interface with the substrate in SEM cross-sections, the modelling of the XRD data with a two-layer model (substrate and surface layer) is possible. The artificial broadening of the interface observed in SIMS and GDOES is dominated by the surface roughening of the interface during sputter depth profiling [26] and should be ignored for the analysis. Thus, the (initial) nitrogen diffusivity can be extrapolated from the decreasing substrate intensity (as a function of grain orientation) if the layer thickness is below 5 µm (where the intensity of the scattered X-rays is attenuated to less than 15% compared to the intensity from the surface). The results of the modelling are shown in Figure 3a, which plots the logarithm of the relative intensity vs. the square root of the nitriding time. Assuming an inverse parabolic layer growth, straight lines are expected—and observed, at least for most of the data. Additionally, errors arise from the fitting procedure of the data, especially the uncertainties in the background subtraction.

As with the qualitative results derived from the SIMS depth profiles, a slightly slower diffusion is observed for 350 °C in steel 904L with the intensity calibration using the known attenuation coefficient [22]. Except for this data point, no variation across the steel grade or the grain orientation is observed. This result could be termed fortuitous, as it is known that variations in the grain size even for values larger than the thickness of the nitrided zone can lead to deviations in the diffusion by 50% [27]. The resulting activation energy of 1.0 ± 0.1 eV is identical for both alloys within the error range, thus indicating identical transport pathways within the grains (as is known from the literature [1]). Additionally, the layer thickness from SIMS is converted to a diffusivity, using the known nitriding times. Here, a reasonable agreement between both datasets for the two steel alloys is observed. It must be noted that the SIMS data result in a diffusivity averaged across the grain orientation, with a slightly higher error due to the inherent roughness increase during sputtering [26].

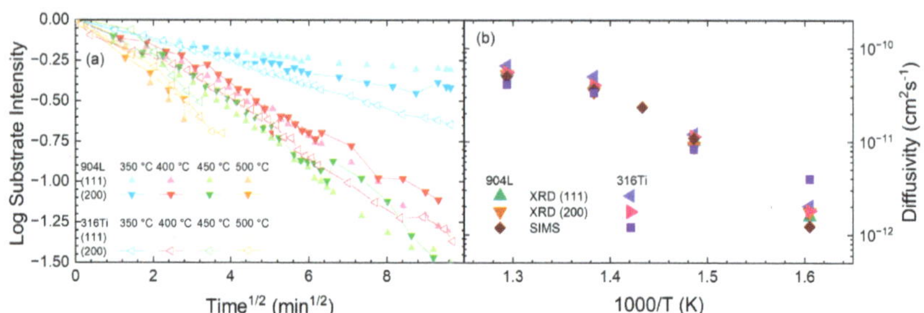

Figure 3. (**a**) Relative intensity of substrate XRD peak in logarithmic scale as a function of square root of process time for 904L and 316Ti steel grade for different process temperatures; (**b**) calculated diffusivity for both type of steels in Arrhenius plot.

Nevertheless, both approaches are complementary: the XRD intensity leads to the initial diffusivity whereas the SIMS measurement calculates an "average" diffusivity across the whole treatment time. As the data are in good agreement, no deviation from the inverse parabolic law can be assumed. The sputtering during nitriding is only a minor corrective term which does not significantly affect the diffusivity of the used short treatments [28]. Conversely, it is difficult or impossible to establish the formation of CrN, which should lead to a different "average diffusivity", as observed in the 5h treatments [29].

3.2. Lattice Expansion

As shown in the previous subsection, the nitrogen content and diffusion are quite similar for both of the investigated alloys. In contrast, the lattice expansion is highly variable between the two alloys, between the process temperatures, and between the grain orientations, as plotted in Figure 4. As a general remark, there is currently no explanation for the evolution of lattice expansion with time; however, existing models of the nitriding of austenitic stainless steel should be able to reproduce this effect [30,31]. With this point in mind, this subsection is more like an empirical data collection where no definitive explanation can yet be provided.

Starting at 350 °C in Figure 4a, despite a similar nitrogen content, a complete reversal of the lattice expansion is seen: 904L, large splitting; 316Ti, small splitting of expansion between orientations; and the (111) expansion is always higher for 316Ti than for 904L during the whole nitriding process.

For longer times, a saturation of the position of all four expanded peaks is apparently obtained. This behavior is mirrored for 400 °C in Figure 4b with a faster saturation (which could be due to a thicker nitrided layer where the interface with the substrate is no longer visible); yet, the (111) expansion is nearly identical for both alloys.

For 450 °C, a different behavior is observed in Figure 4c as the expansion for 316Ti is now lower for both orientations. Still, the (111) expansion saturated quite fast, while the (200) shows a jump for 904L and a gradual increase for 316Ti. This could be due to the formation of a layered structure with two lattice expansions [32], which is investigated in detail below.

For 500 °C in Figure 4d, yet another different behavior is observed: for 904L, the time evolution is very similar to that for 450 °C, whereas for 316Ti a strongly reduced expansion for the (200) oriented grains, even smaller than for the (111) oriented grains at lower temperatures, is measured. At the same time, the formation of CrN precipitates, as indicated by the decreasing intensity of the expanded peak beyond 1 h, is not the cause of this effect. As shown below in Section 3.4, there is still a layered structure, with the CrN-containing layer on top of the expanded austenite and the substrate on the bottom.

At the same time, the Fe-Ni matrix depleted by Cr is converted to a nanoscale ferrite structure [33], which does not contribute to the diffraction intensity at the observed angle of the expanded austenite.

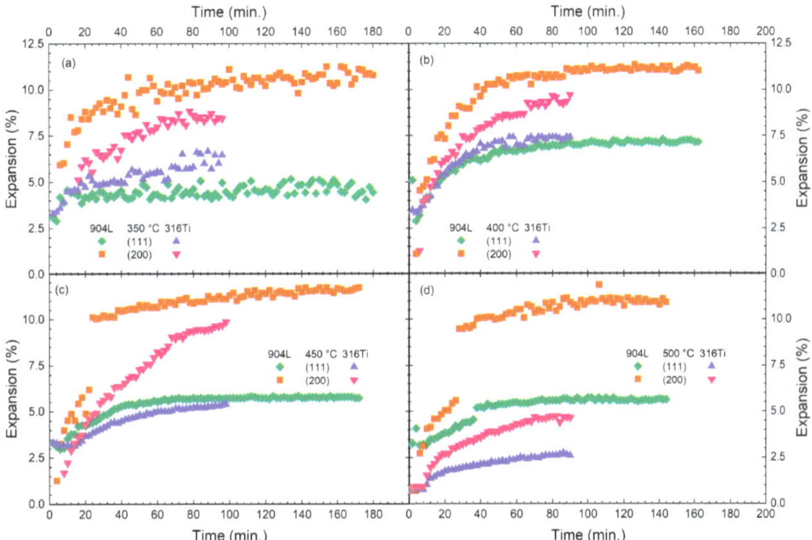

Figure 4. Lattice expansion evolution with increasing time for (111) and (200) orientation for 904L and 316T steel grade for nitriding at (**a**) 350 °C, (**b**) 400 °C, (**c**) 450 °C, and (**d**) 500 °C.

A summary of this behavior is plotted in Figure 5 showing the ratio of the lattice expansion of the (200) oriented vs. the (111) grains as a function of alloy composition, nitriding temperature, and time. The main point here is that there is a huge variation in the ratio of the lattice expansion for the two investigated grain orientations.

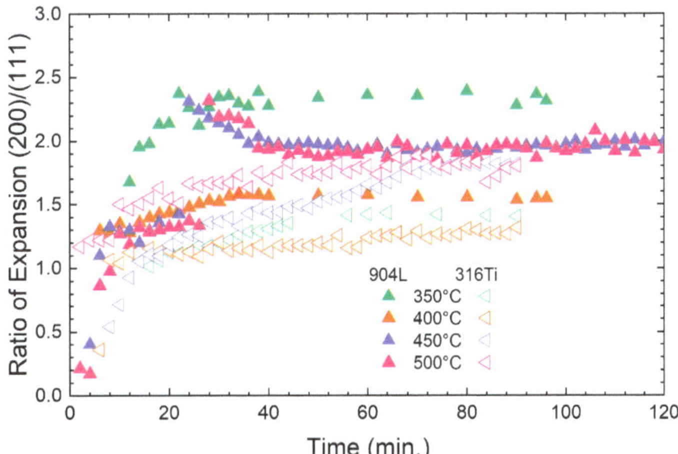

Figure 5. Evolution of ratio of lattice expansion for (111) and (200) orientations with increasing number of diffractograms (respectively, with process time) for 904L and 316Ti steel grades for 4 different process temperatures.

Nevertheless, the first 5–10 min should be treated with caution, as the peak intensity for the expanded peak and the absolute expansion are still small. This introduces an additional error in the fitting routine next to the dominant substrate peak (cf. Figure 2). However, for longer times the plotted values of the lattice expansion are very reliable. In particular, there is only a small concentration gradient of the incorporated nitrogen, and the information depth is heavily skewed towards the surface. The probed depth for the expanded lattice (normal to the surface) is dominated by the first few hundreds of nanometers.

In general, the lattice expansion for the {200} oriented grains is larger than for {111} grains with the absolute values varying between 10 and 150%; larger values are nearly always present for steel 904L compared to 316Ti. Actually, this could be related to the mechanical properties of the underlying steel (or the formed expanded austenite)—and is discussed in more detail in the following paragraphs. However, a big constraint must be inserted: there is a huge variability as a function of processing temperature and processing time—the latter effect would never be visible in ex situ measurements taken after the experiments. For the current in situ experiments, there is nearly no deviation between the last diffractogram measured at the nitriding temperature and the final diffractogram at room temperature. It is negligible beyond the shift due to the thermal expansion; thus, no diffusion or change in the nitrogen distribution occurs during the cooling of the samples.

It was observed before that there are dynamic processes active during nitriding which increase the "scattering intensity" for longer times, especially at intermediate temperatures around 400 °C [23]: the total reflected intensity from the expanded phase is not constant but highly variable. At the same time, the peak width even after correcting for the finite information depth and the nitrogen concentration gradient is also a function of processing temperature [23]. All in all, there are several, mostly unidentified, processes occurring in parallel to the nitrogen insertion and the lattice expansion, including plastic flow due to a stress beyond the yield strength [34], the formation of dislocations [34], or grain rotation [35]. Yet, these phenomena are only the result of underlying processes within the grains and on an atomic level.

Returning to the underlying elastic properties, especially the anisotropy of the elastic modulus in fcc metals, it was proposed that uniaxial stress, together with compositional strain from the isotropic introduction of nitrogen into the lattice [34], can lead to the observed differences in the lattice expansion for differently oriented grains. Yet, there should be a tendency towards a universally valid equation yielding an "anisotropy factor"—similar to the constant which allows the conversion of the lattice parameter into a nitrogen content [34]. Nevertheless, the literature shows that such a simple model, even after correcting for the stacking fault density [36], does not exist. However, variations in the stacking fault energy can affect deformation mechanisms, including dislocation activities and deformation twinning [35].

The only drawback of the in situ method is the limited range in 2θ ending at 54°; thus, only the (111) and (200) oriented grains can be investigated. Here, additional ex situ measurements were performed for the 904L steel samples, and the results are plotted in Figure 6. For the low-orientation reflections—(111) and (200)—no additional information compared to the in situ data is obtained: broad expanded peaks for 350 °C reflect the dominating influence of the gradient within the small layer thickness, compared to the information depth, on the resulting data.

Due to the strong {111} fiber texture, even the higher-order reflections now visible between 60° and 120° in 2θ yield only sparse information. Most of the expanded peaks, broadened for geometrical reasons, are barely visible beyond the noise level. In general, a lattice expansion between 6 and 13% is calculated. This is the same range as already defined by the (111) and (200) reflections and is within the theoretical expectations based on the lattice anisotropy [34]. Thus, even though the ex situ data give additional information on differently oriented grains, no additional insights into the time evolution of the lattice expansion are available.

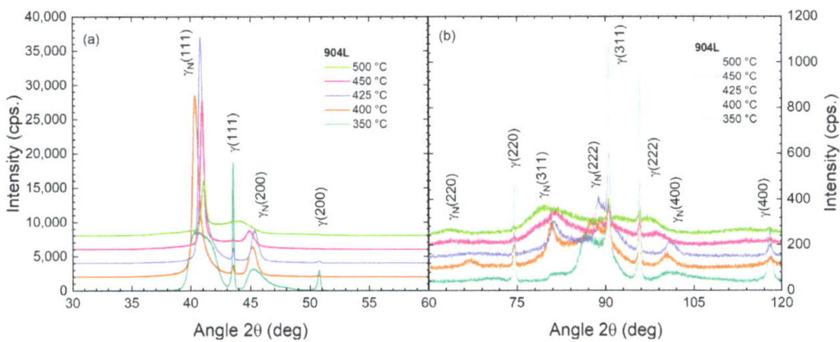

Figure 6. Ex situ measured X-ray diffractograms for 904L steel samples nitrided at different process temperatures, showing the angular range of 2θ from 30° to 120° in two separate panels: (**a**) 30–60° for the (111) reflection and the (200) reflections; (**b**) 60–120° for the region between (220) and (400) reflections.

3.3. Peak Splitting

Another point which is sometimes observed in the literature is peak splitting, which is related to the simultaneous observance of two different expanded peaks, especially for high-alloyed compounds [6]. There, it was argued that a higher Ni content prevents a larger build-up of nitrogen in the surface layers while more Fe is effective in retaining more nitrogen. At the same time, it was proposed that residual stress may induce a sublayer advancement in front of the expanded austenite layer [6].

In previous work, this splitting has been sometimes observed for medium-alloyed steel, e.g., 304 or 316Ti, but only for the (200) oriented grains [30]. There, it was traced to a layered structure with the "low expansion phase" being situated near the interface and the "high expansion phase" near the surface of the sample. While the transition occurred around 1 h after the beginning of the nitriding, at the end of the experiment, after nearly 3 h, only the high-expansion phase was visible in XRD. Subsequent ion etching revealed a surface layer of about 600 nm thickness with the higher expansion (but only in the {200} oriented grain), while no pronounced discontinuity was observed in the nitrogen depth profile [32].

Returning to the current results, both samples implanted here at 450 °C show this (selective) transition in a similar manner, but only for the (200) reflections, as depicted in Figure 7 and evident from Figure 4c. Despite a very similar nitrogen concentration and gradient, the transition occurs earlier in 904L than in 316Ti, and it leads to a thinner layer with a lower lattice expansion. This is evident in the contour plots taken during in situ XRD while sputter-etching the expanded layer, as shown in Figure 7b,d.

Even more pronounced is the effect in Figure 4d for the 904L sample nitrided at 500 °C, where the transition is visible for both orientations. However, there appears to be a slight time delay for the {111} oriented grains. In Figure 4, only the position of the expanded peak with the highest intensity is plotted. When looking at the specific diffractograms, however, the apparent "jump" can be resolved as a function of time (but not depth as the XRD information depth of a few micrometers is still close to the total layer thickness). The layered structure is derived from the sputter investigations shown in Figure 7.

When looking into the details of this transition, as shown in Figure 8, it becomes apparent that the transition is already starting within the first 20 min. There, a new satellite (γ_{N2}) is appearing on the left side of the expanded peak (γ_{N1}) for both orientations and is clearly distinct in the angular position where the intensity is gradually growing. For longer times—near 40 min—this peak is dominating (slightly shifted even further to the left due to the increasing nitrogen content), and the original peak is disappearing.

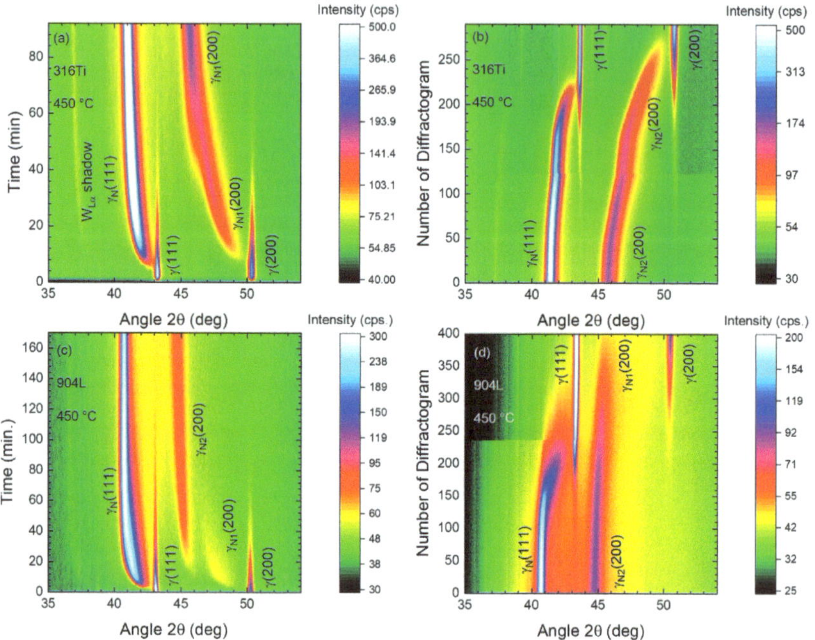

Figure 7. Contour plot of X-ray diffractograms measured during nitriding at 450 °C of (**a**) 316Ti and (**c**) 904L and during sputtering of nitrided samples (**b**) 316Ti and (**d**) 904L. Please note that each ion etching was performed over two days, thus the contour plots show a slight discontinuity. Each diffractogram corresponds to two minutes of sputtering.

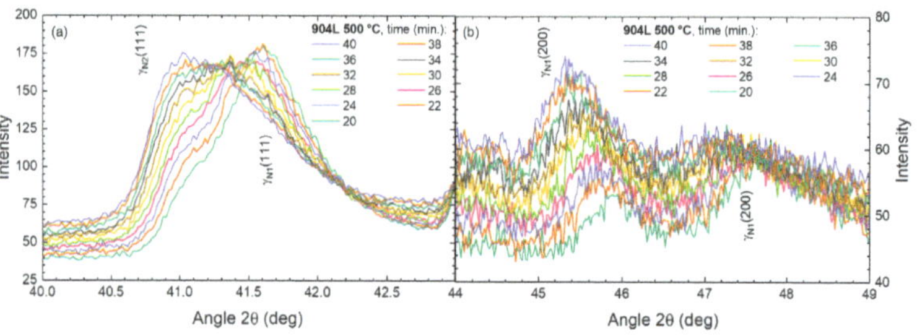

Figure 8. Selected X-ray diffractograms (taken at different times as indicated) measured during nitriding of 904L sample at 500 °C. The diffractograms are vertically shifted for clarity. Please note the different angular scales in both panels: (**a**) (111) reflection and (**b**) (200) reflection.

Thus, it can be concluded that this transition between a low-expansion and a high-expansion phase is observed under some processing conditions during nitriding. As the overlying layer with the high-expansion phase can be quite thick, it often masks the underlying phase in conventional XRD measurement after nitriding. Nevertheless, the transition in the expansion, i.e., the jump ratio, is quite flexible, depending on the temperature, alloy composition, and even the grain orientation. Furthermore, using glancing angle measurements will complicate the analysis as different classes of grains are probed when increasing the angle 2θ [37,38]. At the same time, it has to be kept in mind that only the lattice expansion normal to the surface is probed. Yet, concurrent in-plane measurements yield no unequivocal results about the three-dimensional strain [32].

Hence, a general explanation for the layered expansion is currently not available. One possible candidate is a reduced symmetry, e.g., from fcc to fct symmetry [32,39], with a certain depth or stress necessary to induce the transition. Similarly, work for Ni-based Ni-Ti alloys indicate an origin in the strong depth dependency of the level of residual stress [40].

3.4. Stability of Expanded Austenite

The stability of expanded austenite is crucial for the corrosion resistance as the precipitation of chromium nitride is supposed to prevent the transport of chromium towards the surface [11,24]. Thus, the formation of a protective Cr_2O_3 surface layer is no longer possible at higher temperatures. As the nitrogen is inserted from the surface, the regions closer to the surface experience the nitrogen for a longer time; thus, the CrN formation starts from the surface and progresses towards the bulk [33]. During nitriding, this process is manifested as a decrease in the intensity of the expanded phase after a certain time (cf. Figure 9a beyond 60 min) as the total scattering volume of this phase is decreasing when the CrN precipitates are forming, together with the formation of an Fe-Ni matrix [33]. Initially, the grain size of both phases is too small to be observed by XRD. However, at the end of nitriding, a broad peak at low intensity can be attributed to this nanocrystalline Fe-Ni matrix.

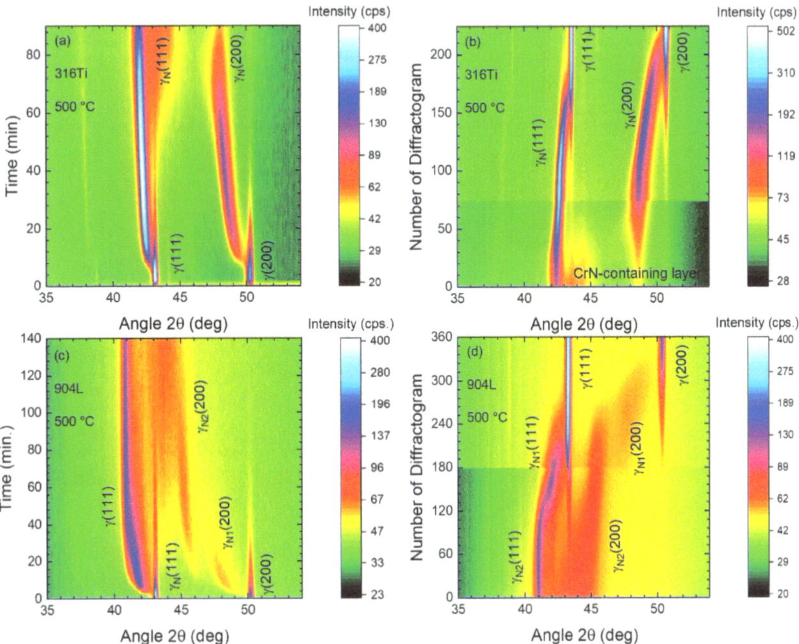

Figure 9. Contour plot of X-ray diffractograms measured during nitriding at 500 °C of steel grade (**a**) 316Ti and (**c**) 904L and during sputtering of nitrided samples of (**b**) 316Ti and (**d**) 904L. Please note that each ion etching was performed over two days; thus, the contour plots show a slight discontinuity. Each diffractogram corresponds to two minutes of sputtering.

As the in situ measurements during nitriding represent the time evolution, integrated over the complete information depth, no actual depth information is obtained there. However, the subsequent sputtering of this layer reveals the structure at the end of nitriding: for 316Ti, a CrN-containing layer on top of expanded austenite followed by the substrate is obtained, whereas for 904L there are two expanded layers on top of each other ($\gamma_{N2}/\gamma_{N1}/\gamma$). This is different from that observed during nitriding as that process is thermally activated—and no depth information is available during nitriding. This layered structure is shown in Figure 9b for the steel 316Ti, nitrided at 500 °C. In contrast, no such effect is observed for steel 904L nitrided at the same temperature (shown in Figure 9c,d). Thus, it appears that expanded austenite in steel 904L is more stable than in steel 316Ti. However, this would contradict the existing literature where CrN precipitates have been identified in 904L at 500 °C using TEM [17,18].

Here, ToF-SIMS presents an alternative method to probe the atomic environment when looking at the relative intensities of cluster ions. It has been shown that the random arrangement of atoms in steel leads to a certain probability of $FeCr^+$, $FeNi^+$, $CrNi^+$, Fe_2^+, Ni_2^+, and Cr_2^+ molecular cluster ions, with the $Cr_2^+/FeCr^+$ being especially sensitive to the formation of CrN precipitates at the nanoscale [33,41]. Here, atomic rearrangements lead to a preferential ordering [42], where chromium atoms are more likely to be surrounded by nitrogen and other chromium atoms and less likely to be found near iron or nickel atoms.

The corresponding results for the currently investigated samples are shown in Figure 10, exemplifying the contrasting behavior of both alloys: in Figure 10b—for 316Ti—the formation of CrN precipitates is only observed at 500 °C—and then is limited to the first 1.5–2 μm of the nitrided layer. This result confirms the information from the in situ XRD measurements shown in Figure 9b. In contrast, the onset of CrN formation in 904L in Figure 10a is already visible at 450 °C—but not at 425 °C (the intensity ratio of the Cr_2^+ and $FeCr^+$ clusters in SIMS at 425 °C is still identical to that at 350 and 400 °C)—where a (slight) deviation of the $Cr_2^+/FeCr^+$ cluster ion ratio from the straight line, which is characteristic for the lower temperatures, is observed.

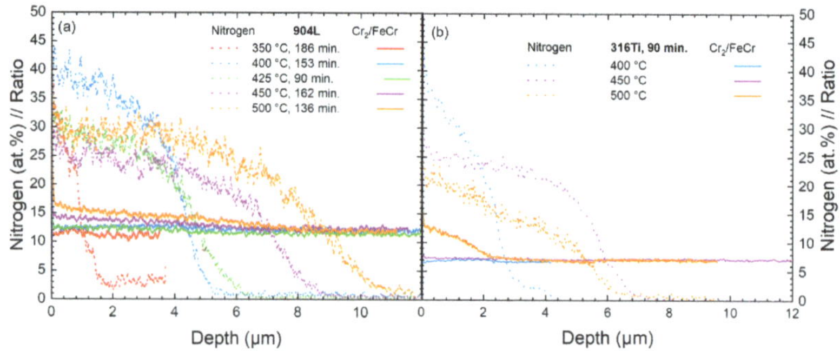

Figure 10. Ratio of intensities for Cr_2 and FeCr cluster ions in ToF-SIMS vs. depth for both steel alloys (steel grade (**a**) 904L and (**b**) 316Ti) at different process temperatures. Additionally, the nitrogen depth profiles after nitriding are shown as dotted lines.

However, the CrN formation is visible in steel 904L across the whole nitrided layer, but at a very low intensity, decreasing from the surface towards the interface. These refined measurements, which are complementary to the in situ XRD data, confirm the observation of CrN precipitates after the nitriding of super austenitic stainless steel in the literature [17,18]. Nevertheless, several open questions remain for further discussion and investigation: why does the nucleation of CrN precipitates start at a much lower temperature in 904L—or at earlier times—during nitriding than in 316Ti? The alloy composition cannot be the only reason as the onset of CrN formation after nitrocarburizing is shifted by about 10 K towards higher temperatures for the higher-alloyed 316L compared to the alloy 304 [29].

At the same time, the growth of these precipitates in 904L is strongly reduced compared to steel 316Ti (or 304 [33]). The scattering intensity of the expanded austenite in steel 904L after nitriding is nearly undisturbed during nitriding as well as ion etching (cf. Figure 9). Thus, the volume fraction of the precipitates, as well as the fraction of a nitrogen- and chromium-free Fe-Ni phase, must be below a few percent of the total volume.

Furthermore, nitriding at 400 °C in an N_2/H_2 atmosphere for extremely long periods, i.e., 99 h, shows that the expanded austenite in 904L is still present, in contrast to 304L, where CrN is dominant [43]. The higher Ni content is supposed to stabilize the austenite phase and prevent the formation of CrN precipitates. Nevertheless, these are visible, but only in the immediate vicinity of the surface, as shown by the glancing angle measurements at $\theta = 3°$ [43]. Prolonged annealing between 450 and 600 °C for 24 to 168 h shows another peculiar behavior as the lattice constant of expanded austenite decreases while no CrN becomes visible (in contrast to 304L, where CrN is visible in XRD) [44]. Localized nitride precipitation (without growth) is assumed to occur.

From an application viewpoint, it would be interesting to investigate the corrosion resistance of super austenitic stainless steel nitrided at higher temperatures (Ref. [18] only compares 300–400 °C) and to establish whether nitriding at even higher temperatures would lead to the growth of CrN precipitates or just a higher nucleation density. At the same time, this could either reduce the processing time or result in a thicker layer, which would be more beneficial for industrial applications.

Here, the nitriding of a Ni-based superalloy (Ni-20Cr) showed a similar effect: an expanded phase with CrN precipitation within this phase while the corrosion resistance was actually slightly improved [45]. Yet, a lattice expansion of 2.3% and 1.0% for the (200) and (111) reflections was measured there, respectively. This was taken as an indication that only a small amount of nitrogen (about 4–10 at.%) was dissolved into the Ni-based superalloy, while the rest of the nitrogen must reside at the grain boundaries or defects. Similarly, a local nitrogen uptake depending on the actual phases present in the material was observed for the nitriding of different Ni-based superalloys with various microstructures [46]. Nevertheless, for the current case of high nitrogen content coupled with a large lattice expansion, the formation of very small CrN precipitates is peculiar and cannot be explained at present.

4. Discussion

For (standard) austenitic stainless steel, no influence of the alloy composition on the nitriding behavior, i.e., nitrogen diffusion, the formation of expanded austenite, and the stability of the expanded phase has been observed [1,4,6]. However, ferritic stainless steel and Ni-based alloys are different in this respect. Ferritic stainless steel shows an expanded phase, sometimes together with an expanded martensite [47]. For austenitic Fe-Ni alloys, in contrast, Fe-rich samples show thicker layers with high N content while Ni-rich samples lead to thinner layers and lower nitrogen content [6].

Looking at the current alloys, 316Ti and 904L, the miniscule addition of Ti (at less than 1 at.%) does not modify the formation of expanded austenite—i.e., nitrogen content and nitrogen diffusion. The formation of TiN instead of CrN has not been observed in TEM investigations [33]. The major difference between the two alloys is the Ni content—12 vs. 25 at.%—and this could be the reason for the difference in the stability of the expanded austenite. According to [6], 904L is still "Fe-rich"; thus, the very similar diffusion and nitrogen content is not surprising; yet, differences in the stacking fault energy do exist [36].

On a microscopic level, dislocation characteristics and their associated strain fields modify the CrN formation [48]. While the insertion of dislocations (due to a stress beyond the yield strength) destroys existing Cr-N chemical pairs, additional nitrogen segregates to the formed dislocations. Thus, segregation of Cr towards nitrogen via the attractive chemical interaction becomes possible. The open question is still whether the different alloy composition, especially the higher Ni content in super austenitic steel, is enough to allow the segregation of isolated Cr-N precipitates, but not the agglomeration of Cr-N clusters.

While the excellent mechanical properties of expanded austenite are well known, and nearly independent of the base material and the nitriding temperature, the corrosion behavior of 904L nitrided at higher temperatures (where CrN clusters, in contrast to CrN precipitates in 316Ti, are formed) is a new topic to be explored in further work.

5. Summary and Conclusions

Using in situ XRD during nitriding as well as ion etching leads to a plethora of information, which is time-resolved but depth-integrated during nitriding, but depth-resolved for the final situation after nitriding at elevated temperatures. While the nitriding of low-alloyed austenitic stainless steel is a common topic, there is nearly no information about super austenitic alloys.

Here, it was shown that the nitrogen uptake and diffusion were very similar for the steels 316Ti and 904L. Between 350 and 500 °C, the diffusion follows an inverse parabolic law with the nitrogen surface concentration between 25 and 40 at.%. No significant difference in the diffusivity was observed when comparing the initial rate, from the decreasing substrate intensity, with the final layer thickness as derived from the SIMS measurements.

In contrast, the lattice expansion (normal to the surface) exhibits a complex behavior depending on temperature, alloy composition, nitriding time, and grain orientation. The main point appears to be that looking at isolated grains is too simplistic; defect formation, grain rotation, and dynamic annealing processes are additionally active and modify the information content of the XRD experiments.

The layered structure with two different lattice expansions (high near the surface and low near the interface—typically around 10% vs. 5%), which is known from the literature, was observed for some conditions, especially at higher nitriding temperatures. For longer times, the surface layer became too thick to see remnants of the interface layer in XRD. A definitive cause could not be identified; strong stress gradients, as proposed in the literature, are a likely candidate.

While not all these results are completely new, the diverging behavior of 316Ti and 904L concerning the formation of CrN precipitates, is unexpected and seen in detail here: 316Ti starts to form CrN precipitates beyond 450 °C, initiated at the surface and progressing towards the interface with a complete decay and the segregation into CrN and an Fe-Ni phase. In contrast, 904L already forms nanoscopic CrN clusters at 450 °C—throughout the whole nitrogen-containing layer, not only at the surface, without subsequent growth or coalescence. At the same time, the original expanded austenite phase is completely undisturbed by this precipitation process. The chemical compositions of the current 904L and the 254 SMO reported in the literature [17,18] are quite similar (X1NiCrMoCu25-20-5 vs. X1CrNiMoCuN20-18-7). Thus, a generalization towards all super austenitic stainless steels is currently not possible, but it could be likely, as the main driving factor of the behavior appears to be the varying Ni content between austenite and super austenite.

The austenitic Fe-Cr-Ni phase is still providing new and unexpected results as soon as the composition deviates from the usual alloys encountered in classical applications of austenitic stainless steel. Unfortunately, we are still relegated to an empirical description of several perplexing effects found during these investigations.

Author Contributions: Conceptualization, D.M.; methodology, D.M.; formal analysis, S.M. and D.M.; writing—original draft preparation, S.M. and D.M.; writing, S.M. and D.M.; funding acquisition, D.M. All authors have read and agreed to the published version of the manuscript.

Funding: This research was funded by Deutsche Forschungsgemeinschaft DFG, grant number MA 4426/5-2.

Data Availability Statement: The raw data supporting the conclusions of this article will be made available by the authors on request.

Acknowledgments: The authors are greatly indebted to H. Neumann and F. Scholze (from IOM) for support during the conception and assembling of the experiment. J.W. Gerlach (from IOM) provided the ex situ XRD measurements. Technical support by M. Müller and R. Woyciechowski (also from IOM) is gratefully acknowledged. The authors thank D. Spemann (from IOM) and J.W. Gerlach for fruitful discussions. M.N. Le, Technische Universität Bergakademie Freiberg, is acknowledged for providing the cross-sectional micrograph.

Conflicts of Interest: The authors declare no conflicts of interest.

References

1. Borgioli, F. From Austenitic Stainless Steel to Expanded Austenite-S Phase. *Metals* **2020**, *10*, 187. [CrossRef]
2. Ichii, K.; Fujimura, K.; Takase, T. Microstructure, corrosion resistance, and hardness of the surface layer in ion nitrided 18-8 stainless steel. *Netsu Shori* **1985**, *25*, 191–195.
3. Somers, M.A.J.; Christiansen, T.L. Low temperature surface hardening of stainless steel. In *Thermochemical Surface Engineering of Steels*; Mittemeijer, E.J., Somers, M.A.J., Eds.; Woodhead Publishing: Oxford, UK, 2015; pp. 557–579.
4. Czerwiec, T.; Renevier, N.; Michel, H. Low-temperature plasma-assisted nitriding. *Surf. Coat. Technol.* **2000**, *131*, 267–277. [CrossRef]
5. Bell, T. Current Status of Supersaturated Surface Engineered S-Phase Materials. *Key Eng. Mater.* **2008**, *373–374*, 289–295. [CrossRef]
6. Williamson, D.L.; Davis, J.A.; Wilbur, P.J. Effect of austenitic stainless steel composition on low-energy, high-flux, nitrogen ion beam processing. *Surf. Coat. Technol.* **1998**, *103–104*, 178–184. [CrossRef]
7. Tao, X.; Matthews, A.; Leyland, A. On the Nitrogen-Induced Lattice Expansion of a Non-stainless Austenitic Steel, Invar 36®, Under Triode Plasma Nitriding. *Metall. Mater. Trans. A* **2020**, *51*, 436–447. [CrossRef]
8. Tang, D.; Zhang, C.; Zhan, H.; Huang, W.; Ding, Z.; Chen, D.; Cui, G. High-Efficient Gas Nitridation of AISI 316L Austenitic Stainless Steel by a Novel Critical Temperature Nitriding Process. *Coatings* **2023**, *13*, 1708. [CrossRef]
9. Luiz, L.A.; Kurelo, B.C.E.S.; de Souza, G.B.; de Andrade, J.; Marino, C.E.B. Effect of nitrogen plasma immersion ion implantation on the corrosion protection mechanisms of different stainless steels. *Mater. Today Commun.* **2021**, *28*, 102655. [CrossRef]
10. Christiansen, T.; Somers, M.A.J. Controlled dissolution of colossal quantities of nitrogen in stainless steel. *Met. Mater. Trans. A* **2006**, *37*, 675–682. [CrossRef]
11. Borgioli, F.; Fossati, A.; Galvanetto, E.; Bacci, T.; Pradelli, G. Glow discharge nitriding of AISI 316L austenitic stainless steel: Influence of treatment pressure. *Surf. Coat. Technol.* **2006**, *200*, 5505–5513. [CrossRef]
12. Drouet, M.; Le Bourhis, E. Low Temperature Nitriding of Metal Alloys for Surface Mechanical Performance. *Materials* **2023**, *16*, 4704. [CrossRef] [PubMed]
13. Czerwiec, T.; Andrieux, A.; Marcos, G.; Michel, H.; Bauer, P. Is "expanded austenite" really a solid solution? Mössbauer observation of an annealed AISI 316L nitrided sample. *J. Alloys Comp.* **2019**, *811*, 151972. [CrossRef]
14. Mittemeijer, E.J.; Somers, M.A.J. Thermodynamics, kinetics, and process control of nitriding. *Surf. Eng.* **1997**, *13*, 483–497. [CrossRef]
15. Borgioli, F. The "Expanded" Phases in the Low-Temperature Treated Stainless Steels: A Review. *Metals* **2022**, *12*, 331. [CrossRef]
16. Totten, G.E.; Casteletti, L.C.; Fernandes, F.A.P.; Gallego, J. Microstructural Characterization of Layers Produced by Plasma Nitriding on Austenitic and Superaustenitic Stainless Steel Grades. *J. ASTM Internat.* **2012**, *9*, 103564. [CrossRef]
17. Fernandes, F.A.P.; Casteletti, L.C.; Gallego, J. Microstructure of nitrided and nitrocarburized layers produced on a superaustenitic stainless steel. *J. Mater. Res. Technol.* **2013**, *2*, 158–164. [CrossRef]
18. Schibicheski Kurelo, B.C.E.; de Souza, G.B.; Serbena, F.C.; Lepienski, C.M.; Chuproski, R.F.; Borges, P.C. Improved saline corrosion and hydrogen embrittlement resistances of superaustenitic stainless steel by PIII nitriding. *J. Mater. Res. Technol.* **2022**, *18*, 1717–1731. [CrossRef]
19. Manova, D.; Bergmann, A.; Mändl, S.; Neumann, H.; Rauschenbach, B. Integration of a broad ion source with a high-temperature vacuum XRD chamber. *Rev. Sci. Instrum.* **2012**, *83*, 113901. [CrossRef] [PubMed]
20. Kaufman, H.R.; Robinson, R.S. Ion Source Design for Industrial Applications. *Am. Inst. Aeronaut. Astronaut. J.* **1982**, *20*, 745–760. [CrossRef]
21. Zeuner, M.; Meichsner, J.; Neumann, H.; Scholze, F.; Bigl, F. Design of ion energy distributions by a broad beam ion source. *J. Appl. Phys.* **1996**, *80*, 611–622. [CrossRef]
22. Henke, B.L.; Gullikson, E.M.; Davis, J.C. X-ray Interactions: Photoabsorption, Scattering, Transmission, and Reflection at E = 50–30,000 eV, Z = 1–92. *At. Data Nucl. Data Tables* **1993**, *54*, 181–342. [CrossRef]
23. Manova, D.; Schlenz, P.; Mändl, S. A combination of ion beam sputtering and in-situ X-ray diffraction as a method for depth resolved phase analysis using nitrogen implanted austenitic stainless steel as an example. *J. Appl. Phys.* **2022**, *131*, 025306. [CrossRef]
24. Williamson, D.L.; Ozturk, O.; Wei, R.; Wilbur, P.J. Metastable phase formation and enhanced diffusion in f.c.c. alloys under high dose, high flux nitrogen implantation at high and low ion energies. *Surf. Coat. Technol.* **1994**, *65*, 15–23. [CrossRef]
25. *DIN EN ISO 17025*; Allgemeine Anforderungen an die Kompetenz von Prüf- und Kalibrierlaboratorien. DIN Media GmbH: Berlin, Germany, 2018. [CrossRef]
26. Manova, D.; Díaz, C.; Pichon, L.; Abrasonis, G.; Mändl, S. Comparability and Accuracy of Nitrogen Depth Profiling in Nitrided Austenitic Stainless Steel. *Nucl. Instrum. Meth. B* **2015**, *349*, 106–113. [CrossRef]
27. Manova, D.; Mändl, S.; Neumann, H.; Rauschenbach, B. Influence of Grain Size on Nitrogen Diffusivity in Austenitic Stainless Steel. *Surf. Coat. Technol.* **2007**, *201*, 6686–6689. [CrossRef]
28. Mändl, S. Nitriding of Stainless Steel: PIII or Low Energy Nitriding? *Plasma Proc. Polym.* **2007**, *4*, 239–245. [CrossRef]
29. Jafarpour, S.M.; Martin, S.; Schimpf, C.; Dalke, A.; Biermann, H.; Leineweber, A. Comparative Plasma Nitrocarburizing of AISI 316L and AISI 304 Steels Using a Solid Carbon Active Screen: Differences in the Developing Microstructures. *Metall. Mater. Trans. A* **2024**, *55*, 1588–1599. [CrossRef]

30. Christiansen, T.; Dahl, K.V.; Somers, M.A.J. Nitrogen diffusion and nitrogen depth profiles in expanded austenite: Experimental assessment, numerical simulation and role of stress. *Mater. Sci. Technol.* **2008**, *24*, 159–167. [CrossRef]
31. Pranevicius, L.; Templier, C.; Rivière, J.-P.; Méheust, P.; Pranevicius, L.L.; Abrasonis, G. On the mechanism of ion nitriding of an austenitic stainless steel. *Surf. Coat. Technol.* **2001**, *135*, 250–257. [CrossRef]
32. Manova, D.; Schlenz, P.; Gerlach, J.W.; Mändl, S. Depth-Resolved Phase Analysis of Expanded Austenite formed in Austenitic Stainless Steel. *Coatings* **2020**, *10*, 1250. [CrossRef]
33. Manova, D.; Lotnyk, A.; Mändl, S.; Neumann, H.; Rauschenbach, B. CrN Precipitation and Elemental Segregation during the Decay of Expanded Austenite. *Mater. Res. Express* **2016**, *3*, 066502. [CrossRef]
34. Czerwiec, T.; He, H.; Marcos, G.; Thiriet, T.; Weber, S.; Michel, H. Fundamental and Innovations in Plasma Assisted Diffusion of Nitrogen and Carbon in Austenitic Stainless Steels and Related Alloys. *Plasma Proc. Polym.* **2009**, *6*, 401–409. [CrossRef]
35. Templier, C.; Stinville, J.C.; Villechaise, P.; Renault, P.O.; Abrasonis, G.; Rivière, J.P.; Martinavičius, A.; Drouet, M. On lattice plane rotation and crystallographic structure of the expanded austenite in plasma nitrided AISI 316L steel. *Surf. Coat. Technol.* **2010**, *204*, 2551–2558. [CrossRef]
36. Wagner, C.; Laplanche, G. Effects of stacking fault energy and temperature on grain boundary strengthening, intrinsic lattice strength and deformation mechanisms in CrMnFeCoNi high-entropy alloys with different Cr/Ni ratios. *Acta Mater.* **2023**, *244*, 118541. [CrossRef]
37. Lutz, J.; Manova, D.; Gerlach, J.W.; Störmer, M.; Mändl, S. Interpretation of Glancing Angle and Bragg-Brentano XRD Measurements for CoCr Alloy and Austenitic Stainless Steel after PIII Nitriding. *IEEE Trans. Plasma Sci.* **2011**, *39*, 3056–3060. [CrossRef]
38. Feugeas, J.; Gómez, B.; Craievich, A. Ion nitriding of stainless steels. Real time surface characterization by synchrotron X-ray diffraction. *Surf. Coat. Technol.* **2002**, *154*, 167–175. [CrossRef]
39. Fewell, M.P.; Mitchell, D.R.G.; Priest, J.M.; Short, K.T.; Collins, G.A. The nature of expanded austenite. *Surf. Coat. Technol.* **2000**, *131*, 300–306. [CrossRef]
40. Fonovic, M.; Leineweber, A.; Robach, O.; Jägle, E.A.; Mittemeijer, E.J. The Nature and Origin of "Double Expanded Austenite" in Ni-Based Ni-Ti Alloys Developing Upon Low Temperature Gaseous Nitriding. *Met. Mater. Trans. A* **2015**, *46*, 4863–4874. [CrossRef]
41. Manova, D.; Gerlach, J.W.; Mändl, S. Nitrogen isotope marker experiments in austenitic stainless steel for identification of trapping/detrapping processes at different temperatures. *Surf. Coat. Technol.* **2023**, *472*, 129952. [CrossRef]
42. Che, H.L.; Yang, X.; Lei, M.K.; Somers, M.A.J. Co-existence of $\gamma'N$ phase and γN phase on nitrided austenitic Fe-Cr-Ni alloys—III. An investigation of the evolution of long-range ordered domains. *Acta Mater.* **2023**, *253*, 118971. [CrossRef]
43. Cao, Y.; Maistro, G.; Norell, M.; Pérez-García, S.A.; Nyborg, L. Multi-technique characterization of low-temperature plasma nitrided austenitic AISI 304L and AISI 904L stainless steel. *Surf. Interface Anal.* **2014**, *46*, 856–860. [CrossRef]
44. Maistro, G.; Pérez-García, S.A.; Norell, M.; Nyborg, L.; Cao, Y. Thermal decomposition of N-expanded austenite in 304L and 904L steels. *Surf. Eng.* **2016**, *33*, 319–326. [CrossRef]
45. Xu, X.L.; Yu, Z.W.; Cui, L.Y. Microstructure and properties of plasma nitrided layers on Ni-based superalloy Ni-20Cr. *Mater. Charact.* **2019**, *155*, 109798. [CrossRef]
46. Chollet, S.; Pichon, L.; Cormier, J.; Dubois, J.B.; Villechaise, P.; Drouet, M.; Declemy, A.; Templier, C. Plasma assisted nitriding of Ni-based superalloys with various microstructures. *Surf. Coat. Technol.* **2013**, *235*, 318–325. [CrossRef]
47. Borgioli, F. Formation of expanded phases in ferritic stainless steel nitrided at low temperatures. *Surf. Coat. Technol.* **2024**, *477*, 130309. [CrossRef]
48. Kawahara, Y.; Kobatake, S.; Kaneko, K.; Sasaki, T.; Ohkubo, T.; Takushima, C.; Hamada, J. Combined effect of interstitial-substitutional elements on dislocation dynamics in nitrogen-added austenitic stainless steels. *Sci. Rep.* **2024**, *14*, 4360. [CrossRef]

Article

Formation and Properties of Nitrocarburizing S-Phase on AISI 316L Stainless Steel-Based WC Composite Layers by Low-Temperature Plasma Nitriding

Shinichiro Adachi * , Takuto Yamaguchi and Nobuhiro Ueda

Research Division of Metal Finishing and Analysis, Osaka Research Institute of Industrial Science and Technology, 2-7-1 Ayumino, Izumi, Osaka 594-1157, Japan; t_yamaguchi@tri-osaka.jp (T.Y.); ueda@tri-osaka.jp (N.U.)
* Correspondence: shinadachi@tri-osaka.jp; Tel.: +81-725-51-2648; Fax: +81-725-51-2749

Abstract: Stainless steel-based WC composite layers fabricated by a laser cladding technique, have strong mechanical strength. However, the wear resistance of WC composite layers is not sufficient for use in severe friction and wear environments, and the corrosion resistance is significantly reduced by the formation of secondary carbides. Low-temperature plasma nitriding and carburizing of austenitic stainless steels, treated at temperatures of less than 450 °C, can produce a supersaturated solid solution of nitrogen or carbon, known as the S-phase. The combined treatment of nitriding and carburizing can form a nitrocarburizing S-phase, which is characterized by a thick layer and superior cross-sectional hardness distribution. During the laser cladding process, free carbon was produced by the decomposition of WC particles. To achieve excellent wear and corrosion resistance, we attempted to use this free carbon to form a nitrocarburizing S-phase on AISI 316 L stainless steel-based WC composite layers by plasma nitriding alone. As a result, the thick nitrocarburizing S-phase was formed. The Vickers hardness of the S-phase ranged from 1200 to 1400 HV, and the hardness depth distribution became smoother. The corrosion resistance was also improved through increasing the pitting resistance equivalent numbers due to the nitrogen that dissolved in the AISI 316 L steel matrix.

Keywords: plasma nitriding; laser cladding; stainless steel; tungsten carbide; hardness; corrosion resistance

Citation: Adachi, S.; Yamaguchi, T.; Ueda, N. Formation and Properties of Nitrocarburizing S-Phase on AISI 316L Stainless Steel-Based WC Composite Layers by Low-Temperature Plasma Nitriding. *Metals* **2021**, *11*, 1538. https://doi.org/10.3390/met11101538

Academic Editor: Andrea Di Schino

Received: 26 August 2021
Accepted: 22 September 2021
Published: 27 September 2021

Publisher's Note: MDPI stays neutral with regard to jurisdictional claims in published maps and institutional affiliations.

1. Introduction

Laser cladding is an effective method for surface modification, such as improving the wear and corrosion resistance of metal substrates, and has many advantages, such as high deposition speed with rapid cooling, dense metallography, strong bonding strength between the cladding layer and substrate, low heat input, and distortion of the substrate [1]. In recent years, the fabrication of three-dimensional objects by additive manufacturing (AM) using laser as a heat source has attracted much attention. There are many metal AM technologies, for example, laser-based powder bed fusion (PBF) methods have two types of selective laser melting (SLM) and selective laser sintering (SLS), in which a laser is irradiated on top of the stacked powders [2–4]. Meanwhile, directed energy deposition (DED) is based on laser cladding with a coaxial nozzle system for feeding powder in the axial direction of the laser beam and has been developed not only as a coating for surface modification, but also as a manufacturing process for various engineering materials and objects. DED achieves a near-perfect net-like shape appearance with faster processing speed and without mold tooling, which can be applied for manufacturing, part repair, and rapid prototyping. DED also enables the reduction in manufacturing costs and has been increasingly used in recent years [5–9].

Among the classes of austenitic stainless steels for laser cladding, AISI 316 L stainless steel is heavily used owing to its excellent corrosion property, high versatility, and the easy of obtaining steel powder [10]. However, deposited AISI 316 L stainless steel layers have a low hardness and poor wear resistance properties, and therefore are difficult to use in severe wear conditions. Several researchers have reported methods to improve the wear resistance and strength of deposited steel layers, and it is concluded that the addition of carbide particles as a reinforcement is highly effective. When the hard carbide particles were dispersed in the steel matrix, the steel structure was modified to one akin to sintered metal matrix composite materials. Consequently, the wear resistance is significantly improved [11–13].

It is required that reinforcement carbides have high hardness, chemical affinity to steel, and high melting temperature [14,15]. Tungsten carbide (WC) is a hard ceramic that can be uniformly dispersed in a steel matrix with good wettability; therefore, WC particles have been considered as excellent candidates for reinforcement. Steel-WC composite layers are currently the most widely applied for hardness and strength characteristics [16,17]. As for AISI 316 L stainless steel layers, it has been reported that the properties of the cladding layers were improved by compositing with SiC, TiC and titanium carbonitride [18–20]. As a matter of course, the WC particles composite for AISI 316 L stainless steel layers obtained by the laser cladding process also leads to improved mechanical properties, in particular, surface hardness and wear resistance [21–23]. During the heating of the laser cladding process, the WC particles decompose and solidify in the AISI 316 L steel matrix, and consequently, the steel matrix structure modifies to form carbide networks as a dendritic structure [21,22,24]. In addition, WC, W_2C phases, $M_{23}C_6$, and $(Fe,W)_3C$ complex carbide phases form, and tungsten and carbon elements solidify [25].

In this way, the wear resistance of WC composite layers can be improved by the hardness of WC particles themselves and also by the formation of carbides in the stainless-steel matrix. However, the hardness of stainless-steel matrix is not sufficient compared with that of hard materials such as ceramics. In addition, when a higher amount of WC particles is added, cracks will appear in the composite layer [26]. Therefore, there is a limit to the amount of WC particles, and the effect of improving wear resistance is also limited. Recently, the fabrication of coatings and objects by laser cladding and DED has attracted considerable attention, especially because of its high functionality for use in severe environments. Accordingly, in addition to the composite of WC particles, further enhancement of wear resistance by other methods should be investigated.

Low-temperature plasma nitriding and carburizing at temperatures less than 450 °C for austenitic stainless steels can improve their wear resistance. Rather than producing a nitride or carbide, this treatment produces supersaturated solid solution of nitrogen or carbon in the face-centered cubic lattice, which is known as the S-phase (or the expanded austenite) [27–33]. Accordingly, the corrosion resistance does not degrade because the ability to form a passive film is maintained after plasma treatment [34,35]. In addition, a combined treatment of nitriding and carburizing has advantages of thickening and improving the toughness of the S-phase compared with the individually nitriding and carburizing [36–39].

These plasma treatments have been applied not only to steel plates but also to thermal spray coatings and laser cladding of AISI 316 L stainless steel, and their effectiveness has been recognized by the authors [27–30]. In addition, the effects of alloy elements in stainless steel plates on the formation and properties of the S-phase have also been examined, and several papers have reported the role of chromium, nickel and molybdenum [32,40–42]. The increase or decrease in these alloy elements causes a variation in the thickness of the S-phase layer and influences the wear and corrosion properties. Furthermore, the nickel element of stainless-steel layers fabricated by laser cladding was reported by the authors [30]. However, the effects of tungsten and tungsten carbides on the S-phase have not yet been reported for both steel plates and cladding layers.

The WC composite layers contain free carbon produced by the decomposition of the WC particles during laser cladding process. In general, a nitrocarburizing S-phase requires two processes of nitriding and carburizing continuously or simultaneously. Through the use of this free carbon, there is a possibility that a carburizing S-phase is formed by nitriding process alone. In this study, AISI 316 L stainless steel-based WC particle composite layers were fabricated using laser cladding, subsequently performed by low-temperature plasma nitriding. Consequently, a thick nitrocarburizing S-phase had formed. Therefore, the metallurgical changes and carbide formation of the WC composite layers during laser cladding process were examined in detail. The formation mechanism of the nitrocarburizing S-phase by low-temperature plasma nitriding were investigated, and the properties of the hardness and corrosion resistance with the WC content were discussed.

2. Experiments

AISI 316L stainless steel-based WC composite layers were obtained using direct laser metal deposition. The mixed powder of AISI 316 L stainless-steel powder with 20 wt.% or 40 wt.% of tungsten carbide (WC) powder was used as a material for the laser cladding. The powder diameter was $-212/+63$ μm of AISI 316 L stainless-steel powder, and $-180/+53$ μm of WC powder.

The laser cladding system consisted of a continuous-wave diode laser beam machine of LDM-2000-60 (Laserline, GmbH, Mulheim-Karlich, Germany) with an attached a coaxial powder feed head of COAX12 (Fraunhofer, Munchen, Germany). The condition of the laser beam was a wavelength of 940 nm and an output power of 1.4 kW. The laser head was scanned one-layer deposition on AISI 304 stainless steel plates (size of 65 mm $\times$ 65 mm $\times$ 10 mm) with a traverse speed of 4 mm s^{-1} and length of 50 mm at 3 mm interval. The mixed powder was delivered in the coaxial direction of the laser beam through the COAX12 laser head with mass flow rates of 17–19 g min^{-1}, and using argon as the carrier gas.

To remove the surface oxidation film and flatten the surface profile such as a mirror, the deposited AISI 316 L stainless steel-based WC composite layers were polished, and the final polishing was treated using a 1 μm diamond paste. After the polishing, the thickness of the deposited layers was over 1 mm.

DC plasma ion treatment machine of FECH-1N (Fuji Electronics Industry, Osaka, Japan) was used in low-temperature plasma nitriding. The AISI 316L stainless steel-based WC composite layers on the AISI 304 stainless steel plate were connected to a cathode electrode in the glass chamber, and the copper plate was connected to an anode electrode. DC current was applied between the anode and cathode electrode at a voltage of 300 V. The plasma gas with a composition ratio of N_2:H_2 = 80:20 flowed at a rate of 1 L min^{-1} in the chamber to keep the pressure at 667 Pa. The nitriding temperatures were at 400 °C, 425 °C and 450 °C, and the processing time was 4 h after reaching the nitriding temperatures.

The crystal structure of the WC composite layers was examined using an X-ray diffraction measurement of SmartLab (Rigaku, Tokyo, Japan) in the conventional θ–2θ scan of a Cu-Kα radiation with 40 kV and 150 mA.

Cross-sectional micrographs of the WC composite layers were obtained using an optical microscope of ECLIPSE MA100N (Nikon, Tokyo, Japan). The thickness of the nitride layers was measured from the cross-sectional micrographs at approximately 60 points for each sample.

The compositions of the WC composite layers were investigated by SEM with an energy-dispersive X-ray (EDX) analysis, SEM of ERA-8900FE (Elionix, Tokyo, Japan), and EDX of ApolloX (AMETEK, Berwyn, IL, USA), and the operating SEM voltage was 20 kV. In addition, the nitrogen and carbon depth distribution profiles were obtained using a glow-discharge optical emission spectroscopy (GDOES) of GDA750 system (Rigaku, Tokyo, Japan).

The hardness depth profiles in the cross-section of the nitrided layers were measured using a nanoindentation tester of ENT-1100a (Elionix, Tokyo, Japan) with a test load of 1.5 mN. The measurements for each sample were repeated thrice. The surface hardness of the WC composite layers was measured using a Vickers tester of HM-220D (Mitutoyo, Tokyo, Japan) with a test load of 0.245 N. The Vickers tests were repeated for each sample at least eight times, and the maximum and minimum values were rejected.

The corrosion resistance of the WC composite layers was estimated from anodic polarization curves. The anodic polarization curves were obtained using a potentiostat of HSV-110 (Hokuto Denko Corporation, Tokyo, Japan). The reference and counter electrode were an Ag/AgCl and a platinum plate, respectively. The measurement voltage was applied at a range of -0.6 V to 1 V with a scan rate of 1 mV s^{-1}. The experiments were carried out in a 3.5 wt.% NaCl solution at a temperature of 30 °C. Before the measurements, the dissolved oxygen in the NaCl solution was degassed by bubbling with N$_2$ gas for at least 0.5 h. After the measurements, the corrosion morphologies of the surfaces were obtained using a digital microscope of Dino-Lite (AnMo Electronics Corporation, Taipei, Taiwan).

3. Results and Discussion

3.1. Carbide Formation of AISI 316 L Stainless Steel with WC Particles during Laser Cladding

Figure 1 shows an overall view of the cross-section of 20 wt.% and 40 wt.% of WC particle composite layers. Note that the WC compositions in the text are the content of the feeding powder material used for the laser cladding, and the compositions of the deposited layers should be varied during the laser cladding process. The WC particles were dispersed in the AISI 316 L steel matrix, similarly to the metal matrix composite materials. However, cracks perpendicular to the substrate were observed in some areas of the WC 40 wt.% layers.

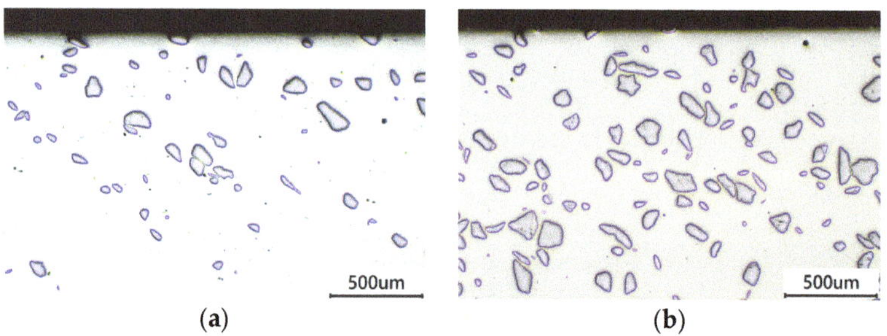

Figure 1. Cross-sectional micrographs of the as-deposited AISI 316 L steel-based WC composite layers with various WC compositions: (**a**) WC 20 wt.%, and (**b**) WC 40 wt.%.

The XRD patterns in Figure 2 present that the as-deposited WC composite layers contained the austenite phase of AISI 316 L stainless steel, the carbides of WC, W$_2$C and M$_6$C, and M$_{12}$C in only the 40 wt.% layer. It has been reported that WC, W$_2$C phases, M$_{23}$C$_6$, and (Fe,W)$_3$C complex phases are formed simultaneously in Fe/WC metal matrix composite coatings produced using the Yb: YAG laser [25].

Furthermore, W$_2$C carbide is produced by the decomposition of WC, the metastable W$_2$C decomposes to M$_6$C of Fe$_3$W$_3$C [43,44], and the Eta-carbides of M$_6$C and M$_{12}$C are formed in the quaternary system Fe-W-C-Cr [45]. In this examination, the WC particles in the deposited layers were decomposed to W$_2$C; subsequently, W$_2$C decomposed and reacted with the AISI 316 L steel. Finally, the secondary carbides of M$_6$C and M$_{12}$C were synthesized. In Figure 3 of the cross-sectional microstructures after the etching treatment with Marble's reagent (5 mL HCl, 5 mL H$_2$O, and 1 g CuSO$_4$), the gray contrasting areas with a net-like appearance in the AISI 316 L steel matrix indicate the precipitation of M$_6$C and M$_{12}$C as secondary carbides.

The magnified image by SEM in Figure 4 also showed the presence of eutectic carbides between the austenite in the WC 40 wt.% layer. In addition, elemental analysis of SEM-EDX showed that tungsten was dissolved in the austenite, and the chemical compositions of W ranged from 10 to 15 wt.%.

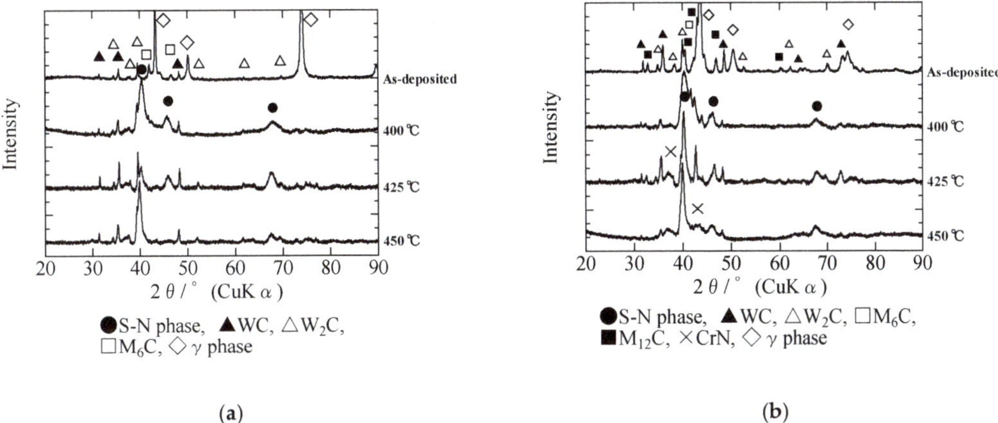

Figure 2. X-ray diffraction patterns of the AISI 316 L steel-based WC composite layers with various WC compositions and nitriding temperatures: (**a**) WC 20 wt.%, and (**b**) WC 40 wt.%.

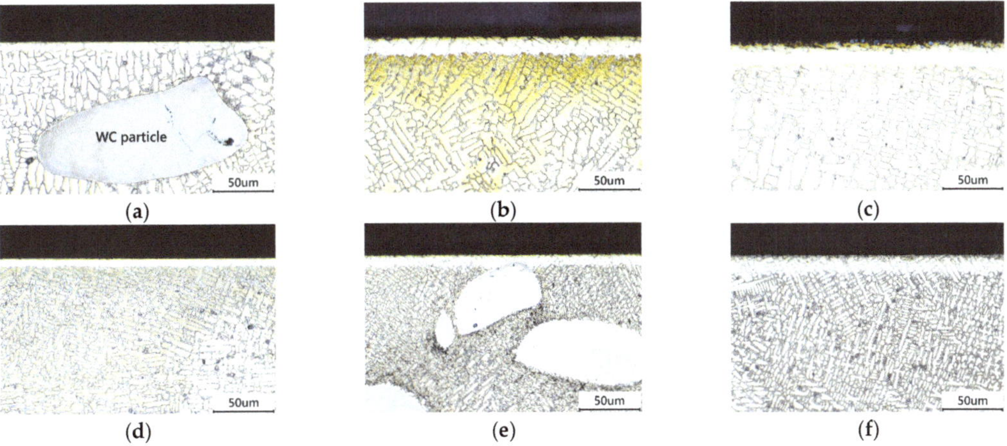

Figure 3. Cross-sectional micrographs of the AISI 316 L steel-based WC composite layers with various WC compositions and nitriding temperatures: (**a**) WC 20 wt.% at 400 °C, (**b**) WC 20 wt.% at 425 °C, (**c**) WC 20 wt.% at 450 °C, (**d**) WC 40 wt.% at 400 °C, (**e**) WC 40 wt.% at 425 °C, and (**f**) WC 40 wt.% at 450 °C.

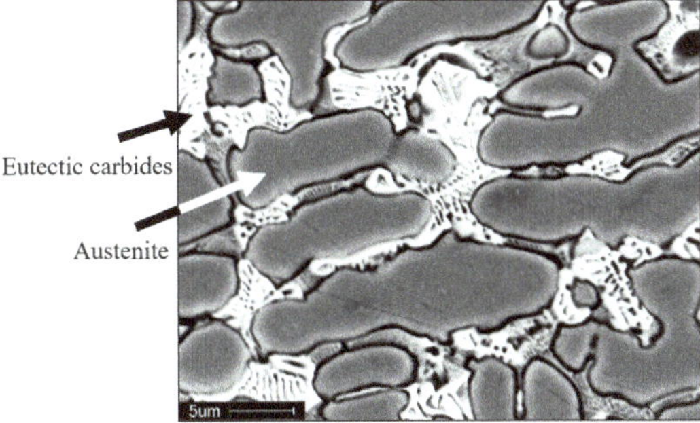

Figure 4. Cross-sectional SEM image of the AISI 316 L steel-based WC 40 wt.% composite layer.

3.2. Formation Mechanism of Nitrocarburizing S-Phase by Plasma Nitriding

The XRD of the nitrided WC composite layers in Figure 2 show that the (111), (200) and (220) planes of the γ-phase shifted towards the lower angles than those of the as-deposited layers. These shifts of the γ-phase peaks are caused by expanding the fcc lattice constants due to the dissolution of nitrogen. In addition, the cross-sectional microstructures with the etching treatment in Figure 3 indicate that the bright contrast layers are present on the surface. These XRD peak shifts and presence of the bright contrast layers suggest the formation of the S-phase.

Figure 5 shows the thickness of the S-phase measured from the bright contrast layers in the cross-sectional microstructures (Figure 3). The average thickness of the S-phase increased with increasing nitriding temperature. WC 20 wt.% layers were 6.7 μm at the nitriding temperature of 400 °C, 13.5 μm at 425 °C, and 20.6 μm at 450 °C. WC 40 wt.% layers were 5.6 μm at 400 °C, 9.8 μm at 425 °C, and 15.7 μm at 450 °C.

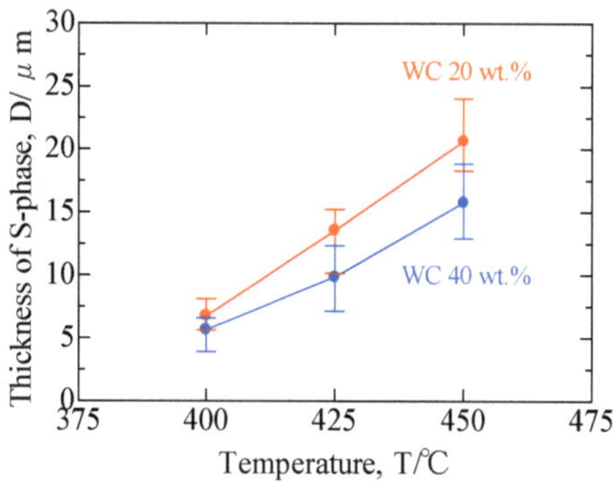

Figure 5. Thickness of the S-phase layers with various WC compositions and nitriding temperatures.

The nitrogen depth distributions of the S-phase obtained by the GDOES measurements in Figure 6, show that nitrogen's distribution is plateau and step-like, rather than a smooth decrease which would obey Fick's law. Plateau and step-like distributions have been reported due to the dissolved chromium atoms binding to nitrogen, and it is considered that these distributions are unique to the S-phase [46–48]. Meanwhile, the carbon distributions show that the carbon peaks are present at a greater depth than where the nitrogen diffusion had completed. These distributions of carbon are very similar to that obtained by continuously or simultaneously nitriding and carburizing treatments [36–38]. Furthermore, the thickness of the S-phase, measured from the bright contrast layers in the cross-sectional microstructures (Figure 5), was thicker than the diffusion depth of nitrogen in GDOES. The thickness was almost equal to the sum of the diffusion depth of the nitrogen and the depth of the peak position of carbon. These results suggest that a carburizing S-phase presents where the carbon peaks exist.

The source of carbon in the carburizing S-phase seems to be free carbon produced by the decomposition of WC to W_2C during the laser cladding process, since the possibility of decomposition of carbides by nitriding is low from the point of view of thermodynamics. Furthermore, some WC particles may have dissolved due to the local temperature increase caused by the laser scanning [49]. It is also possible that free carbon derived from the formation of the secondary carbides, M_6C and $M_{12}C$. It has been reported that the diffusion rate of carbon in austenitic stainless steels is fast compared with that of nitrogen [36]; therefore, the free carbon could penetrate ahead of nitrogen due to the nitrogen pushing the free carbon inward during the plasma nitriding, finally the carburizing S-phase would be formed. Consequently, the S-phase became a dual layer of nitrocarburizing; the nitriding S-phase sited at the outside, and the carburizing S-phase at the inside.

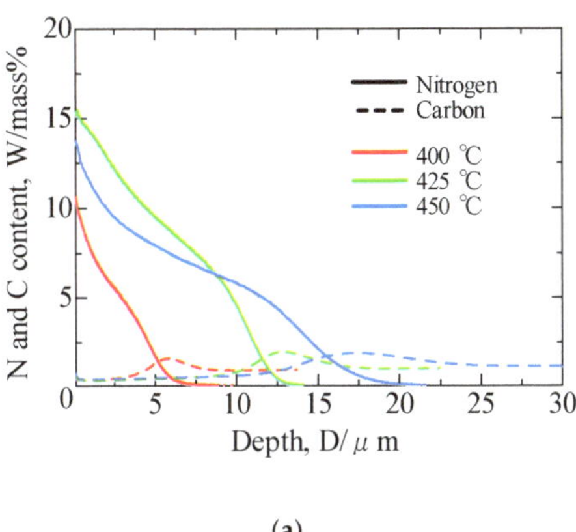
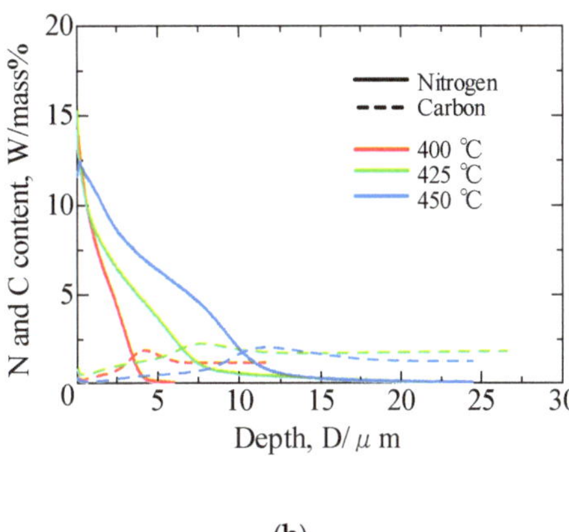

(**a**)   (**b**)

Figure 6. Nitrogen and carbon depth distributions of the AISI 316 L steel-based WC composite layers with various WC compositions and nitriding temperatures: (**a**) WC 20 wt.%, and (**b**) WC 40 wt.%.

3.3. Effects of WC Content on Thickness of Nitrocarburizing S-Phase

The thickness of the S-phase of the WC composite layers was greater than that of 8.2 μm of the single AISI 316 L steel layer without WC at 450 °C [30]. The formation of the carburizing layer as well as the nitriding layer has resulted in the thickening of the S-phase. The GDS result for WC 20 wt.% at 450 °C in Figure 6 presents the nitrogen plateau region reaching up to 12 μm depth. Even excluding the carburizing S-phase, the nitriding S-phase was thicker than that of the single AISI 316 L steel layer. As shown in the previous section, the compositions of the solid solution of tungsten in the austenite ranged from 10 to 15 wt.%. Therefore, the solid solution of tungsten may promote nitrogen diffusion.

Meanwhile, comparing the nitrogen diffusion depth between the 20 wt.% and 40 wt.% WC layers in GDOES (Figure 6), the WC 40 wt.% layers diffused at a much shallower rate. The WC 40 wt.% layers had dense secondary carbide precipitations of a net-like appearance (Figure 3), resulting in the decrease in the crystal grain size. These dense carbide precipitations and small grain size would prevent the nitrogen diffusion. In contrast, the thickness of the carburizing S-phase (width of the carbon peaks in Figure 6) was almost the same in both layers. The WC 40 wt.% layers had a higher volume of the free carbon. Therefore, the shallow diffusion of nitrogen would not change the thickness of the carburizing S-phase layers.

3.4. Hardness Depth Profiles and Vickers Hardness of Nitrocarburizing S-Phase

Figure 7 shows the nanoindentation hardness-depth profiles of the cross-section at a nitriding temperature of 450 °C. The hardness gradually decreased from the surface to the inside, and hardening was observed up to a depth of approximately 20 μm for the 20 wt.% layer and 15 μm for the WC 40 wt.% layer. These hardening depths are almost identical to the thickness of the S-phase (Figure 5). In general, hardness depth profiles formed by nitriding alone have been reported to abruptly decrease at the boundary between the S-phase layer and the base metal. In contrast, a combined treatment of nitriding and carburizing decreases gradually from the surface to the inside [50]. Therefore, these hardness depth profiles also prove the formation of the carburizing S-phase. The variations in hardness in the deep region where the S-phase was not formed are likely due to the hardness of the secondary carbides. From these variations, the hardness of the secondary carbides was derived to be 10–20 GPa, which is lower than the hardness of the surfaces of the nitriding S-phase layers.

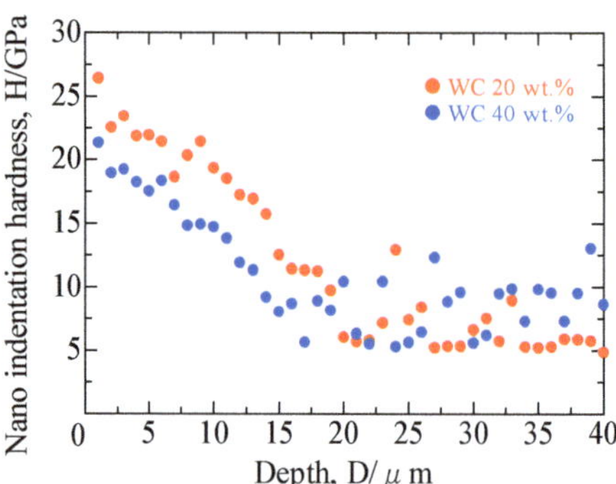

Figure 7. Hardness depth profiles of the S-phase layers at the nitriding temperature of 450 °C with various WC compositions.

The Vickers hardness values of the surfaces of the WC composite layers are shown in Figure 8. The Vickers indentations were aimed at the surfaces of the AISI 316 L steel matrix, except for the WC particle location. The average hardness of the as-deposited 20 wt.% and 40 wt.% WC layers was 540 HV and 654 HV, respectively. Meanwhile, the single AISI 316 L steel layer without WC was 205 HV [30]. Therefore, the composite of WC particles improved the hardness of the as-deposited WC composite layers by more than 300 HV. The reason for this improvement owes to the formation of secondary carbides and the dissolved solid solution of tungsten. In addition, the hardness of the as-deposited 40 wt.% layer was 100 HV harder than that of the WC 20 wt.% layer because more secondary carbides existed.

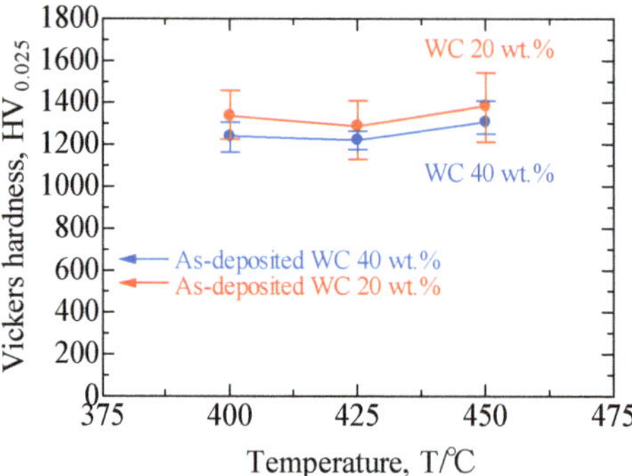

Figure 8. Vickers hardness of the surfaces of the as-deposited and nitrided layers with various WC compositions and nitriding temperatures.

In contrast, the average hardness of the S-phase at a nitriding temperature of 450 °C was 1384 HV for the 20 wt.% layer and 1308 HV for the WC 40 wt.% layer. This hardness was almost the same as that of the S-phase on the single AISI 316 L steel layer without WC [30]. As mentioned above, the hardness of the secondary carbides was lower than that of the nitriding S-phase. Therefore, after plasma nitriding, the secondary carbides and solid solution of tungsten are not effective in improving the hardness. On the contrary, the WC 40 wt.% layers had a low hardness compared to the WC 20 wt.% layers, suggesting that an increase in WC content decreases the S-phase hardness because the many secondary carbides with low hardness are present.

In consequence, the composite of WC particles only improved the hardness of as-deposited AISI 316 L steel matrix, not improved the S-phase hardness. However, the nitrocarburizing S-phase was formed by the composite of the WC particles, that has the advantage of smoothing the hardness depth distribution, which is expected to restrict the delamination of the S-phase layer from the substrate when an external force is applied [50].

3.5. Corrosion Resistance of Nitrocarburizing S-Phase

Figure 9 shows the anodic polarization curves of the as-deposited and nitrided layers. The as-deposited WC 20 wt.% layer had a narrower passive region, and the WC 40 wt.% layer had no passive region.

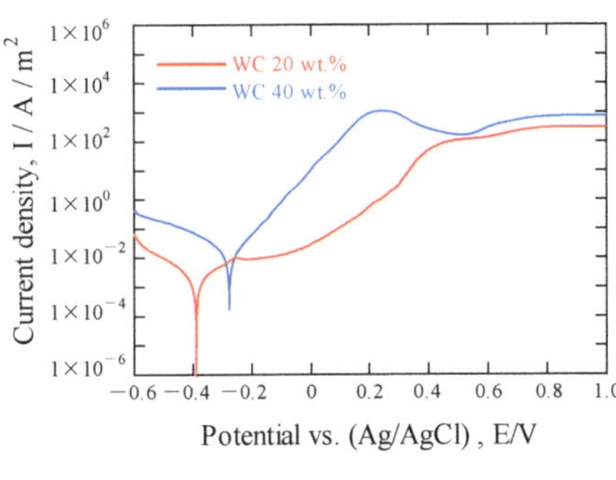

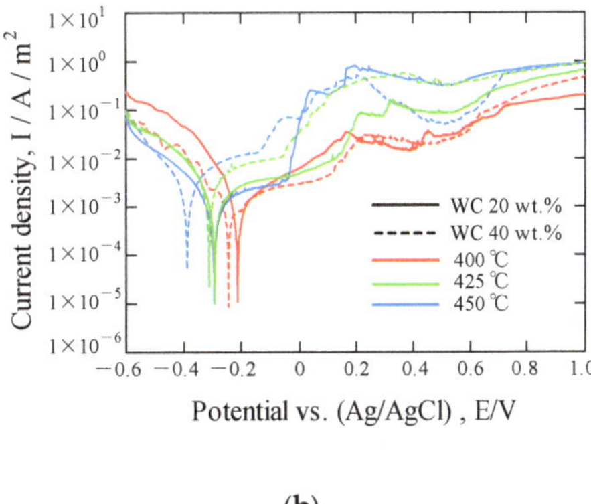

Figure 9. Anodic polarization curves in 3.5 mass% NaCl solution of the as-deposited and nitrided layers with various WC compositions and nitriding temperatures: (**a**) of as-deposited layers, and (**b**) of nitrided layers.

The absence of a passive region implies a low corrosion resistance. These results indicate that the WC composite layers were inferior to the single AISI 316 L steel layer without WC [30]. In general, the effect of alloying elements on the pitting resistance of various types of stainless steels reported that Cr, Mo, W, and N strongly influence the pitting resistance equivalent numbers (PREN) as per Equation (1) [51].

$$PREN = \%Cr + 3.3 \times (\%Mo + 0.5\%W) + 16 \times \%N \tag{1}$$

The solid solution of tungsten in the AISI 316 L steel matrix can improve the PREN. However, elemental analysis of SEM-EDX indicated that the compositions of chromium were approximately 11–13 wt.% of both the 20 wt.% and 40 wt.% WC layers, which showed a decrease compared to 17 wt.% of the feeding AISI 316 L steel powder. Passive films on austenitic stainless steels are composed of oxy-hydroxides of iron, chromium, and molybdenum species [52]; in particular, chromium oxy-hydroxides play an important role. Therefore, the decrease in the solid solution of chromium would degrade the passive film. Furthermore, secondary carbides should also prevent the uniform formation of the passive film on the AISI 316 L steel matrix. Therefore, the as-deposited WC composite layers exhibited a poor corrosion resistance.

In contrast, the nitrided WC composite layers had a passive region, and the corrosion current density was generally lower than that of the as-deposited layers. Figure 10 shows that the surfaces of the as-deposited layers exhibited pitting corrosion around the boundary of the tested area (5 mm × 5 mm), while the nitrided layers were discolored slightly, and no pitting corrosion was observed. Note that the black dots visible on the nitrided layers were not pitting corrosion, but WC particles. The dissolved nitrogen greatly improves the PREN; therefore, the formation of the S-phase improved the corrosion resistance.

The corrosion current density of the 40 wt.% layers was higher than that of the WC 20 wt.% layers because containing more secondary carbides would increase the corrosion current. In addition, the XRD pattern in Figure 2b indicates small CrN peaks in WC 40 wt.% layers treated at nitriding temperatures of 425 °C and 450 °C. The formation of CrN can also decrease the solid solution of chromium to deteriorate the corrosion resistance. Therefore, increasing the WC content would deteriorate the corrosion resistance. For the same reason, the S-phase of the WC composite layers was inferior to that of the single AISI 316 L steel layer without WC [30].

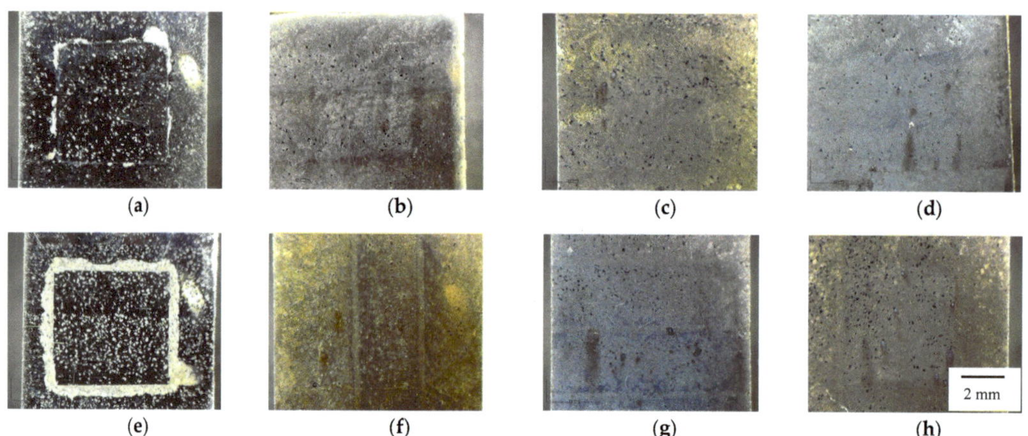

Figure 10. Surface morphologies after anodic polarization measurements of the as-deposited and nitrided layers with various WC compositions and nitriding temperatures: (**a**) WC 20 wt.% with as-deposited, (**b**) WC 20 wt.% at 400 °C, (**c**) WC 20 wt.% at 425 °C, (**d**) WC 20 wt.% at 450 °C, (**e**) WC 40 wt.% with as-deposited, (**f**) WC 40 wt.% at 400 °C, (**g**) WC 40 wt.% at 425 °C, and (**h**) WC 40 wt.% at 450 °C.

3.6. Summary and Perspectives

Nitriding as heat treatments has been applied to steel materials and has been highly effective in improving their properties. On the other hand, several papers about nitriding for stainless steel and other steel materials produced by laser cladding, selective laser melting and powder bed fusion, have been published on the improvement of wear and other properties [53–56]. In this study, it was found for the first time that the WC composite stainless steel layers have dendritic secondary carbides and dissolved tungsten, but the S-phase is formed by low-temperature plasma nitriding. In addition, not only a nitriding S-phase was formed, but also a carburizing S-phase was simultaneously formed by free carbon decomposed from WC particles. Owing to the nitrocarburizing, the S-phase thickened, and the hardness depth profiles were improved. This nitrocarburizing S-phase formed by nitriding alone is a unique phenomenon.

Recently, laser cladding and DED can be applied for an additive manufacturing as well as coatings. In these applications, there is a high demand for improved properties such as wear and corrosion resistance. As for steel materials, various heat treatment technologies have been developed and put into practical use to improve the characteristics. It is expected that heat treatment technologies will be applied to laser cladding and DED, contributing to the industry by expanding its applications.

4. Conclusions

In this study, we attempted to investigate whether low-temperature plasma nitriding applied to improving the hardness and corrosion resistance of AISI 316 L stainless steel-based WC composite layers fabricated by laser cladding. The effects of tungsten and tungsten carbides on the S-phase, which has not yet been clarified, were investigated. During laser cladding process, free carbon was produced by the decomposition of WC particles. Then, we attempted to form a nitrocarburizing S-phase by plasma nitriding alone using this free carbon. For these purposes, the metallurgical structure changes in the AISI 316 L stainless steel by WC particle dissolution during laser cladding, were examined in detail. The formation mechanism of the S-phase was discussed, and the hardness and corrosion resistance were evaluated.

1. W_2C, M_6C, and $M_{12}C$ were synthesized in the as-deposited layers during the laser cladding process because the WC particles decomposed and reacted with the AISI 316 L steel matrix. The AISI 316 L steel matrix became a dendritic structure, and secondary carbides were distributed in a network. In addition, tungsten was dissolved in the austenite of the steel matrix;
2. Free carbon was also produced during the laser cladding process. By low-temperature plasma nitriding, this free carbon was pushed inward by the diffused nitrogen, forming a carburizing S-phase. Consequently, the S-phase became a dual layer, the nitriding layer sited at the outside, and the carburizing layer at the inside;
3. Owing to the nitrocarburizing, the thickness of the S-phase was greater than that of the single AISI 316 L steel layer without WC. In the WC 40 wt.% layers, secondary carbides densely precipitated, and the grain size was small, which would prevent the nitrogen diffusion. As a result, the WC 20 wt.% layers had thick S-phase compared with the WC 40 wt.% layers;
4. The Vickers hardness of the surfaces was improved to the range of 1200 to 1400 HV by low-temperature plasma nitriding. This hardness was almost the same as that of the S-phase on the single AISI 316 L steel layer without WC. The WC 40 wt.% layers were of a low hardness compared with the WC 20 wt.% layers, because the many secondary carbides with a low hardness were present. Meanwhile, the hardness depth profiles were improved to gradually decrease in the depth direction by the nitrocarburizing S-phase. This is expected to enhance the wear resistance and prevent the delamination of the S-phase layer from the substrate;
5. The corrosion resistance of the as-deposited layers was degraded due to the WC particle composite. In contrast, the nitrided layers exhibited a low current density, and those surfaces after the anodic polarization measurements showed a slight color change, with no major damage such as pitting corrosion. Accordingly, low-temperature plasma nitriding of the WC composite layers improved the corrosion resistance to a generally excellent degree. The corrosion current density of the 40 wt.% layers was higher than that of the WC 20 wt.% layers because containing more secondary carbides and CrN.

Author Contributions: Conceptualization, S.A.; methodology, S.A.; investigation, S.A., T.Y. and N.U.; resources, S.A.; data curation, S.A.; formal analysis, S.A.; writing—original draft preparation, S.A.; writing—review and editing, S.A.; visualization, S.A.; supervision, S.A.; project administration, S.A.; funding acquisition, S.A. All authors have read and agreed to the published version of the manuscript.

Funding: This research was funded by JSPS KAKENHI, grant number 18K04792.

Conflicts of Interest: The authors declare no conflict of interest.

References

1. Desale, G.R.; Paul, C.P.; Gandhi, B.K.; Jain, S.C. Erosion wear behavior of laser clad surfaces of low carbon austenitic steel. *Wear* **2009**, *266*, 975–987. [CrossRef]
2. Bandyopadhyay, A.; Zhang, Y.; Bose, S. Recent developments in metal additive manufacturing. *Curr. Opin. Chem. Eng.* **2020**, *28*, 96–104. [CrossRef] [PubMed]
3. DebRoy, T.; Wei, H.L.; Zuback, J.S.; Mukherjee, T.; Elmer, J.W.; Milewski, J.O.; Beese, A.M.; Wilson-Heid, A.; De, A.; Zhang, W. Additive manufacturing of metallic components—Process, structure and properties. *Prog. Mater. Sci.* **2018**, *92*, 112–224. [CrossRef]
4. Smelov, V.G.; Sotov, A.V.; Murzin, S.P. Particularly selective sintering of metal powders by pulsed laser radiation. *Key Eng. Mater.* **2016**, *685*, 403–407. [CrossRef]
5. Toyserkani, E.; Khajepour, A.; Corbin, S. 3-D finite element modeling of laser cladding by powder injection: Effects of laser pulse shaping on the process. *Opt. Lasers Eng.* **2004**, *41*, 849–867. [CrossRef]
6. El Cheikh, H.; Courant, B.; Branchu, S.; Huang, X.; Hascot, J.Y.; Guillen, R. Direct Laser Fabrication process with coaxial powder projection of 316L steel. Geometrical characteristics and microstructure characterization of wall structure. *Opt. Lasers Eng.* **2012**, *50*, 1779–1784. [CrossRef]

7.   Thompson, S.M.; Bian, L.; Shamsaei, N.; Yadollahi, A. An overview of Direct Laser Deposition for additive manufacturing; Part I: Transport phenomena, modeling and diagnostics. *Addit. Manuf.* **2015**, *8*, 36–62. [CrossRef]

8.   Moradi, M.; Ashoori, A.; Hasani, A. Additive manufacturing of stellite 6 superalloy by direct laser metal deposition—Part 1: Effects of laser power and focal plane position. *Opt. Laser Technol.* **2020**, *131*, 106328. [CrossRef]

9.   Zhu, L.; Xue, P.; Lan, Q.; Meng, G.; Ren, Y.; Yang, Z.; Xu, P.; Liu, Z. Recent research and development status of laser cladding: A review. *Opt. Laser Technol.* **2021**, *138*, 106915. [CrossRef]

10. Majumdar, J.D.; Pinkerton, A.; Liu, Z.; Manna, I.; Li, L. Mechanical and electrochemical properties of multiple-layer diode laser cladding of 316L stainless steel. *Appl. Surf. Sci.* **2005**, *247*, 373–377. [CrossRef]

11. Bartkowski, D.; Młynarczak, A.; Piasecki, A.; Dudziak, B.; Gościański, M.; Bartkowska, A. Microstructure, microhardness and corrosion resistance of Stellite-6 coatings reinforced with WC particles using laser cladding. *Opt. Laser Technol.* **2015**, *68*, 191–201. [CrossRef]

12. Guo, C.; Zhou, J.; Chen, J.; Zhao, J.; Yu, Y.; Zhou, H. High temperature wear resistance of laser cladding NiCrBSi and NiCrBSi/WC-Ni composite coating. *Wear* **2011**, *270*, 492–498. [CrossRef]

13. Zhou, S.; Dai, X.; Zheng, H. Microstructure and wear resistance of Fe-based WC coating by multi-track over lapping laser induction hybrid rapid cladding. *Opt. Laser Technol.* **2012**, *44*, 190–197. [CrossRef]

14. Mertens, A.; L'Hoest, T.; Magnien, J.; Carrus, R.; Lecomte-Beckers, J. On the elaboration of metal-ceramic composite coatings by laser cladding. *Mater. Sci. Forum* **2017**, *879*, 1288–1293. [CrossRef]

15. Mertens, A.I.; Lecomte-Beckers, J. On the role of interfacial reactions, dissolution and secondary precipitation during the laser additive manufacturing of metal matrix composites: A review. *New Trends 3D Print.* **2016**, 187–213. [CrossRef]

16. Wang, Z.; Tan, M.; Wang, J.; Zeng, J.; Zhao, F.; Xiao, X.; Xu, S.; Liu, B.; Gong, L.; Sui, Q.; et al. Core-shell structural iron based metal matrix composite powder for laser cladding. *J. Alloys Compd.* **2021**, *878*, 160127. [CrossRef]

17. Xiao, Q.; Sun, W.; Yang, K.; Xing, X.; Chen, Z.; Zhou, H.; Lu, J. Wear mechanisms and micro-evaluation on WC particles investigation of WC-Fe composite coatings fabricated by laser cladding. *Surf. Coat. Technol.* **2021**, *420*, 127341. [CrossRef]

18. Majumdar, J.D.; Kumar, A.; Li, L. Direct laser cladding of SiC dispersed AISI 316L stainless steel. *Tribol. Int.* **2009**, *42*, 750–753. [CrossRef]

19. Wu, C.L.; Zhang, S.; Zhang, C.H.; Zhang, J.B.; Liu, Y. Formation mechanism and phase evolution of in situ synthesizing TiC-reinforced 316L stainless steel matrix composites by laser melting deposition. *Mater. Lett.* **2018**, *217*, 304–307. [CrossRef]

20. Pejakovic, V.; Berger, L.M.; Thiele, S.; Rojacz, H.; Ripoll, M.R. Fine grained titanium carbonitride reinforcements for laser deposition processes of 316L boost tribocorrosion resistance in marine environments. *Mater. Des.* **2021**, *207*, 109847. [CrossRef]

21. Fetni, S.; Enrici, T.M.; Niccolini, T.; Tran, S.H.; Dedry, O.; Jardin, R.; Duchene, L.; Mertens, A.; Habraken, A.M. 2D thermal finite element analysis of laser cladding of 316L + WC composite coatings. *Procedia Manuf.* **2020**, *50*, 86–92. [CrossRef]

22. Enrici, T.M.; Dedry, O.; Boschini, F.; Tchuindjang, J.T.; Mertens, A. Microstructural and thermal characterization of 316L + WC composite coatings obtained by laser cladding. *Adv. Eng. Mater.* **2020**, *22*, 2000291. [CrossRef]

23. Fetni, S.; Enrici, T.M.; Niccolini, T. Thermal model for the directed energy deposition of composite coatings of 316L stainless steel enriched with tungsten carbides. *Mater. Des.* **2021**, *204*, 109661. [CrossRef]

24. Lu, J.Z.; Caa, J.; Lu, H.F.; Zhang, L.Y.; Luo, K.Y. Wear properties and microstructural analyses of Fe-based coatings with various WC contents on H13 die steel by laser cladding. *Surf. Coat. Technol.* **2019**, *369*, 228–237. [CrossRef]

25. Bartkowski, D.; Bartkowska, A.; Jurci, P. Laser cladding process of Fe/WC metal matrix composite coatings on low carbon steel using Yb: YAG disk laser. *Opt. Laser Technol.* **2021**, *136*, 106784. [CrossRef]

26. Wang, J.; Li, L.; Tao, W. Crack initiation and propagation behavior of WC particles reinforced Fe-based metal matrix composite produced by laser melting deposition. *Opt. Laser Technol.* **2016**, *82*, 170–182. [CrossRef]

27. Adachi, S.; Ueda, N. Formation of S-phase layer on plasma sprayed AISI 316L stainless steel coating by plasma nitriding at low temperature. *Thin Solid Films* **2012**, *523*, 11–14. [CrossRef]

28. Adachi, S.; Ueda, N. Surface hardness improvement of plasma-sprayed AISI 316L stainless steel coating by low-temperature plasma carburizing. *Adv. Powder Technol.* **2013**, *24*, 818–823. [CrossRef]

29. Adachi, S.; Ueda, N. Formation of expanded austenite on a cold-sprayed AISI 316L coating by low-temperature plasma nitriding. *J. Therm. Spray Technol.* **2015**, *24*, 1399–1407. [CrossRef]

30. Adachi, S.; Egawa, M.; Yamaguchi, T.; Ueda, N. Low-temperature plasma nitriding for austenitic stainless steel layers with various nickel contents fabricated via direct laser metal deposition. *Coatings* **2020**, *10*, 365. [CrossRef]

31. Borgioli, F. From austenitic stainless steel to expanded austenite-S phase: Formation, characteristics and properties of an elusive metastable phase. *Metals* **2020**, *10*, 187. [CrossRef]

32. Borgioli, F.; Galvanetto, E.; Bacci, T. Corrosion behaviour of low temperature nitrided nickel-free, AISI 200 and AISI 300 series austenitic stainless steels in NaCl solution. *Corros. Sci.* **2018**, *136*, 352–365. [CrossRef]

33. Huang, Z.; Guo, Z.X.; Liu, L.; Guo, Y.Y.; Chen, J.; Zhang, Z.; Li, J.L.; Li, Y.; Zhou, Y.W.; Liang, Y.S. Structure and corrosion behavior of ultra-thick nitrided layer produced by plasma nitriding of austenitic stainless steel. *Surf. Coat. Technol.* **2021**, *405*, 126689. [CrossRef]

34. Liu, H.Y.; Che, H.L.; Li, G.B.; Lei, M.K. Low-pressure hollow cathode plasma source carburizing technique at low temperature. *Surf. Coat. Technol.* **2021**, *422*, 127511. [CrossRef]

35. Li, L.; Yan, J.; Xiao, J.; Sun, L.; Fan, H.; Wang, J. A comparative study of corrosion behavior of S-phase with AISI 304 austenitic stainless steel in $H_2S/CO_2/Cl$- media. *Corros. Sci.* **2021**, *187*, 109472. [CrossRef]
36. Sun, Y. Hybrid plasma surface alloying of austenitic stainless steels with nitrogen and carbon. *Mater. Sci. Eng. A* **2005**, *404*, 124–129. [CrossRef]
37. Adachi, S.; Ueda, N. Combined plasma carburizing and nitriding of sprayed AISI 316L steel coating for improved wear resistance. *Surf. Coat. Technol.* **2014**, *259*, 44–49. [CrossRef]
38. Adachi, S.; Ueda, N. Wear and corrosion properties of cold-sprayed AISI 316L coatings treated by combined plasma carburizing and nitriding at low temperature. *Coatings* **2018**, *8*, 456. [CrossRef]
39. Fernandes, F.A.P.; Casteletti, L.C.; Gallego, J. Microstructure of nitrided and nitrocarburized layers produced on a superaustenitic stainless steel. *J. Mater. Res. Technol.* **2013**, *2*, 158–164. [CrossRef]
40. Menthe, E.; Rie, K.-T.; Schultze, J.W.; Simson, S. Structure and properties of plasma-nitrided stainless steel. *Surf. Coat. Technol.* **1995**, *74*, 412–416. [CrossRef]
41. Buhagiar, J.; Li, X.; Dong, H. Formation and microstructural characterisation of S-phase layers in Ni-free austenitic stainless steels by low-temperature plasma surface alloying. *Surf. Coat. Technol.* **2009**, *204*, 330–335. [CrossRef]
42. Wydorska, K.M.; Kabulska, F.I.; Flis, J. Corrosion of low-temperature nitrided molybdenum-bearing stainless steels. *Corros. Sci.* **2011**, *53*, 1762–1769. [CrossRef]
43. Lee, E.S.; Park, W.J.; Jung, J.Y.; Ahn, S. Solidification microstructure and M2C carbide decomposition in a spray-formed high-speed steel. *Metall. Mater. Trans. A* **1998**, *29*, 1395–1404. [CrossRef]
44. Zhou, S.; Xu, T.; Hu, C.; Wu, H.; Liu, H.; Ma, X. A comparative study of tungsten carbide and carbon nanotubes reinforced Inconel 625 composite coatings fabricated by laser cladding. *Opt. Laser Technol.* **2021**, *140*, 106967. [CrossRef]
45. Bergstrom, M. The eta-carbides in the quaternary system Fe-W-C-Cr at 1250 °C. *Mater. Sci. Eng.* **1977**, *27*, 271–286. [CrossRef]
46. Moller, W.; Parascandola, S.; Telbizova, T.; Gunzel, R.; Richter, E. Surface processes and diffusion mechanisms of ion nitriding of stainless steel and aluminium. *Surf. Coat. Technol.* **2001**, *136*, 73–79. [CrossRef]
47. Riviere, J.P.; Meheust, P.; Villain, J.P.; Templier, C.; Cahoreau, M.; Abrasonis, G.; Pranevicius, L. High current density nitrogen implantation of an austenitic stainless steel. *Surf. Coat. Technol.* **2002**, *158–159*, 99–104. [CrossRef]
48. Czerwiec, T.; Andrieux, A.; Marcos, G.; Michel, H.; Bauer, P. Is "expanded austenite" really a solid solution? Mossbauer observation of an annealed AISI 316L nitrided sample. *J. Alloys Compd.* **2019**, *811*, 151972–151983. [CrossRef]
49. Martínez, S.; Lamikiz, A.; Ukar, E.; Calleja, A.; Arrizubieta, J.A.; Lopez de Lacalle, L.N. Analysis of the regimes in the scanner-based laser hardening process. *Opt Lasers Eng.* **2017**, *90*, 72–80. [CrossRef]
50. Sun, Y.; Haruman, E. Effect of carbon addition on low-temperature plasma nitriding characteristics of austenitic stainless steel. *Vacuum* **2006**, *81*, 114–119. [CrossRef]
51. Okamoto, H. The effect of tungsten and molybdenum on the perfomance of super duplex stainless steels. In Proceedings of the Applications of Stainless Steel '92 Conference, Stockholm, Sweden, 9–11 June 1992; Volume 1, pp. 360–369.
52. Langberg, M.; Ornek, C.; Zhang, F.; Cheng, J.; Liu, M.; Granas, E.; Wiemann, C.; Gloskovskii, A.; Matveyev, Y.; Kulkarni, S. Characterization of native oxide and passive film on Austenite/Ferrite phases of duplex stainless steel using synchrotron HAXPEEM. *J. Electrochem. Soc.* **2019**, *166*, C3336–C3340. [CrossRef]
53. Godec, M.; Donik, Č.; Kocijan, A.; Podgornik, B.; Skobir Balantič, D.A. Effect of post-treated low-temperature plasma nitriding on the wear and corrosion resistance of 316L stainless steel manufactured by laser powder-bed fusion. *Addit. Manuf.* **2020**, *32*, 101000–101008. [CrossRef]
54. Dai, S.; Zuo, D.; Fang, C.; Zhu, L.; Cheng, H.; Gao, Y.-X.; Li, W.-W. Characterization of laser cladded Fe-Mn-Cr alloy coatings modied by plasma nitriding. *Mater. Trans.* **2016**, *57*, 539–543. [CrossRef]
55. Hong, Y.; Dong, D.D.; Lin, S.S.; Wang, W.C.; Tang, M.; Kuang, T.C.; Dai, M.J. Improving surface mechanical properties of the selective laser melted 18Ni300 maraging steel via plasma nitriding. *Surf. Coat. Technol.* **2021**, *406*, 126675. [CrossRef]
56. Li, B.; Zhu, H.; Qiu, C.; Gong, X. Laser cladding and in-situ nitriding of martensitic stainless steel coating with striking performance. *Mater. Lett.* **2020**, *259*, 126829. [CrossRef]

Article

Cold Gas Spraying of Solution-Hardened 316L Grade Stainless Steel Powder

Thomas Lindner [1],*, Martin Löbel [1], Maximilian Grimm [1] and Jochen Fiebig [2]

[1] Materials and Surface Engineering Group, Institute of Materials Science and Engineering, Chemnitz University of Technology, D-09107 Chemnitz, Germany; martin.loebel.ww@gmail.com (M.L.); maximilian.grimm@mb.tu-chemnitz.de (M.G.)

[2] Forschungszentrum Jülich, Institute of Energy and Climate Research (IEK-1), D-52428 Julich, Germany; j.fiebig@fz-juelich.de

* Correspondence: th.lindner@mb.tu-chemnitz.de

Abstract: Austenitic steels are characterized by their outstanding corrosion resistance. They are therefore suitable for a wide range of surface protection requirements. The application potential of these stainless steels is often limited by their poor wear resistance. In the field of wrought alloys, interstitial surface hardening has become established for simultaneously acting surface stresses. This approach also offers great potential for improvement in the field of coating technology. The hardening of powder feedstock materials promises an advantage in the treatment of large components and also as a repair technology. In this work, the surface hardening of AISI 316L powder and its processing by thermal spraying is presented. A partial formation of the metastable expanded austenitic phase was observed for the powder particles by low-temperature gas nitrocarburizing. The successful deposition was demonstrated by cold gas spraying. The amount of expanded austenitic phase within the coating structure strongly depends on the processing conditions. Microstructure, corrosion and wear behavior were studied. Process diagnostic methods were used to validate the results.

Keywords: thermal spray; cold gas spraying; cold spray; thermochemical treatment; powder; austenitic stainless steel; expanded austenite; S-phase

check for updates

Citation: Lindner, T.; Löbel, M.; Grimm, M.; Fiebig, J. Cold Gas Spraying of Solution-Hardened 316L Grade Stainless Steel Powder. *Metals* **2022**, *12*, 30. https://doi.org/10.3390/met12010030

Academic Editors: Pasquale Cavaliere and Wei Zhou

Received: 19 November 2021
Accepted: 22 December 2021
Published: 24 December 2021

Publisher's Note: MDPI stays neutral with regard to jurisdictional claims in published maps and institutional affiliations.

1. Introduction

The excellent corrosion resistance of stainless steels contributes to several wide-ranging surface protection applications. Nickel stabilizes the paramagnetic austenitic grades that offer the best performance. Because of the comparatively soft and ductile matrix, similar material pairings have to be avoided due to the risk of cold welding. In addition, tribological loads often cause short service lives of the components. In order to provide adequate surface protection under superimposed stresses, the property profile of these materials can be improved by solution hardening. For applications where an adequate corrosion resistance is required, interstitial hardening is preferred over precipitation hardening. Carbon or nitrogen diffuses into the interstices, while chromium is still dissolved in the solid solution [1,2]. Hence, the corrosion resistance is maintained. The elemental enrichment of the surface layer results in strong compressive residual stresses [3]. Additionally, a significant increase in hardness and wear resistance is achieved. A successful enrichment has already been demonstrated in wrought alloys [1–3]. The first results on precipitation hardening by thermochemical treatment of thermal spray coatings were obtained by Nestler et al., while Wielage et al. demonstrated interstitial hardening of coating systems [4,5]. Additional studies demonstrate the general feasibility of coating treatment by thermochemical processes in the low-temperature range [6–12]. While the success of diffusion treatment generally depends on the surface activation, Lindner et al. demonstrated that this is not required in the case of thermally sprayed AISI 316L coatings. The heterogeneous formation of the passive layer is stated as a possible reason [11]. The expanded austenitic layer in porous coating systems can exceed the thickness achieved for wrought material. This is related to the penetration of gaseous treatment media [12]. With regard to the lattice parameters, a similar lattice expansion was demonstrated for thermally sprayed coatings compared to wrought alloys.

An alternative approach for the treatment of components is the solution hardening of feedstock materials for the coating process. Particle diameters below 60 μm allow for a complete interstitial diffusion enrichment. The successful powder treatment and its processing promises solutions in the field of repair of worn components with an expanded austenitic surface layer. For large components, an economical opportunity for partial or complete coating can be developed. Various approaches to nitride steel powder have been investigated [13,14]. Lindner et al. successfully demonstrated S-phase formation in austenitic steel powder. In order to retain the metastable phase, further processing of the powder requires a sufficiently low process temperature. The subsequent powder processing by thermal spraying revealed the transfer of the expanded austenite to the coating. In comparison with atmospheric-plasma-sprayed and high-velocity-flame-sprayed AISI 316L coatings, it was shown that increasing thermal energy input reduces the lattice expansion [15]. However, an expanded austenitic phase was still detected for both coating systems. A suitable method for the further reduction in the thermal load of the feedstock material is cold gas spraying (CGS). A process gas, typically nitrogen, helium or a mixture of both, with a temperature of several hundred degrees and a pressure of a few MPa is used to accelerate the particles to velocities between 400 and 1200 m/s. In contrast to other thermal spray techniques, the powder particles possess a high kinetic energy but moderate temperatures and are solid during the impact on the substrate. The relatively low process temperature offers several advantages: (i) in-flight oxidation of powder particles is avoided; (ii) phase transformation, e.g., the formation of brittle phases, is hindered; (iii) low thermal stresses are present in cold-sprayed coatings [16–18]. CGS coating formation depends on the deformation capacity of the powder feedstock material. A widely accepted explanation for the bonding mechanism in cold gas spray is the adiabatic shear instability as described by Assadi et al. [19] and Grujicic et al. [20]. For a successful deposition, the particles have to exceed a material-dependent critical velocity. A generalized deposition window, determined by critical and erosion velocity, was introduced by Schmidt et al. [21,22] based on fluid dynamics simulations, finite element analysis and experimental validation. Analytical descriptions of for the material- and particle-size-dependent critical velocities were developed and allow us to estimate the deposition window for a variety of metals and alloys [19,21–23]. Austenitic steels are ideally suited due to their ductile behavior. Several authors [8,24,25] have studied cold gas spraying of AISI 316L steel, and different aspects were investigated. Al-Mangour et al. analyzed the influence of heat treatments on the microstructure and mechanical properties of cold-sprayed AISI 316L steel, which demonstrated an increase in tensile strength and ductility after heat treatments [24]. Furthermore, the authors found dense coatings with low porosity when helium was used as a propellant gas. Xie et al. found an increase in the adhesion strength by increasing the substrate pre-heating temperature [25]. The effects of gas temperature and gas pressure of nitrogen on the deposition behavior of AISI 316L coatings were studied by Adachi and Ueda. An increase in gas temperature and gas pressure significantly reduces coating porosity, but at high gas temperatures slightly reduces the coating hardness [8].

In the presented study, precipitation-free hardening by interstitial solvation of carbon and nitrogen was performed for the feedstock material in combination with coating production by CGS. This work contributes to a better understanding of the requirements of surface-hardened powders for CGS by linking the processing and coating properties.

2. Materials and Methods

Gas-atomized AISI 316L powder (GTV Verschleissschutz GmbH, Luckenbach, Germany) with a specified particle size of −53 + 20 μm was used. The particle size distribution was examined by laser diffraction analysis in a Cilas 930 device (Cilas, Orléans, France). In order to obtain an expanded austenitic phase modification by interstitial diffusion enrichment, an industrial low-temperature gas nitrocarburization treatment was conducted at a temperature below 450 °C and a duration time of 30 h. Because of the low weight of the powder particles, the risk of them whirling within the gas flow increased. Therefore, the powder was added to a vessel up to a fill level of 50 mm. The coatings were deposited on stainless steel AISI 316L sheets with a thickness of 2 mm. The substrates were ground with SiC paper up to a 1200 mesh and subsequently polished with a diamond solution in the final step with 3 μm diamond solution. Subsequently, the substrates were cleaned in an ultrasonic ethanol bath for 10 min. Coating deposition was conducted using an Impact 5/11 high-pressure cold gas spray system (Impact Innovations GmbH, Rattenkirchen, Germany) equipped with the standard water-cooled D-24 de-Laval-type converging–diverging nozzle. Nitrogen and helium were used as propellant gas. The inlet gas pressure and the temperature were varied in the range of 4–5 MPa and 850–1100 °C, respectively. The standoff distance from the nozzle exit to the substrate surface was 30 mm. The coating was applied in five passes with a spray angle of 90°. Variations in process parameters are summarized in Table 1. The parameter selection is based on publications on cold gas spraying of AISI 316L steel [8,24,25].

Table 1. Cold gas spraying (CGS) parameters of AISI 316L with Impact 5/11 system.

Temperature (°C)	Pressure (Bar)	N_2/He Ratio (%/%)
850	40	100/0
950	40	100/0
1100	50	100/0
1100	50	75/25

A Cold Spray Meter (CMS) system (Tecnar Automation Ltée, Saint-Bruno-de-Montarville, QC, Canada) was used to determine the state variables of the feedstock during processing. The particles are irradiated in-flight by a laser beam with a wavelength of 790 nm and a power of 3.3 W. An optical sensor detects the light reflected by the particle surface. In front of the optical sensor, a double-slit mask is placed. If a particle passes in front of the mask, the reflected laser light will result in a double-peak signal. The time of flight (TOF) of the particles is the time passed between the maxima of the two peaks. With the known dimensions of the mask, namely the distance between the middle of the two slits, the magnification of the lens and the TOF of the particle velocity can be calculated. Further details can be found in literature [26–28]. For a stand-off distance of 30 mm, the analysis of the particle velocity was performed on the basis of 10,000 individual particles for all process parameters listed in Table 1. The particle velocities were filtered between 400 and 1200 m/s. In addition, the thermal and kinematic states during impact were determined as a function of particle size, using the KSS software.

The surface of the coated samples was ground up to mesh 1000, resulting in a final coating thickness of approximately 200 μm. Metallographic cross-sections were prepared according to standard metallographic procedures. For the visualization of different microstructural domains, the cross-sections were etched using a Beraha-II color etchant. The resulting contrast enables the evaluation of diffusion-enriched areas regarding their presence and distribution. For the metallographic investigations, an optical microscope GX51 equipped with a SC50 camera (Olympus, Shinjuku, Japan) was used.

A phase-selective hardness measurement was carried out by nanoindentation. For this, a Fischerscope HM 2000 XYp (Helmut Fischer GmbH, Sindelfingen, Germany) with a Vickers tip was used for progressive measurement with a load of 50 mN. Borosilicate glass BK7 was used as the calibration standard. Phase analyses were conducted by X-ray diffraction (XRD) using a D8 DISCOVER diffractometer (Bruker AXS, Billerica, MA, USA) with Co-K$_\alpha$ radiation (tube voltage: 40 kV; tube current: 40 mA). The diffractometer was equipped with polycap optics for beam shaping, a 1 mm pinhole collimator and a 1D Lynxeye XE (Bruker AXS, Billerica, MA, USA) detector. All of the diffractograms were measured in a diffraction angle (2θ) range from 20° to 120°, with a step size of 0.01° and

1.5 s/step. Because of the use of the 1D detector, this value corresponds to 288 s/step. The powder diffraction file (PDF) database from 2014 was applied for the phase identification and for the determination of the lattice parameters.

In order to investigate tribological behavior under abrasive wear conditions, scratch tests were carried out using a CSM Revetest-RST device (CSM Instruments SA, Peseux, Switzerland). The applied parameters are summarized in Table 2. Tactile measurements using the truncated diamond cone tip of the wear-testing device with a load of 0.9 N were performed in order to measure the wear marks.

Table 2. Scratch test parameters.

Mode	Force	Speed	Length	Tip	Radius
progressive	1–200 N	2.5 mm/min	5 mm	truncated diamond cone	200 μm

The corrosion behavior was investigated by polarization curves, with the aim of studying the effects of different coating microstructures on corrosion current density. A round surface with a diameter of 10 mm was tested with a three-electrode arrangement. A newly developed 0.6 M NaCl gel electrolyte with a custom designed corrosion cell was used for the measurements. A platinum sheet with dimensions (2×2 cm^2) was used as the counter electrode and a platinum wire was used as the pseudo-reference electrode. After a 30 min measurement of the open-circuit potential (OCP), the polarization curves were examined at a scanning speed of 1 mV/s in the potential range from -100 mV to $+500$ mV based on the OCP. The experiments were performed at room temperature and were recorded with a potentiometer Zennium (Zahner-Elektrik GmbH & Co. KG, Kornach, Germany). The polarization curves were evaluated by an MatLab script using the Butler–Volmer equation to determine i_{corr} and E_{corr}. Due to the immobility of the electrolyte, the corrosion behavior in the immediate surface area was able to be characterized. This is beneficial for a more accurate analysis of real corrosion processes, such as those under biological films, as penetration of the corrosion medium into the coating and contacting of the substrate material can be avoided compared to investigations with liquid electrolytes.

3. Results

Because of interstitial solution hardening, the corrosion behavior of the material changes. Hence, applying a Beraha II etching to the cross-section of the solution-hardened AISI 316L powder reveals the detection of the expanded austenitic phase. Figure 1 shows the etched cross-section of the solution-hardened AISI 316L powder. A multiphase state with phase fractions of the metastable expanded the austenitic phase and the initial fcc phase is detected. A deposition of color only occurs at the initial phase. While some particles exhibit complete interstitial hardening, an enrichment limited to the surface area is detectable in the majority of the powder. This can be attributed to the lack of circulation of the enrichment medium during gas nitrocarburizing in the vessel. In addition, deposits of carbon can be partially found on the particle surface, which further impedes circulation and enrichment. The treatment of the powders caused an increase in the characteristic values of the particle size distribution with $d_{10} = 29$ μm, $d_{50} = 62$ μm and $d_{90} = 43$ μm. The determined hardness values confirm the heterogeneous state of the powder. The results show that the direct transfer of existing treatment routines to powder materials is impossible. Nevertheless, an enrichment of the powder is achieved, which is suitable for investigating the basic suitability for cold gas spraying. A coating deposition could be achieved with all investigated process parameters.

Figure 1. Optical-microscopic image of Beraha-II-etched AISI 316L powder feedstock after gas nitrocarburization.

For example, for the particle sizes of 15 μm and 35 μm, the mean values of the measurement results of the kinematic and thermal state variables are summarized in Table 3. A steady increase in particle temperature and velocity with increasing process temperature can be seen. By adding 25% He to the process gas, the particle temperature at impact can be lowered. Hence, the particle temperatures exceed the stability limit of the expanded austenitic phase for long-term stress. Due to the very short residence times within process gas, no impairment of the phase stability can be assumed.

Table 3. Impact temperature and velocity of the AISI 316L particles depending on the coating parameters during cold gas spraying. The values were calculated with the KSS software. For the experiments with the cold spray meter, the average particle velocities and standard deviations are listed.

Parameter Setting	Impact Speed (m/s)		CSM	Impact Temperature (°C)	
	15 μm	35 μm		15 μm	35 μm
850 °C	706	612	717 ± 106	422	618
950 °C	734	629	775 ± 114	487	699
1100 °C	782	680	796 ± 114	575	814
1100 °C, 25% He	860	726	863 ± 121	489	749

The measured and calculated particle velocities differ significantly from each other, Figure 2. Furthermore, all CSM measurements lead to a standard deviation of more than 100 m/s, see Table 1. Several factors could be responsible for the enhanced particle velocities. Firstly, during the CSM experiments no substrate was placed in front of the nozzle exit. Due to this setup, the bow shock effect, which mainly affects small and light particles at short stand-off distances [29], is avoided and the measured particle velocities should be compared to those in a free gas stream. The effect is small for the used feedstock, since most particles are larger than 29 μm. Furthermore, several studies [30–33] have demonstrated that particle measurements with DPV systems (including CSM) typically lead to higher average velocities than CFD simulations. Increases of up to 8% were reported, corresponding to a velocity enhancement of 40 to 60 m/s for the experiments carried out in this work. Nevertheless, the deviations in the current study are significantly higher, which might be related to the non-spherical shape of a portion of the feedstock, as it can be seen from the cross-section images in Figure 1 and in Figure 2 [15]. A recent study by Özdemir et al. [34] has demonstrated the importance of considering the particle shape when comparing particle velocities obtained by CFD simulations with experimental results from velocimetry. For non-spherical particles, other models for the drag coefficient are incorporated in the calculations, resulting in higher particle velocities. Özdemir et al.

replaced the drag coefficient given by Schiller–Naumann [35] with the drag coefficient of Haider and Levenspiel [36]. This model includes a shape factor, called sphericity $\psi = s/S$. (s = Surface Area of a Spherical Particle of Equivalent Volume)/S = Surface Area of the Particle Under Investigation), which was introduced by Wadell [37]. For a sphere the sphericity parameter is equal to 1, and for non-spherical particles the parameter is smaller than 1.

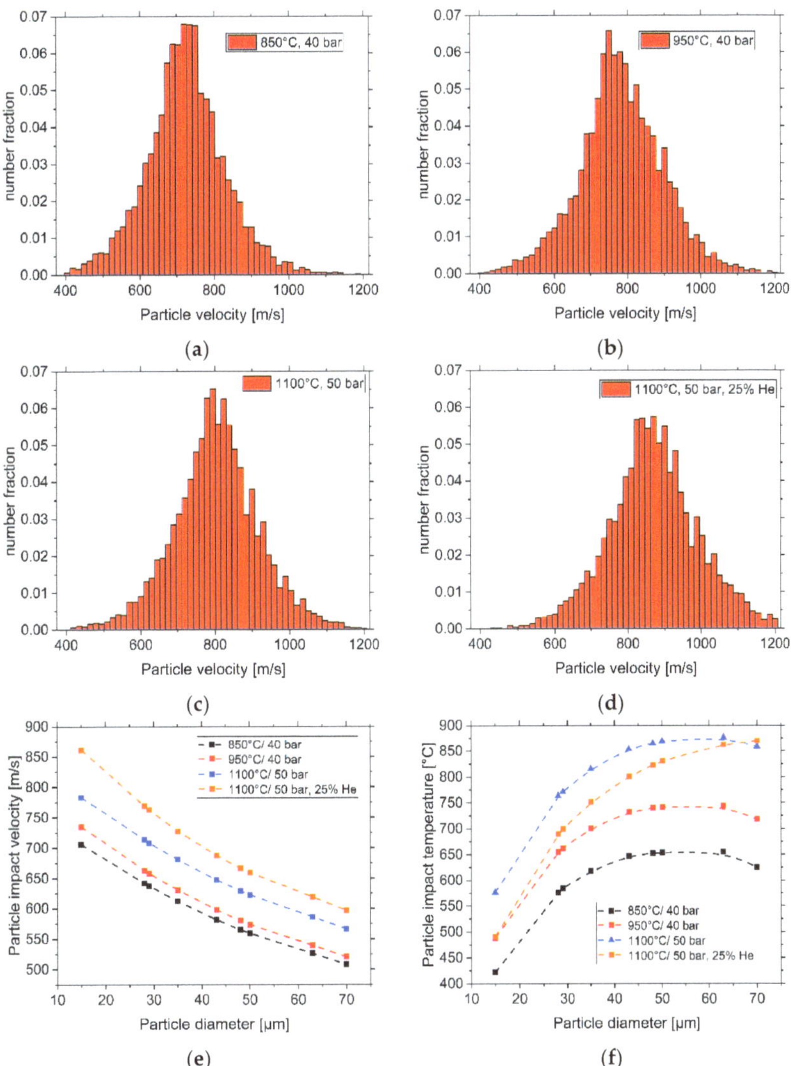

Figure 2. Particle velocity distributions measured with a cold spray meter: (**a**) 850 °C, (**b**) 950 °C, (**c**) 1100 °C and (**d**) 1100 °C and 25% He. The detailed spray parameters can be found in Table 1. The particle impact velocities and particle impact temperatures depending on particle diameter and spray parameters are presented in (**e**,**f**), respectively. For the calculations, KSS software was used. For certain particle diameters, the values can be found in Table 3.

The particle velocities in Table 3 or Figure 2 were calculated with KSS software, assuming spherical particles. For a more detailed understanding of velocimetry data, it is therefore necessary to experimentally determine the sphericity and to use the corresponding drag coefficient for the calculations. Figure 3 shows examples of cross-sections of coating systems produced with different process parameters. A clear contrast between the color-etched austenitic matrix phase and solution-hardened areas without color deposits is still present. While for the coating produced with a process temperature of 850 °C, the phase proportions roughly correspond to those of the powder, a decrease in the proportion of the expanded austenitic phase in the coating structure was observed with an increase in process speed. This can be assumed to be influenced by the different hardness levels in the feedstock powder. The high hardness of particles with the expanded austenitic phase contributes to an increase in the proportion of reflected particles.

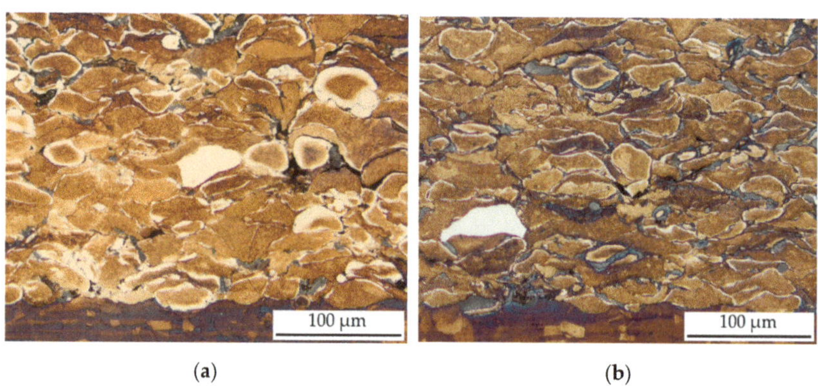

(**a**)   (**b**)

Figure 3. Optical-microscopic image of a Beraha-II-etched AISI 316L CGS coating using (**a**) 850 °C and 40 bar and (**b**) 1100 °C and 50 bar with 25% He in cross-section.

At the same time, the proportion of impurity incorporation of the starting powders in the coating structure increases for a deposition with a process temperature of 1100 °C. This can be explained by the improved formability and low hardness. In contrast to the results for HVOF and APS coatings in Ref. [15], no changes in the phase domains within the individual particles can be observed due to the low thermal load in the process. The nanoindentation measurements showed an increase in hardness for the austenitic phase with increasing process speed. Likewise, the average hardness value increased from 410 HV0.005 for 850 °C, to 460 HV0.005 for 1100 °C with 25% He. While only minor variations were observed, a strong gradient was detectable within the expanded austenitic phase. Accordingly, the hardness increased to values between 510–850 HV0.005.

The results of the XRD measurements are shown in Figure 4. Interstitial enrichment is always accompanied by lattice expansion. Due to the heterogeneous treatment condition, the strongest peak occurs for the fcc phase. Characteristic diffraction patterns of precipitates are not detectable.

Due to the heterogeneous phase state, a broad range with no distinct peak appears at lower diffraction angles. This can be assigned to the expanded austenitic phase. Due to the deformation of the particles during processing, the stress states within the grains change. This results in a broadening of the peaks. Due to the higher process speed, the broadening was at its maximum for the coatings deposited with a process temperature of 1100 °C. Although the intensity in the area of the expanded austenitic phase is relatively low, a decrease with increasing process speed can still be observed. This confirms the assumption that the volume fraction of the expanded austenitic phase decreases compared to the initial powder.

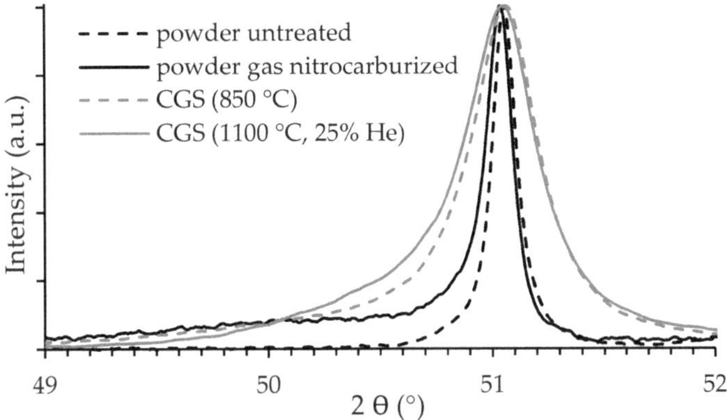

Figure 4. Sections of diffractograms of the untreated and gas-nitrocarburized AISI 316L powders and the CGS coating using 850 °C as well as 1100 °C and 25% He for the main fcc lattice peak.

The wear resistance of the CGS coatings was characterized under abrasive conditions in the scratch tests, Figure 5. With increasing temperature and pressure, an increase in wear resistance can be observed. The best results were obtained using 1100 °C and 50 bar with the addition of 25% He. However, the wear resistance of this coating did not exceed that of the HVOF coating [15]. This can be explained by the decrease in the hardness level of the solution-hardened particles. Nevertheless, the results confirm the potential for the improvement of thermally sprayed austenitic steel coatings by using solution-hardened powder feedstock.

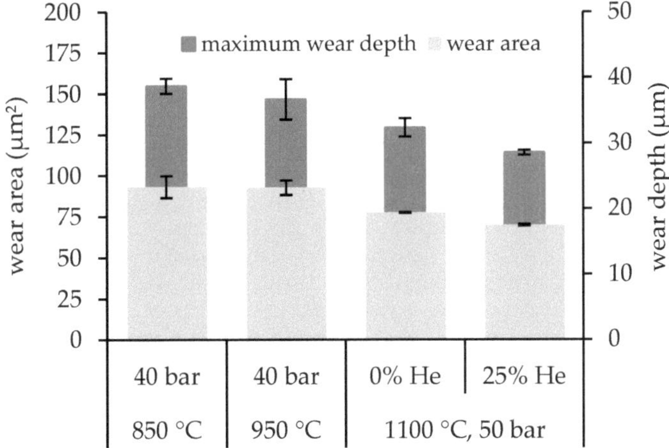

Figure 5. Comparison of the results for wear area and wear depth in scratch tests of CGS coatings in dependence of the process parameters.

The polarization curves also showed a dependence of the spraying parameters on the corrosion behavior, shown in Figure 6. Accordingly, the highest corrosion current densities could be detected in the coating systems that were coated with a process temperature of 850 °C. With an increase in process speed, a continuous decrease in the corrosion current densities could be observed. This can be attributed to the decrease in the volume fractions of the expanded austenitic phase, which exhibits lower resistance compared to the austenitic matrix.

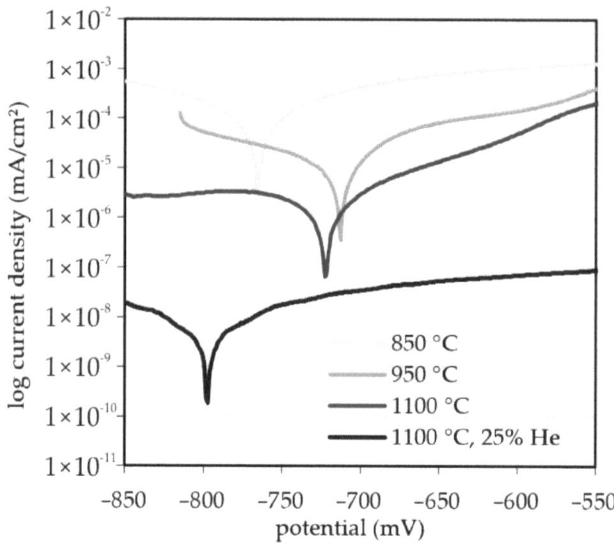

Figure 6. Polarization curves of CGS-sprayed AISI 316L coatings in dependence of the process parameters.

4. Conclusions

Solution-hardened AISI 316L powders were successfully processed by cold gas spraying. The deposition of coatings could be achieved with all investigated process parameters. A heterogeneous hardness level of the different phase fractions could be demonstrated. The hardness of the initial austenitic phase increases with the impact velocity. Due to the low thermal energy input in the coating process, the heterogeneous phase structure of the feedstock could be largely preserved for the coatings, while the phase fraction of expanded austenite decreases with increasing impact velocity of the particles. This indicates a reflection effect of hardened particles. An increased proportion of the austenitic phase is associated with a lower corrosion current density in the polarization measurement. Hence, an impairment of the corrosion resistance by the interstitial solid solution phase can be deduced. With increasing kinetic energy, hardness and wear resistance was improved.

The results of this study confirm the general feasibility of the novel process approach for cold gas spraying. The prerequisite for further investigations is the improvement of the quality of the powder treatment. For this, adjustments to the standard industrial routine are necessary. For example, carbon deposition on the surface can be prevented by pure nitrogen enrichment. The development of suitable cleaning steps is another option. Furthermore, a circulation of the powder or a pressure modulation is suitable to achieve a more uniform enrichment medium within the powder bed.

Author Contributions: Conceptualization, T.L., M.L. and J.F.; methodology, T.L., M.L., M.G. and J.F.; validation and investigation, T.L., M.L., M.G. and J.F.; writing—original draft preparation, T.L. and J.F.; writing—review and editing, M.L. and M.G.; project administration, T.L. and J.F. All authors have read and agreed to the published version of the manuscript.

Funding: No funding was received for this study.

Data Availability Statement: Not applicable.

Acknowledgments: The authors thank Niclas Hanisch for conducting the wear tests and Paul Seidel for support in the metallographic investigation and hardness measurement.

Conflicts of Interest: The authors declare no conflict of interest.

References

1. Zhao, C.; Li, C.X.; Dong, H.; Bell, T. Low temperature plasma nitrocarburising of AISI 316 austenitic stainless steel. *Surf. Coat. Technol.* **2005**, *191*, 195–200. [CrossRef]
2. Bell, T. Current status of supersaturated surface engineered S-phase materials. In *Key Engineering Materials*; Trans Tech Publications: Zurich, Switzerland, 2008; pp. 289–295. [CrossRef]
3. Borgioli, F. From Austenitic Stainless Steel to Expanded Austenite-S Phase: Formation, Characteristics and Properties of an Elusive Metastable Phase. *Metals* **2020**, *10*, 187. [CrossRef]
4. Nestler, M.C.; Spies, H.; Hermann, K. Production of duplex coatings by thermal spraying and nitriding. *Surf. Eng.* **1996**, *12*, 299–302. [CrossRef]
5. Wielage, B.; Rupprecht, C.; Lindner, T.; Hunger, R. Surface modification of austenitic thermal spray coatings by low-temperature carburization. In Proceedings of the International Thermal Spray Conference & Exposition 2011, Hamburg, Germany, 27–29 September 2011; p. 276. Available online: www.researchgate.net/publication/320323004 (accessed on 1 November 2021).
6. Adachi, S.; Ueda, N. Formation of S-phase layer on plasma sprayed AISI 316L stainless steel coating by plasma nitriding at low temperature. *Thin Solid Film* **2012**, *523*, 11–14. [CrossRef]
7. Park, G.; Bae, G.; Moon, K.; Lee, C. Effect of plasma nitriding and nitrocarburinzing on HVOF-sprayed stainless steel coatings. *J. Therm. Spray Technol.* **2013**, *22*, 1366–1373. [CrossRef]
8. Adachi, S.; Ueda, N. Effect of Cold-Spray Conditions Using a Nitrogen Propellant Gas on AISI 316L Stainless Steel-Coating Microstructures. *Coatings* **2017**, *7*, 87. [CrossRef]
9. Adachi, S.; Ueda, N. Wear and Corrosion Properties of Cold-Sprayed AISI 316L Coatings Treated by Combined Plasma Carburizing and Nitriding at Low Temperature. *Coatings* **2018**, *8*, 456. [CrossRef]
10. Lindner, T.; Mehner, T.; Lampke, T. Surface modification of austenitic thermal-spray coatings by low-temperature nitrocarburizing. In *IOP Conference Series: Materials Science and Engineering*; IOP Publishing: Bristol, UK, 2016; Volume 118. [CrossRef]
11. Lindner, T.; Kutschmann, P.; Löbel, M.; Lampke, T. Hardening of HVOF-Sprayed Austenitic Stainless-Steel Coatings by Gas Nitriding. *Coatings* **2018**, *8*, 348. [CrossRef]
12. Kutschmann, P.; Lindner, T.; Börner, K.; Reese, U.; Lampke, T. Effect of Adjusted Gas Nitriding Parameters on Microstructure and Wear Resistance of HVOF-Sprayed AISI 316L Coatings. *Materials* **2019**, *12*, 1760. [CrossRef]
13. Duan, C.; Shen, Y.; Feng, X.; Chen, C.; Zhang, J. Nitriding of Fe–18Cr–11Mn powders using mechanical alloying method through aerating nitrogen circularly. *Mater. Sci. Technol.* **2016**, *32*, 1231–1239. [CrossRef]
14. Gnedovets, A.G.; Ankudinov, A.B.; Zelenskii, V.A.; Kovalev, E.P.; Wisniewska-Weinert, H.; Alymov, M.I. Synthesis of Micron Particles with Fe–Fe4N Core–Shell Structure at Low-Temperature Gaseous Nitriding of Iron Powder in a Stream of Ammonia. *Inorgan. Mater. Appl. Res.* **2016**, *7*, 303–309. [CrossRef]
15. Lindner, T.; Löbel, M.; Lampke, T. Phase Stability and Microstructure Evolution of Solution-Hardened 316L Powder Feedstock for Thermal Spraying. *Metals* **2018**, *8*, 1063. [CrossRef]
16. Grujicic, M.; Zhao, C.L.; Tong, C.; DeRosset, W.S.; Helfricht, D. Analysis of the impact velocity of powder particles in the cold-gas dynamic-spray process. *Mater. Sci. Eng. A* **2004**, *368*, 222–230. [CrossRef]
17. Assadi, H.; Kreye, H.; Gärtner, F.; Klassen, T. Cold spraying—A materials perspective. *Acta Mater.* **2016**, *116*, 382–407. [CrossRef]
18. Rokni, M.R.; Nutt, S.R.; Widener, C.A.; Champagne, V.K.; Hrabde, R.H. Review of Relationship Between Particle Deformation, Coating Microstructure, and Properties in High-Pressure Cold Spray. *J. Therm. Spray Technol.* **2017**, *26*, 1308–1355. [CrossRef]
19. Assadi, H.; Gärtner, F.; Stoltenhoff, T.; Kreye, H. Bonding mechanism in cold gas spraying. *Acta Mater.* **2003**, *51*, 4379–4394. [CrossRef]
20. Grujicic, M.; Zhao, C.L.; DeRosset, W.S.; Helfricht, D. Adiabatic shear instability based mechanism for particles/substrate bonding in the cold-gas dynamic-spray process. *Mater. Des.* **2004**, *25*, 681–688. [CrossRef]
21. Schmidt, T.; Gärtner, F.; Assadi, H.; Kreye, H. Development of a generalized parameter window for cold spray deposition. *Acta Mater.* **2006**, *54*, 729–742. [CrossRef]
22. Schmidt, T.; Assadi, H.; Gärtner, F.; Richter, H.; Stoltenhoff, T.; Kreye, H.; Klassen, T. From Particle Acceleration to Impact and Bonding in Cold Spraying. *J. Therm. Spray Technol.* **2009**, *18*, 794–808. [CrossRef]
23. Assadi, H.; Schmidt, T.; Richter, H.; Kliemann, J.O.; Binder, K.; Gartner, F.; Klassen, T.; Kreye, H. On Parameter Selection in Cold Spraying. *J. Therm. Spray Technol.* **2011**, *20*, 1161–1176. [CrossRef]
24. Al-Mangour, B.; Vo, P.; Mongrain, R.; Irissou, E.; Yue, S. Effect of Heat Treatment on the Microstructure and Mechanical Properties of Stainless Steel 316L Coatings Produced by Cold Spray for Biomedical Applications. *J. Therm. Spray Technol.* **2013**, *23*, 641–652. [CrossRef]
25. Xie, Y.; Planche, M.-P.; Raoelison, R.; Liao, H.; Suo, X.; Hervé, P. Effect of Substrate Preheating on Adhesive Strength of SS 316L Cold Spray Coatings. *J. Therm. Spray Technol.* **2016**, *26*, 123–130. [CrossRef]
26. Huang, R.Z.; Sun, B.; Ohno, N.; Fukanuma, H. Study on the Influences of DPV-2000 Software Parameters on the Measured Results in Cold Spray, in Thermal Spray 2006: Science, Innovation, and Application (ASM International). 2006. Available online: https://www.researchgate.net/publication/268346796 (accessed on 1 November 2021).
27. Fukanuma, H.; Ohno, N.; Sun, B.; Huang, R. In-flight particle velocity measurements with DPV-2000 in cold spray. *Surf. Coat. Technol.* **2006**, *201*, 1935–1941. [CrossRef]

28. Mauer, G.; Singh, R.; Rauwald, K.-H.; Schrüfer, S.; Wilson, S.; Vaßen, R. Diagnostics of Cold-Sprayed Particle Velocities Approaching Critical Deposition Conditions. *J. Therm. Spray Technol.* **2017**, *26*, 1423–1433. [CrossRef]
29. Pattison, J.; Celetto, S.; Khan, A.; O'Neill, W. Standoff distance and bow shock phenomena in the Cold Spray process. *Surf. Coat. Technol.* **2008**, *202*, 1443–1454. [CrossRef]
30. Champagne, V.K.; Helfritch, D.J.; Dinavahi, S.P.G.; Leyman, P.F. Theoretical and Experimental Particle Velocity in Cold Spray. *J. Therm. Spray Technol.* **2011**, *20*, 425–431. [CrossRef]
31. Huang, R.; Fukanuma, H. Study of the Influence of Particle Velocity on Adhesive Strength of Cold Spray Deposits. *J. Therm. Spray Technol.* **2012**, *21*, 541–549. [CrossRef]
32. List, A.; Gärtner, F.; Schmidt, T.; Klassen, T. Impact Conditions for Cold Spraying of Hard Metallic Glasses. *J. Therm. Spray Technol.* **2012**, *21*, 531–540. [CrossRef]
33. Suo, X.; Yin, S.; Planche, M.-P.; Liu, T.; Liao, H. Strong effect of carrier gas species on particle velocity during cold spray processes. *Surf. Coat. Technol.* **2015**, *268*, 90–93. [CrossRef]
34. Özdemir, O.Ç.; Conahan, J.M.; Müftü, S. Particle Velocimetry, CFD, and the Role of Particle Sphericity in Cold Spray. *Coatings* **2020**, *10*, 1254. [CrossRef]
35. Schiller, L.; Naumann, A. Über die grundlegenden Berechnungen bei der Schwerkraftaufbereitung. *Vdi Zeits* **1933**, *77*, 318–320.
36. Haider, A.; Levenspiel, O. Drag Coefficient and Terminal Velocity of Spherical and Nonspherical Particles. *Powder Technol.* **1989**, *58*, 63–70. [CrossRef]
37. Wadell, H. The Coefficient of Resistance as a Function of Renoylds Number for Solids of Various Shapes. *J. Frankl. Inst.* **1934**. [CrossRef]

metals

Article

Tribo-Mechanical Behavior of Films and Modified Layers Produced by Cathodic Cage and Glow Discharge Plasma Nitriding Techniques

Bruna C. E. Schibicheski Kurelo [1,2,*], Gelson Biscaia De Souza [2], Silvio Luiz Rutz Da Silva [2], Carlos Maurício Lepienski [1], Clodomiro Alves Júnior [3], Rafael Fillus Chuproski [2] and Giuseppe Pintaúde [1]

1 Department of Mechanical Engineering, Federal Technological University of Paraná, Campus Ecoville, Curitiba 81280-34, Paraná, Brazil
2 Laboratory of Mechanical Properties and Surfaces, Department of Physics, State University of Ponta Grossa, Campus Uvaranas, Ponta Grossa 84030-900, Paraná, Brazil
3 Integrated Center for Technological Innovation of the Semiarid Region, Federal Rural University of the Semi-Arid, Av. Francisco Mota 572, Mossoró 59625900, Rio Grande do Norte, Brazil
* Correspondence: brunakurelo@utfpr.edu.br or bruna_schibicheski@hotmail.com

Abstract: Two surface modification techniques, the glow discharge plasma nitriding (GDPN) and the cathodic cage plasma nitriding (CCPN), were compared regarding the mechanical and tribological behavior of layers produced on AISI 316 stainless-steel surfaces. The analyses were carried out at the micro/nanoscale using nanoindentation and nanoscratch tests. The nitriding temperature (°C) and time (h) parameters were 350/6, 400/6, and 450/6. Morphology, structure, and microstructure were evaluated by X-ray diffraction, scanning electron and optical interferometry microscopies, and energy-dispersive X-ray spectroscopy. GDPN results in stratified modified surfaces, solidly integrated with the substrate, with a temperature-dependent composition comprising nitrides (γ'-Fe$_4$N, ε-Fe$_{2+x}$N, CrN) and N-solid solution (γ_N phase). The latter prevails for the low treatment temperatures. Hardness increases from ~2.5 GPa (bare surface) to ~15.5 GPa (450 °C). The scratch resistance of the GDPN-modified surfaces presents a strong correlation with the layer composition and thickness, with the result that the 400 °C condition exhibits the highest standards against microwear. In contrast, CCPN results in well-defined dual-layers for any of the temperatures. A top 0.3–0.8 μm-thick nitride film (most ε-phase), brittle and easily removable under scratch with loads as low as 63 mN, covers a γ_N-rich case with hardness of 10 GPa. The thickness of the underneath CCPN layer produced at 450 °C is similar to that from GDPN at 400 °C (3 μm); on the other hand, the average roughness is much lower, comparable to the reference surface (Ra ~10 nm), while the layer formation involves no chromium depletion. Moreover, edge effects are absent across the entire sample´s surface. In conclusion, among the studied conditions, the GDPN 400 °C disclosed the best tribo-mechanical performance, whereas CCPN resulted in superior surface finishing for application purposes.

Keywords: plasma-based nitriding; thermal diffusion; deposition; thin film; iron nitrides; S-phase

Citation: Schibicheski Kurelo, B.C.E.; De Souza, G.B.; Da Silva, S.L.R.; Lepienski, C.M.; Alves Júnior, C.; Chuproski, R.F.; Pintaúde, G. Tribo-Mechanical Behavior of Films and Modified Layers Produced by Cathodic Cage and Glow Discharge Plasma Nitriding Techniques. *Metals* **2023**, *13*, 430. https://doi.org/10.3390/met13020430

Academic Editors: Shinichiro Adachi, Francesca Borgioli and Thomas Lindner

Received: 30 January 2023
Revised: 14 February 2023
Accepted: 16 February 2023
Published: 19 February 2023

1. Introduction

Glow discharge plasma nitriding (GDPN) is a surface treatment widely used to improve the tribo-mechanical behavior of stainless-steels. Then, after the GDPN treatment, steel surfaces are expected to increase in hardness and wear resistance. Furthermore, depending on the microstructure of the formed layer, the corrosion resistance should be maintained or even improved after nitriding. Keeping the corrosion resistance at the same level as stainless-steel depends on the control of the formation of chromium and iron nitrides, by performing treatments at low temperatures and suitable times [1]. The modified layers are generally desirable to be rich in nitrogen-expanded austenite (γ_N) phase, promoting excellent characteristics in stainless-steels [2,3].

There are some limitations associated with the use of GDPN treatment, such as the difficulty in treating samples with complex geometries, and the formation of non-uniform layers or edge effects. Both phenomena are associated with variations of the electric field around the sample in the plasma [4–6]. Furthermore, after GDPN treatments, surfaces undergo micrometric changes in the surface morphology due to sputtering/deposition and orientation-dependent anisotropic swelling. A relatively new technique that avoids these undesirable effects on the surface morphology is the cathodic cage plasma nitriding (CCPN) [5–8].

In the CCPN, a type of Physical Vapor Deposition (PVD) process, the sample is electrically isolated from the cathode and immersed in a floating electrical potential due to the presence of the surrounding and biased cathodic cage. Scientific evidence shows that a Kölbel's-like "sputtering and re-condensation" model, supposedly valid to the GDPN technique, is also applicable in a similar way to CCPN by considering the interaction among species from the atmosphere and adsorption on the surface [9]. In this process, electrons electrically trapped in the holes of the cage successively collide with atoms in the atmosphere, increasing the ion density. In this region, the ions, in turn, bombard the inner walls of the holes, sputtering atoms from the cage that will eventually be deposited on the surface of the sample. Thus, the plasma atmosphere, chemical composition of the cathodic cage, and arrangement of holes are parameters that govern the characteristics of the deposited film. In general, even carbon steel cages result in the deposition of iron nitrides on the electrically isolated substrates [5–7]. Another relevant feature of this technique is forming films that are brittle and thinner than the modified layers obtained by GDPN treatments, when employing the same treatment times [8,10]. However, the CCPN also allows the formation of a modified layer below the film due to nitrogen diffusion in the bulk, which is heated by thermal radiation. To our knowledge, it is not common to evaluate the modified diffusion layer formed below the film in CCPN treatments.

The wear performance of nitride layers depends on two aspects involving deformation. The first one is the level of the strain of the substrate, which can deform during contact between two bodies and correlates with hardness. Pintaude et al. [11] applied a method similar to CCPN, named active screen plasma nitriding (ASPN), for two duplex stainless-steels. They verified, after scratch tests, an occurrence of microcracks much more significant on the softer substrate than the harder one, featuring an eggshell effect. The other aspect is the relation between the load conditions and the layer thickness. For thicker nitrided layers processed onto martensitic stainless by ASPN, Rovani et al. [12] described less occurrence of cracks, as the thickness of the layers increased. In the same work, when the load in the scratch tests was reduced from 15 N to 8 N, cracks were not observed on the worn surfaces of the nitrided layers, emphasizing that the treated surfaces have a critical loading value that is important for the application of these layers [12]. These characteristics are important for films processed by CCPN and will be explored in the present study.

The surface mechanical strengthening of the AISI 316 stainless-steel finds applications in several areas, such as in biomedical devices, making it a good reason to investigate varied methods of nitriding for this alloy. Samanta et al. [13] carried out a detailed study of the mechanical and tribological behavior of nitrided AISI 316L surfaces, however they employed only one technique, the GDPN. Herein, the tribo-mechanical performance of 316 surfaces modified by GDPN, analyzed in micro/nanoscale, were contrasted with those subjected to CCPN. The study aims to contribute to the understanding of how the structure and microstructure of the layers affect the wear performance, as well as to the proper selection of the nitriding method.

2. Materials and Methods

Commercial samples of AISI 316 austenitic stainless-steel (15.7% Cr, 11.6% Ni, 1.7% Mn, 1.9% Mo, 0.3% Si, 0.1% C, 0.03% S, and 0.03% P, with Fe in balance) supplied by the company Villares Metals (São Paulo, SP, Brazil) were mechanically polished with sandpapers (SiC) and diamond pastes up to a 1 μm particle size, and the final polishing was carried out in a 15% volume solution of colloidal silica in H_2O_2. Before nitriding, the samples were cleaned with two ultrasound baths in acetone.

The equipment used for GDPN and CCPN is a custom-made equipament consisting of a vacuum chamber, an exhaust and a gas supply system, a voltage source, a K-type thermocouple and electronic sensors, described in detail in previous reports [6,14,15]. The treatment atmosphere was controlled by a flowmeter (model MKS 1179A, MKS Instruments, Andover, MA, USA), and the temperature by a thermocouple connected to the center of the sample holder (cathode). Voltage and current were monitored by a voltmeter and ammeter , both supplied by the company Minipa (São Paulo, SP, Brazil) and integrated into the voltage source. The chamber's internal pressure was measured by an Active Gauge controller RS 232 Edwards pressure gauge (Edwards Ltd., Burgess Hill, UK).

The samples were subjected to sputtering with H_2 at (150 ± 10) °C and 300 Pa for one hour to remove the oxide surface layer and then nitrided by GDPN. An atmosphere of 1:4 (N_2:H_2) (in volume), resulting in a total pressure of 300 Pa and a total flux of 20 sccm, was used. Both nitriding treatments by CCPN and GDPN were carried out at temperatures of 350 °C, 400 °C, and 450 °C for 6 h.

For CCPN treatments, a cathodic cage made of AISI 1008 steel was placed into the same GDPN equipment. Schematic diagrams for the GDPN and CCPN are shown in Figure 1. The material´s chemical composition was 0.1% C, 0.5% Mn, 0.04% P, and 0.05% S (wt.%), in balance with Fe. The cage dimensions were 112 mm × 25 mm × 0.8 mm (diameter × height × thickness), the diameter of the cage holes was 8 mm, and the distance between the center of adjacent holes was 9.2 mm. The samples were placed on an alumina insulator with a diameter of 55.8 mm. The cage was positioned above the cathode to enclose the alumina insulator assembly and the samples. The samples remained electrically isolated in this arrangement, and the cage was in electrical contact with the cathode. Initially, the cage was sputtered with H_2 at 200 ± 5 °C for one hour at 30 Pa. After sputtering, the samples were nitrided by CCPN in an N_2:H_2 atmosphere of 4:1 (in volume), resulting in a total pressure of the order of 80 Pa and an N_2/H_2 gas flow of 16/4 sccm. These flow and pressure values in CCPN treatments were defined based on optimized sample properties in previous work [6,14,15]. The pressure value was defined as being the hollow cathode pressure the cage uses for that gaseous mixture, which is necessary to maximize the film thickness. The voltage varied between 464 V and 588 V during nitriding, according to the treatment temperature. A temperature of 450 °C was chosen to evaluate the formation of chromium nitrides in the different techniques. The nitriding of AISI 316 steel for this work by GDPN and CCPN was carried out concomitantly with other martensitic steels, whose research was previously published [8,10].

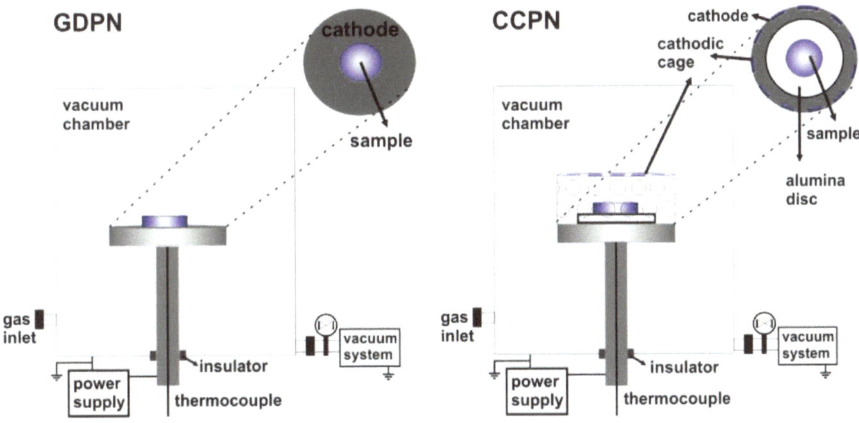

Figure 1. Schematic diagrams for the GDPN and CCPN treatments.

The cross-sections of the samples were mechanically polished and analyzed by field-emission gun scanning electron microscopy (FEG-SEM) model Mira3) (TESCAN, Kohoutovice, Czech Republic) equipped with energy dispersive spectroscopy (EDS) model XMaxN SDD (Oxford, Abingdon, UK) to evaluate the nitrogen uptake in the modified layers. Then, the chemical etching was carried out with Murakami's reagent composed by 10 g KOH and 10 g $K_3Fe(CN)_6$ from the brand Sigma Aldrich (San Luis, MO, USA) and 50 ml H_2O, allowing the thickness of the modified layers to be measured through FEG-SEM analysis. In the samples treated by the CCPN technique, the thickness of the films was determined by optical interferometry (OI) model Talysurf CCI—Lite 3D (Taylor Hobson, Leicester, UK) on the surfaces of the samples, analyzing the height difference between the film surface and the interface with the modified layer in regions where the nitrides film was detached. OI estimated the roughness of all surfaces.

X-ray diffraction (XRD) data were collected by the X-ray diffractometer model Ultima IV (Rigaku, Tokyo, Japan.), with CuKα radiation (λ = 0.15406 nm), Bragg–Brentano geometry (θ–2θ), a 0.02° step, and a counting time of 4 s. The diffraction peaks were identified using powder crystallographic data sheets and literature data to γ_N phase peaks [3,16].

The nanoindentation technique was applied according to the ISO 14577-1 (2002) standard to measure the hardness and elastic modulus. The equipment used in the tests was an Nanoindenter XP (MTS Systems Corporation, Eden Prairie, MN, USA). Here, 25 indentations were performed with a Berkovich indenter arranged in a matrix with 100 μm between the indentations. To obtain profiles of hardness and modulus of elasticity by depth, multiple-loading cyclic nanoindentation was used. The tests were carried out in the central regions of the samples with a maximum load of 400 mN and 8 loading–unloading cycles. For the analysis of the loading curves, the contact stiffness technique was used to minimize the roughness effect [17].

The nanoscratch tests were performed on the same equipment used in the nanoindentation tests. They were carried out with the Berkovich tip, with displacement towards one of the vertices, ramp loading (linear from zero), and a maximum load of 400 mN. The sliding speed was 10 μm/s, and the scratch length was 600 μm. The images of the nanoindentations and scratches were performed with a SEM model JSM-6360LV (JEOL, Peabody, MA, USA).

3. Results and Discussions

3.1. Microstructural and Morphological Analysis

The microstructures of surfaces nitrided by GDPN and CCPN can be seen in Figure 2. All the surfaces were homogeneous after the treatment but, while the CCPN-nitrided surfaces showed small, sharp needle-shaped peaks, characteristic of the deposition process, the GDPN-treated samples disclosed broader peaks and valleys. For all the treatment conditions and at the same temperature, the difference in height between peaks and valleys was always smaller in the CCPN-nitrided surfaces when compared to GDPN-nitrided surfaces. This indicates that the mean roughness of samples treated by GDPN was always higher for the same treatment temperature and time.

Table 1 summarizes the thickness, roughness, and hardness (measured at a 200 nm depth) of layers and films produced by GDPN and CCPN techniques. The compared average roughness values were higher for all the GDPN conditions. Some possible explanations are plasma sputtering and swelling of grains, with the latter being more evident in GDPN than in CCPN. It is a result of the anisotropic retention of nitrogen in solid solution, dependent on the crystal direction of the grain, which is restrained laterally and grows toward the free surface [18]. The sputtering occurs via the non-uniform electric polarization of the sample in the plasma. According to accepted models [19], it plays an important role in the nitriding process since released atoms combine with ions from the plasma and fall with the ion flux toward the surface. This is likely a non-uniform process across the treated area and can result in the observed topography. On the contrary, average roughness values of the electrically isolated CCPN surfaces were similar to the reference one.

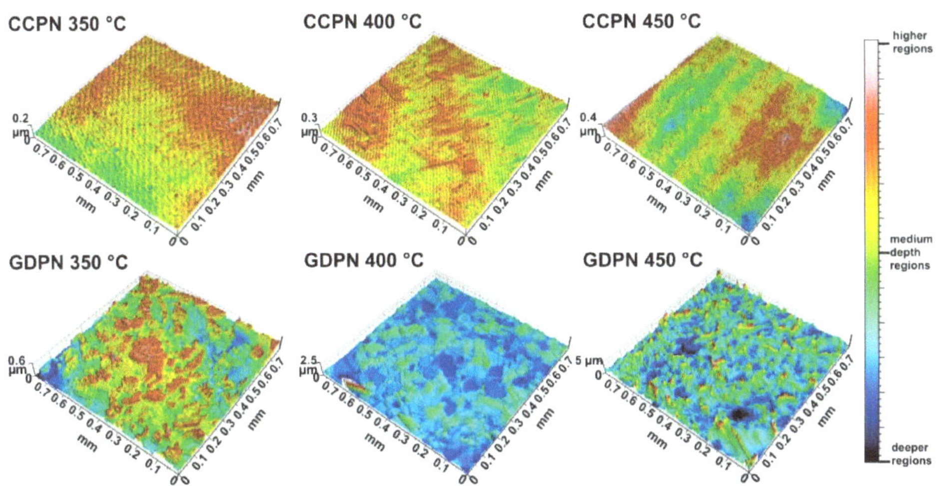

Figure 2. Surface microstructures of samples nitrided by CCPN and GDPN, obtained by OI.

Table 1. The thickness of modified layers and films, roughness, and hardness of the untreated and nitrided samples.

Sample	Layer Thickness Modified Layer (μm)	Layer Thickness Film (μm)	Roughness Ra (nm)	Roughness Rq (nm)	Top Hardness (GPa)
Untreated	-	-	11.6 ± 1.7	14.7 ± 2.4	3.3 ± 0.3
GDPN 350 °C	1.2 ± 0.5	-	48.1 ± 6.5	37.8 ± 4.5	8.5 ± 2.9
GDPN 400 °C	4.3 ± 1.1	-	65.4 ± 7.5	83.6 ± 8.5	12 ± 0.7
GDPN 450 °C	18.4 ± 2.5	-	246.1 ± 30.1	319.1 ± 35.1	15.1 ± 1.6
CCPN 350 °C	0.5 ± 0.1	0.3 ± 0.1	8.7 ± 0.9	10.8 ± 1	3.8 ± 0.3
CCPN 400 °C	1.2 ± 0.1	0.5 ± 0.1	9.5 ± 0.9	11.7 ± 1.3	4.9 ± 0.4
CCPN 450 °C	3.0 ± 0.5	0.8 ± 0.1	18.7 ± 2.1	24.6 ± 2.9	3.6 ± 0.3

Figure 3 shows the cross-section SEM images of surface plasma nitrided by GDPN and CCPN at the indicated temperatures, etched with Murakami's reagent. Modified layers are visible for all the nitriding treatments. At a same temperature, the thickness of the modified layer was smaller for the CCPN technique than for the GDPN technique. This difference increased with temperature, as seen in Table 1.

The nitrogen concentration shown in Figure 4 was obtained by EDS point analysis carried out in several points through the cross-section of the GDPN- and CCPN-treated samples. The N concentration was relatively higher for the GDPN technique than the CCPN one, for the same temperatures. In addition, the concentration profiles were consistent with the layer thicknesses, identified and measured after chemical etching, as seen in Figure 3 and summarized in Table 1.

In Figure 5, the nitride films produced by the CCPN technique were brittle and could be removed from the surface. The regions where the film was detached, such as the one shown in Figure 5a, were convenient to infer the film thickness, by measuring the height difference with the yN-rich surface below it. These height differences are presented in Figure 5b–d.

The average thickness of the films varied between 0.3 μm and 0.8 μm for CCPN 350 °C and CCPN 450 °C samples, respectively, as shown in Table 1. The values are close to those obtained by (Atomic Force Microscope) AFM in martensitic steels treated by the CCPN technique under the same conditions [8]. Thus, even by adding thicknesses of the nitride films with the modified layers produced by the CCPN technique, they were still thinner than those produced by the GDPN technique, in agreement with other studies [8,10,20]. On the other hand, standard deviations of films + layer thicknesses produced by CCPN were smaller, for all the treatment conditions, than the corresponding ones from the GDPN technique. This result corroborates the homogeneity and uniformity of layers produced by the CCPN technique.

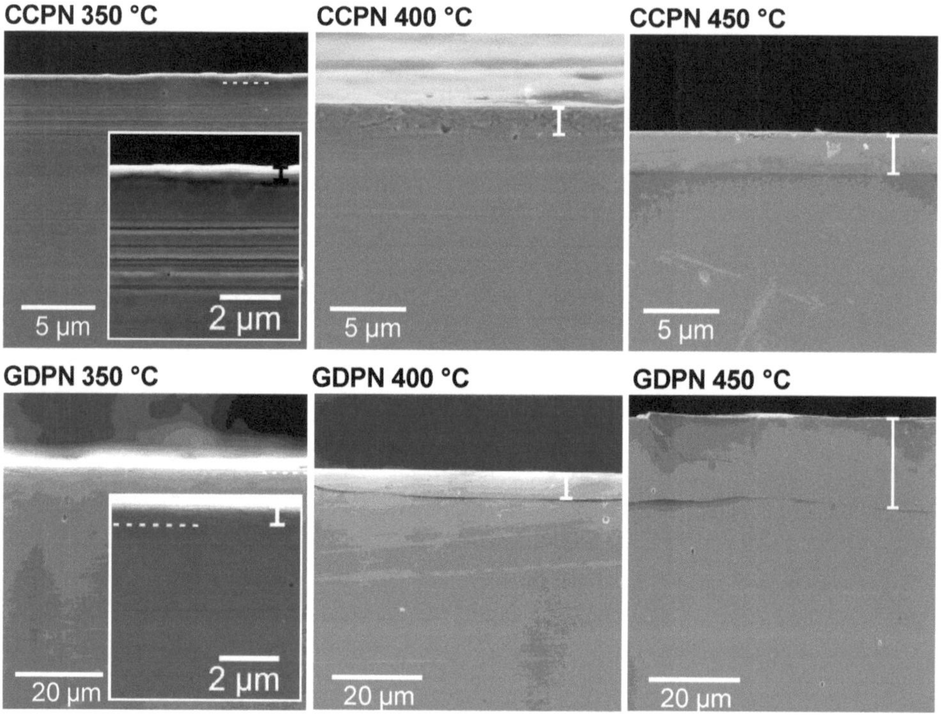

Figure 3. Cross-section SEM images of the plasma-nitrided samples by GDPN and CCPN at the indicated temperatures, etched with Murakami's reagent. Regions indicated correspond to modified layers produced, respectively.

3.2. XRD Analysis

Figure 6 shows the X-ray diffractograms for the untreated sample and the samples nitrided by GDPN. The reference sample presented only the expected austenite peaks (γ). After nitriding, the diffractogram of the GDPN 350 °C sample showed the nitrogen-expanded austenite (γ_N phase), which corresponded to the presence of nitrogen in solid solution in the steel's crystalline structure. The γ_N peak was wide, displaced to greater interplanar distances with regard to the γ phase. This phase corresponded to the superposition of several Fe(N) sub-stoichiometries of the nitrogen in solid solution, consisting in a gradient of lattice micro-strains.

In the sample GDPN 400 °C, the contribution of the γ phase peak considerably decreased compared to the GDPN 350 °C sample due to the increased thickness of the modified layer (Table 1). At this treatment temperature, the γ_N phase peak was displaced to greater interplanar distances, indicating a more significant expansion of the crystal lattice.

In the GDPN 450 °C sample, the diffractogram disclosed the formation of iron nitrides γ'-Fe$_4$N and ε-Fe$_{2+x}$N (with x between 0 and 1), in addition to the drastic reduction of the peak's intensities corresponding to austenite. The formation of the γ' phase indicated that this temperature and treatment time were sufficient to cause the supersaturation of nitrogen in interstitial solid solution, resulting in the precipitation of nitrides [3]. After reaching the supersaturation limit, the formation of nitrides that have a lower enthalpy of formation occurred, such as the γ'-Fe$_4$N phase. With increasing temperature, conditions became more favorable for the formation of nitrides that have a higher enthalpy of formation than the γ' phase, i.e., the ε phase [21]. Then, with the increase in temperature from 400 °C to 450 °C, the replacement of γ_N by γ' and ε is expected. In addition, the increase of the temperature provided more favorable conditions for the diffusion of nitrogen into the steel. Thus, the increase in temperature led to forming a thicker modified layer with more precipitates [2,22,23].

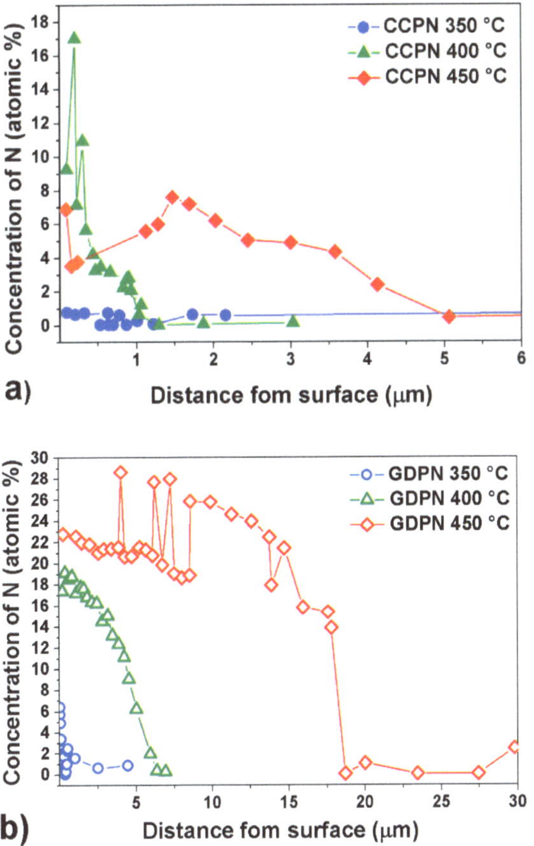

Figure 4. The nitrogen concentration, in at.%, carried out in several points through the cross-section of the (**a**) GDPN- and (**b**) CCPN-treated samples.

At 450 °C, it was also possible to attribute contributions of the chromium nitride CrN in the diffractograms, a result of the chromium depletion in the steel [3]. This phenomenon is undesirable since the presence of Cr in the austenitic matrix is a determining factor for the "stainless" characteristic of the steel.

Figure 7 shows the diffractograms for the untreated and CCPN-nitrided samples. They significantly differed from the GDPN ones. Regardless of the treatment temperature, there was the formation of the ε phase. The analysis carried out from the crystallographic files revealed that this phase had predominantly the Fe2N stoichiometry. The γ'-Fe$_4$N phase became more evident in the diffractogram corresponding to the 400 °C treatment. The expanded austenite phase (γ_N) was also present in all treatment conditions but exhibiting smaller lattice expansions. The peak at ~43° at temperatures of 350 °C and 400 °C and the peak at ~42.5° in the sample nitrided at 450 °C were broad, as they have contributions from the γ and γ_N phases. The identified phases follow the literature results for this austenitic steel treated by CCPN [6,7,15]. In the CCPN process, the deposition phenomenon prevails over diffusion, so iron nitride phases can be formed and deposited regardless of the treatment temperature.

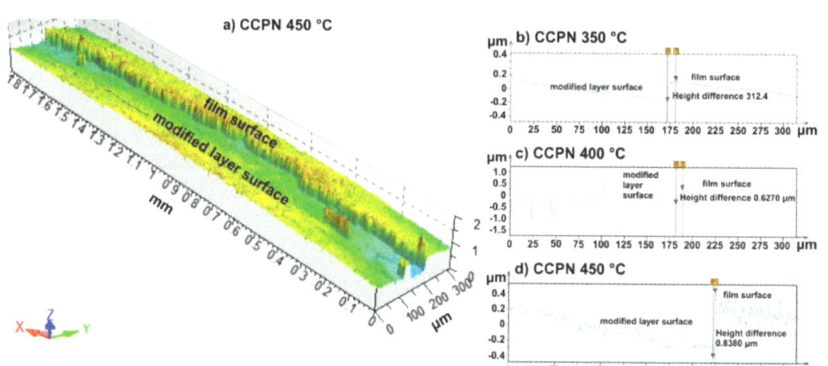

Figure 5. (**a**) Optical interferometry image of a region, on the CCPN 450 °C surface, showing the boundary between the nitride film and the hardened surface below it. (**b–d**) Analyses of the height differences between the film and modified layer surfaces.

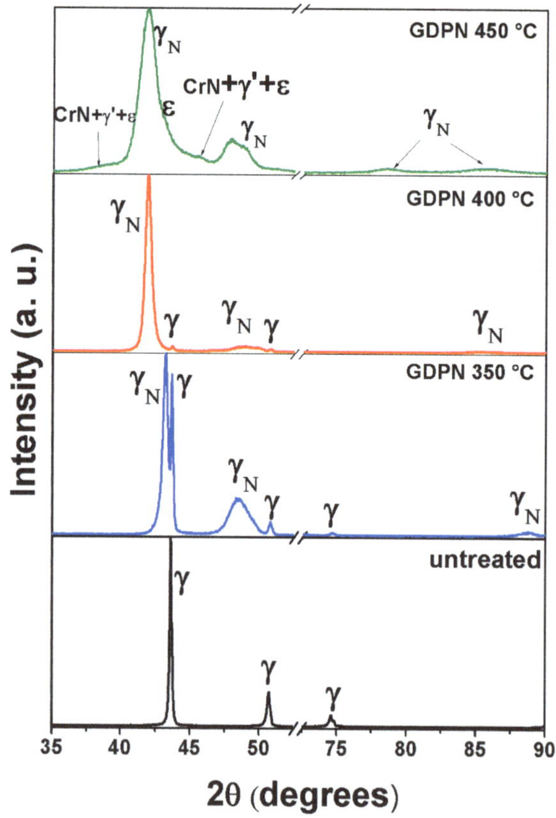

Figure 6. X-ray diffractograms for samples untreated and nitrided by GDPN AISI 316 steel at different temperatures.

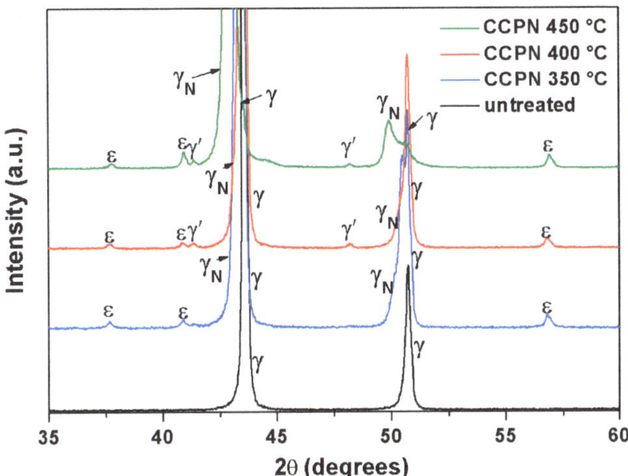

Figure 7. X-ray diffractograms of samples untreated and nitrided by CCPN AISI 316 at different temperatures.

One advantage of the CCPN is a concomitant formation of a diffusion zone beneath the film. After removing the nitride layer from the GDPN 450 °C sample, the γ' phase was no longer identified in the diffractogram (Figure 8), while the predominant phase was γ_N. Hence, the N-solid solution laid mostly below the film, instead of composing it. The small ε phase peaks in this diffractogram may be due to some layer residues that remained adhered to the surface. In the CCPN 450 °C sample, perhaps because the level of nitrogen supersaturation was low in the γ_N region below the film, it did not decay into ε, γ', or chromium nitride phases. This result proves that a modified layer also formed below the film, despite the sample in the CCPN technique being electrically isolated from the cathode. Furthermore, after the formation of the film, it can act as a shield for the diffusion of nitrogen to the bulk. The diffusion of nitrogen into the matrix can be due to the release of N from unstable deposited phases, such as FeN [9].

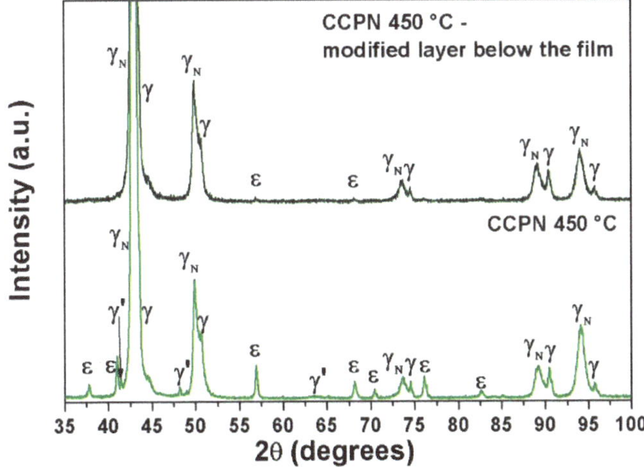

Figure 8. X-ray diffractograms of the CCPN 450 °C sample for the nitrided layer and in the modified layer below the film deposited.

Comparing the techniques studied, GDPN and CCPN, it is relevant to stress that XRD did not identify chromium nitrides neither in the film nor in the modified layer produced by the CCPN technique. It can be attributed to the low nitrogen content in such coating. The absence of chromium nitrides is beneficial to preserve or increase the corrosion resistance of treated materials [24].

3.3. Nanoindentation Measurements

Figure 9 shows SEM micrographs of residual impressions produced by a Berkovich tip on the untreated and GDPN-nitrided AISI 316 steel samples. Lateral cracks were visualized in the GDPN 350 °C sample; however, none of the impressions disclosed radial cracks. The smaller residual impression on the GDPN 450 °C sample was a result of the high hardness of the layer. The surface morphology of the samples treated at 350 °C and 400 °C were similar to that of the untreated sample. In Figure 9d, the vertical structures around the indentation can be attributed to deformation twins.

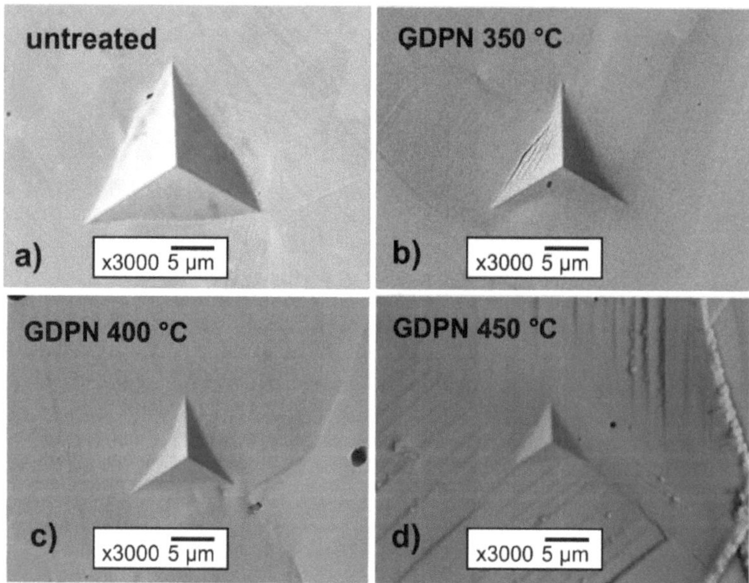

Figure 9. SEM micrographs of residual impressions produced by Berkovich tip indentation on untreated (**a**) and GDPN-nitrided (**b–d**) AISI 316 steel samples.

In CCPN treatments, however, the surface morphologies, shown in Figure 10, differed considerably from the original surface due to the film deposited on it, uniformly covering the surface. The residual impressions presented cracks for all the treatment temperatures, indicating the brittle behavior of these layers. The CCPN layer was not detached since the same surface morphology was also seen inside the indentation imprints. The residual impressions of the nitrided samples presented a reduction in the projected area compared to the impressions on the untreated sample, a conjoined result of the high hardness of CCPN layers and the nitrogen diffusion through the subsurface region (Figure 8).

Figure 11 shows hardness profiles of AISI 316 steel samples without treatment and treated by GDPN at different temperatures. The hardness profile of the reference sample was not constant due to the polishing-induced work hardening: it converged to 2.5 ± 0.2 GPa. The GDPN 350 °C sample presented high hardness only on the top surface, decaying until reaching the substrate values at a depth of approximately 1100 nm, while the GDPN 400 °C sample had an initial "plateau"-shaped profile (values constant or within error bars) to approximately 280 nm, then decayed towards the substrate value. The hardness profile for the GDPN 450 °C sample was approximately constant over the entire depth range analyzed, converging to 15.2 ± 1.5 GPa at a depth of 730 nm. The large error bars for the 350 °C surface may, in addition to roughness effects, be due to the formation of lateral cracks (Figure 9) in the first stages of loading.

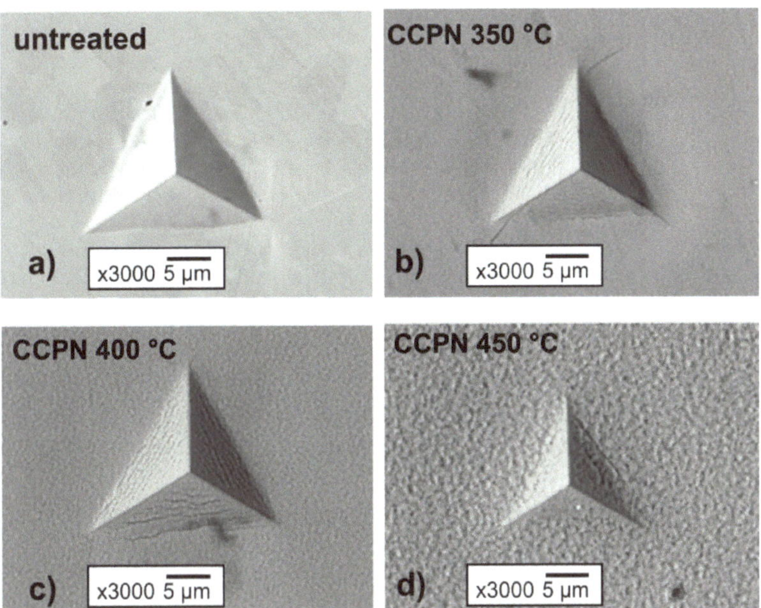

Figure 10. SEM micrographs of residual impressions produced by Berkovich tip indentation on untreated (**a**) and CCPN-nitrided (**b–d**) AISI 316 steel samples.

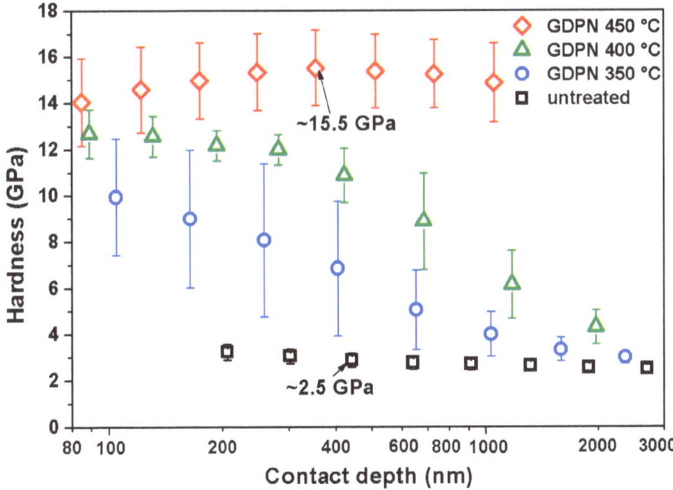

Figure 11. Hardness profiles of untreated and GDPN-nitrided samples.

The hardness profiles agreed with the X-ray diffractograms of these samples (Figure 6). In the samples treated at 350 °C and 400 °C, due to the predominant formation of the γ_N phase, the primary hardening mechanism was solid solution formation, in which interstitial atoms can migrate to sites in the dislocations. This phenomenon, known as the "Cottrell atmosphere," prevents the movement of dislocations and can reduce the number of slip systems, increasing the material's resistance to plastic deformation. The GDPN 450 °C sample may also present the precipitation hardening and dispersion as a contributing mechanism due to the formation of ε and γ' precipitates that act as obstacles to the movement of dislocations, increasing the material's resistance to plastic deformation.

The graph in Figure 12a shows hardness profiles of the untreated and CCPN-nitrided AISI 316 steel surfaces. The behavior of the 350 °C and 450 °C conditions was similar to the untreated one at depths below ~200 nm. The sample treated at 400 °C showed an increase in hardness profiles from 3.3 ± 0.3 GPa to 5.5 ± 0.5 GPa. However, the CCPN 450 °C curve presented an elevation at shallow depths and then a decrease after ~600 nm. This phenomenon may be related to the intense brittleness of the layer produced by CCPN when it is subjected to normal loading, as it can be seen in the indentation imprints shown in Figure 10. The imprints disclosed radial or lateral cracks, which caused errors in the calculated hardness values. In addition, all hardness profiles of the nitrided samples converged to substrate values (untreated sample) at a ~2600 nm depth.

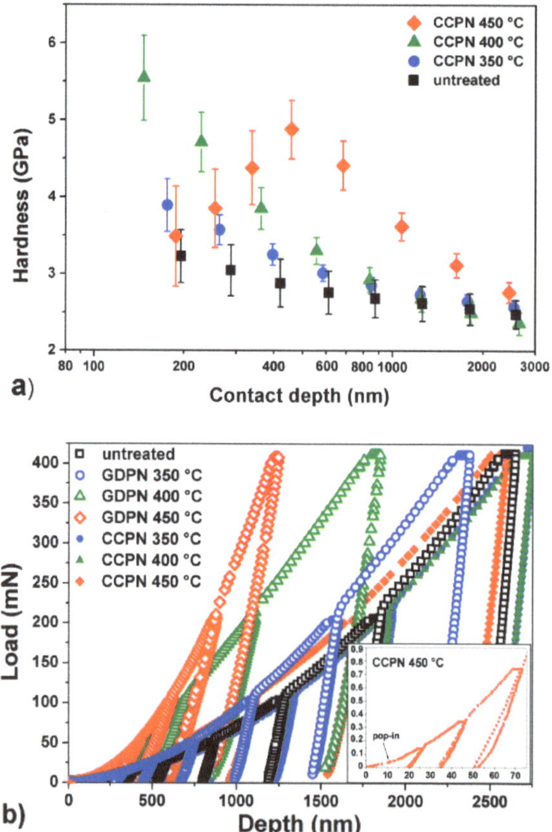

Figure 12. (**a**) Hardness profiles of untreated and CCPN-nitrided samples and (**b**) loading curves for all samples. The inset shows a pop-in event in the loading curve of the CCPN 450 °C sample.

In Figure 12b, loading curves for all samples are presented. Analyzing the loading curves of the sample treated at 450 °C (Figure 12b), it is observed that a significant change in inclination and a pop-in occurred in the first stages of loading which, in this case, can be attributed to the formation of fractures. The film yielded under the indenter, causing incursions into the material without the need of increasing the applied load. Consequently, the algorithm of the equipment underestimated the hardness values obtained by the Oliver and Pharr method (in which $H \propto 1/h_c^2$) [25].

Figure 13 shows the hardness profile obtained after the removal of the film from the CCPN 450 °C sample surface, which revealed a nitrogen diffusion zone, as identified by XRD (Figure 8). The hardness profile no longer presented the anomalous behavior shown in Figure 12a while disclosing a GDPN-like behavior (Figure 11) at this modified layer. The hardness at a depth of ~200 nm was 9.2 ± 0.7 GPa, that is, greater than that (3.6 ± 0.3 GPa) for the nitride film over the same sample. It was also of the same order of values, reported by others, for cross-sectional profiles measured in the same AISI 316 and AISI316L treated by CCPN [6,26]; probably, with the microhardness method, they could not test the thin nitride layer separately. The hardness obtained for the CCPN 450 °C condition, below the nitride film, was lower than that measured on surfaces treated by GDPN at the same temperature, in agreement with other works [8,10,20,26,27]. Moreover, the hardness values were very close to those obtained for the GDPN 350 °C condition (Figure 11), where γ_N was the main phase (Figure 6). In the absence of a plateau in the profiles and considering the 10% rule [28], we can estimate that this modified layer had a thickness of ~1 μm. As already mentioned, the origin of this nitrogen diffusion zone in samples electrically isolated in the plasma chamber may be related to the deposition and subsequent dissociation of unstable nitride species (FeN) and their diffusion by thermal effects [9,29].

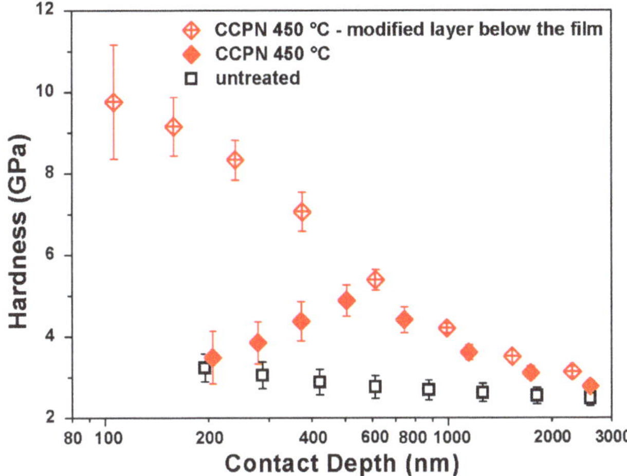

Figure 13. Hardness profiles of the CCPN 450 °C sample in the film and the region below the deposited film compared to the untreated sample.

These results suggest that the anomalous and low hardness profiles (compared to GDPN treatments under similar conditions) shown in Figure 12a are related to the brittle nitride layer produced by the cathodic cage. The true hardness of the layer can even be high, in the order of values observed for conventional plasma treatments, since it is made up of iron nitrides (Figure 7). However, the generation of cracks prevents the correct determination of hardness by normal loading indentation.

3.4. Scratch Tests

Figure 14 presents profiles obtained during and after scratching of untreated and GDPN-nitrided surfaces, with micrographs from central regions of the grooves.

The maximum depth and elastic recovery values are summarized in Table 2. The penetration depth profiles during loading, the residual depth of the nanoscratch tracks, and the track widths were smaller in the nitrided samples than in the reference sample (not shown). These characteristics agree with the mechanical strengthening of the surfaces, which increased with treatment temperatures. On the surface nitrided at 450 °C, the morphology influenced the nanoscratch path so that the tip course deviated near structures that can be swelled grain boundaries. On this surface, a step-like topography in the inner region of the track can also be observed, which is reflected in the "rough" aspect of the scratch profile. In all the nitrided samples, the elastic recovery of the surface after the load removal was more significant than in the untreated condition. The maximum scratch depth in samples treated at 400 °C and 450 °C was within the range evaluated for layer thicknesses ($\geq$3.2 μm). However, the overall behavior of the scratch profiles did not significantly differ from the untreated sample, despite the ~4-fold increase in hardness. This fact may be associated with the abrasive action of hard nitride precipitates that were removed and displaced by the tangential movement of the tip. The samples GDPN 350 °C and GDPN 450 °C had elastic recovery close to those of the reference sample, and the sample GDPN 400 °C had a maximum scratch penetration depth of approximately half the value of the untreated sample.

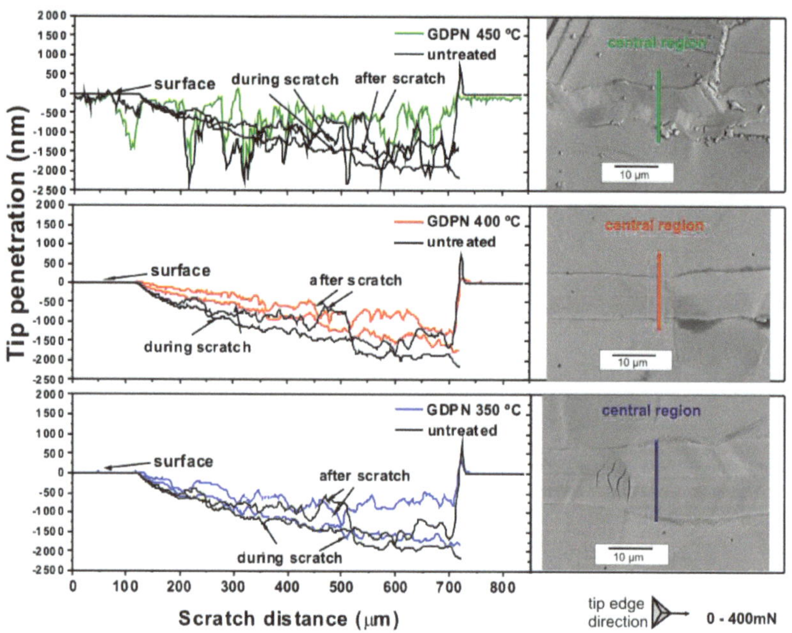

Figure 14. Nanoscratch profiles of untreated and GDPN-treated AISI 316 steel samples and micrographs obtained by backscattered SEM.

Table 2. Critical load, maximum depth, and elastic recovery (under 200 mN load) of untreated and nitrided samples.

Sample	Critical Load (mN)	Maximum Depth (nm)	Elastic Recovery (%)
Untreated	-	1375 ± 64	22 ± 8
GDPN 350 °C	-	1281 ± 66	33 ± 5
GDPN 400 °C	-	698 ± 67	55 ± 6
GDPN 450 °C	-	1254 ± 35	35 ± 3
CCPN 350 °C	9 ± 1	1434 ± 39	-
CCPN 400 °C	63 ± 13	1225 ± 24	-
CCPN 450 °C	24 ± 2	1575 ± 22	-

Figure 15 shows electron micrographs of the scratch path produced on the untreated and CCPN-treated samples. It is possible to observe regions where the film was delaminated and fractured, while in the rest of the track the scratching caused fragmentation and detachment of the nitride layer.

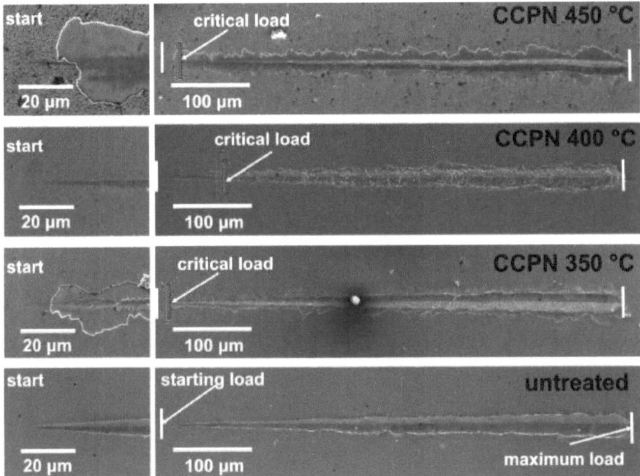

Figure 15. Nanoscratch micrographs of untreated and CCPN-treated AISI 316 steel samples obtained by SEM.

The critical load of scratching at which the film remained adhered to the substrate can be taken as a comparison measurement of the performance among the different CCPN surfaces. Figure 16 shows the profile graph and nanoscratch micrograph of the AISI 316 steel sample treated at 400 °C for 6 h. In Figure 16, in the backscattered SEM image, it is possible to identify by contrast (due to the different chemical compositions) the nitride film and regions where it was chipped. Therefore, the lighter region corresponds to the substrate exposed after removing the nitride film (the darker region). The distance from which the film chipping occurred was approximately 112 µm from the scratch beginning. At this distance, the penetration curve changed in slope and roughness during loading; from this position on, the removal of the film became more pronounced. The critical load for scratch resistance at this point was 63 mN. With this methodology, we obtained the critical load values for the scratch resistance of the CCPN-nitrided surfaces, which are summarized in Table 2. Among the treatments with a cathode cage, the CCPN 400 °C sample showed the highest critical load value for scratch resistance, with a mean value of (63 ± 13) mN. A similar behavior was observed in treatments performed on martensitic steels nitrided under the same conditions [8].

Steels that undergo nitriding processes are typically employed under severe wear and abrasion conditions. Therefore, the film deposited for these applications should withstand loads higher than those used in this test (400 mN = 40 gf); however, the maximum critical load of the formed films was only ~63 mN. In practice, this brittle and nitride-rich film can act as a sacrificial layer, where wear and corrosion simultaneously occur. In addition, according to the literature, the CCPN treatments can enhance the corrosion resistance of austenitic steels [30]. After the rupture of this top layer, a modified and homogeneous surface where γ_N is the predominant phase will be exposed to the working conditions. It is expected that in this condition, the sample will show good tribological performance associated with good corrosion resistance. Therefore, additional tribological and corrosion resistance studies are necessary to evaluate if the nitrides and γ_N layer may act together as a complex system.

Regarding the maximum depths, the lowest value was observed for the GDPN 400 °C sample. In this situation, there was a ~50% reduction in the maximum depth compared to the untreated sample. In contrast, the maximum depth reached by the CCPN-treated samples was very close to or higher than that of the untreated sample, possibly due to film breakage and incorporation of debris in the abrasion process within the wear track.

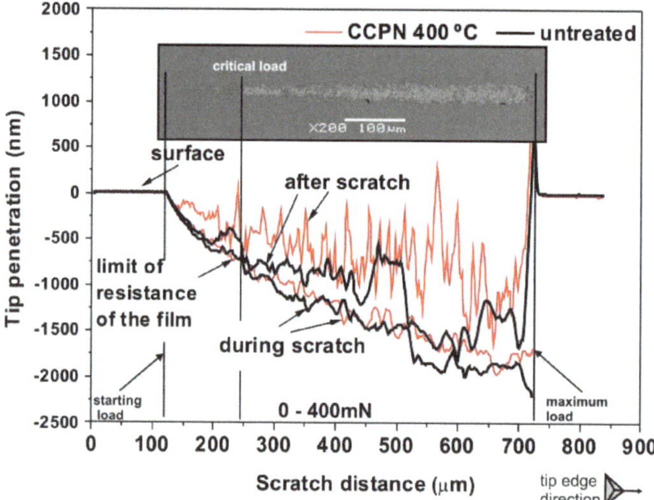

Figure 16. Profile graph and nanoscratch micrograph obtained by backscattered SEM of the AISI 316 steel sample treated at 400 °C for 6 h by CCPN.

Figure 17 presents the cross-section topographies of grooves produced on surfaces nitrided at 400 °C by CCPN and GDPN, which were the ones that presented the best tribological performance in each of the nitriding techniques.

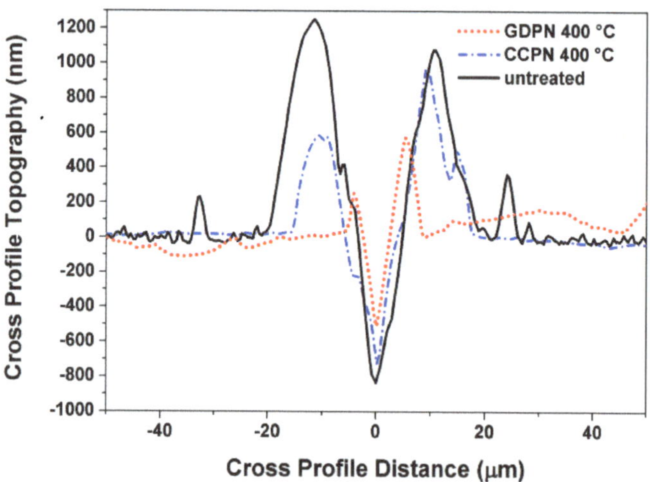

Figure 17. Cross-section topographies taken at the grooves' middle regions, corresponding to 200 mN of applied load.

In agreement with the previous analyses, the grooves' profiles of the GDPN 400 °C surface were consistent with the best wear resistance, as evidenced by the reduction in the residual depth (analyzed after removing the load) and a considerably lower pile-up.

4. General Remarks

To sum up, the results obtained for the GDPN 400 °C sample were superior to those obtained for the samples treated by CCPN, and they also corroborated the best performance of this GDPN treatment, among the conditions studied here, for improving the tribological behavior of the AISI 316L steel at typical applications.

On the other hand, below a brittle thin layer, the CCPN resulted in graded γ_N-rich cases, which were also well-established with the substrate and (an important feature) free from the undesirable edge effect. A possible solution to achieve the best of both processes, avoiding the brittleness of nitride films, is to keep the material under treatment in the cathodic cage but electrically biased, instead of in a floating potential. This condition is expected to prioritize the formation of the modified layer to the detriment of the film deposition process, therefore reducing the edge effect. Some researchers have reported good results in this treatment condition [31], which is planned to be investigated by nanoindentation and nanoscratch in the future.

5. Conclusions

At temperatures between 350 °C and 450 °C, the surface modification by CCPN produced 0.3–0.8 μm-thick nitride films with low roughness (8.7 nm < Ra < 18.7 nm), laying above and presenting a well-defined interface with a 0.5–3 μm-thick nitrogen solid solution (γ_N) region. The GDPN treatments produced thicker modified layers (1.2–18 μm), composed of γ_N phase and nitrides, with higher roughness (48 nm < Ra < 246 nm) than CCPN.

In the GDPN technique, different phases (γ_N, γ'-Fe$_4$N, ε-Fe$_{2+x}$N, CrN) were formed with a predominance of the γ_N phase in treatments carried out at low temperatures (up to 400 °C). In the CCPN treatments, the formation of the ε phase occurred independently of the treatment temperature due to the deposition process. High temperatures (from 450 °C) in CCPN treatments favored the formation of a modified layer rich in γ_N phase below the film due to the greater diffusion of nitrogen to the substrate; however, chromium nitrides were absent from both layers, which is an interesting feature for the corrosion resistance.

The GDPN 450 °C sample presented the highest hardness values (15.5 GPa) among the studied conditions, while in the CCPN 450 °C sample, below the brittle nitrides layer, hardness was 10 GPa, 35% smaller than the GDPN 450 °C sample.

The scratch resistance of the GDPN-nitrided surfaces significantly increased as compared to the untreated sample. The film from the CCPN-treated samples showed less scratch resistance due to delamination and cracking of the films during the analyses. The highest critical load for scratch resistance was 63 ± 13 mN.

The layers produced by the GDPN technique were more mechanical-resistant than those produced by the CCPN technique. However, depending on the application, the corrosion resistance of these layers must be evaluated. In addition, due to uniformity, low roughness, and absence of chromium depletion, surfaces modified by CCPN may perform better than the ones subjected to the GDPN technique, providing that the top brittle nitride layer can be removed or is irrelevant for application purposes.

Author Contributions: Conceptualization, B.C.E.S.K., G.B.D.S. and S.L.R.D.S.; methodology, B.C.E.S.K., G.B.D.S. and S.L.R.D.S.; formal analysis, B.C.E.S.K., G.B.D.S., S.L.R.D.S., C.M.L., C.A.J., R.F.C. and G.P.; investigation, B.C.E.S.K. and R.F.C.; data curation, B.C.E.S.K.; writing—original draft preparation, B.C.E.S.K.; writing—review and editing, G.B.D.S., S.L.R.D.S., C.M.L., C.A.J., R.F.C. and G.P. All authors have read and agreed to the published version of the manuscript.

Funding: Bruna Kurelo was supported by a postdoctoral research scholarship PNPD granted by the Coordination for the Improvement of Higher Education Personnel during the completion of this research.

Institutional Review Board Statement: Not applicable.

Informed Consent Statement: Not applicable.

Data Availability Statement: The data used in this research are available in the article.

Acknowledgments: The authors thank the C-LABMU/UEPG, CMCM/UTFPR, LabNano/UFPR and CME/UFPR for the use of characterization facilities.

Conflicts of Interest: The authors declare no conflict of interest.

References

1. Olzon-Dionysio, M.; de Souza, S.D.; Basso, R.L.O.; de Souza, S. Application of Mössbauer Spectroscopy to the Study of Corrosion Resistance in NaCl Solution of Plasma Nitrided AISI 316L Stainless Steel. *Surf. Coat. Technol.* **2008**, *202*, 3607–3614. [CrossRef]
2. Nascimento, F.C.; Foerster, C.E.; Da Silva, S.L.R.; Lepienski, C.M.; Siqueira, C.J.D.M.; Junior, C.A. A Comparative Study of Mechanical and Tribological Properties of AISI-304 and AISI-316 Submitted to Glow Discharge Nitriding. *Mater. Res.* **2009**, *12*, 173–180. [CrossRef]
3. Dong, H. S-Phase Surface Engineering of Fe-Cr, Co-Cr and Ni-Cr Alloys. *Int. Mater. Rev.* **2010**, *55*, 65–98. [CrossRef]
4. Olzon-Dionysio, M.; Campos, M.; Kapp, M.; de Souza, S.; de Souza, S.D. Influences of Plasma Nitriding Edge Effect on Properties of 316L Stainless Steel. *Surf. Coat. Technol.* **2010**, *204*, 3623–3628. [CrossRef]
5. Junior, C.A.; Freitas, N.V.D.S.; De Morais, P.B.; Vitoriano, J.D.O. Changing the Characteristics of the Nitrided Layer Using Three Different Plasma Configurations. *Rev. Mater.* **2019**, *24*, 1–8. [CrossRef]
6. De Sousa, R.R.M.; De Araújo, F.O.; Gontijo, L.C.; Da Costa, J.A.P.; Alves, C. Cathodic Cage Plasma Nitriding (CCPN) of Austenitic Stainless Steel (AISI 316): Influence of the Different Ratios of the (N_2/H_2) on the Nitrided Layers Properties. *Vacuum* **2012**, *86*, 2048–2053. [CrossRef]
7. Alves, C.; de Araújo, F.O.; Ribeiro, K.J.B.; da Costa, J.A.P.; Sousa, R.R.M.; de Sousa, R.S. Use of Cathodic Cage in Plasma Nitriding. *Surf. Coat. Technol.* **2006**, *201*, 2450–2454. [CrossRef]
8. Kurelo, B.C.; de Souza, G.B.; da Silva, S.L.R.; Daudt, N.D.F.; Alves, C.; Torres, R.D.; Serbena, F.C. Tribo-Mechanical Features of Nitride Coatings and Diffusion Layers Produced by Cathodic Cage Technique on Martensitic and Supermartensitic Stainless Steels. *Surf. Coat. Technol.* **2015**, *275*, 41–50. [CrossRef]
9. Saeed, A.; Khan, A.W.; Jan, F.; Abrar, M.; Khalid, M.; Zakaullah, M. Validity of "Sputtering and Re-Condensation" Model in Active Screen Cage Plasma Nitriding Process. *Appl. Surf. Sci.* **2013**, *273*, 173–178. [CrossRef]
10. Kurelo, B.C.E.S.; De Souza, G.B.; Da Silva, S.L.R.; Serbena, F.C.; Foerster, C.E.; Alves, C. Plasma Nitriding of HP13Cr Supermartensitic Stainless Steel. *Appl. Surf. Sci.* **2015**, *349*, 403–414. [CrossRef]
11. Pintaude, G.; Rovani, A.C.; das Neves, J.C.K.; Lagoeiro, L.E.; Li, X.; Dong, H.S. Wear and Corrosion Resistances of Active Screen Plasma-Nitrided Duplex Stainless Steels. *J. Mater. Eng. Perform.* **2019**, *28*, 3673–3682. [CrossRef]
12. Rovani, A.C.; Breganon, R.; de Souza, G.S.; Brunatto, S.F.; Pintaúde, G. Scratch Resistance of Low-Temperature Plasma Nitrided and Carburized Martensitic Stainless Steel. *Wear* **2017**, *376–377*, 70–76. [CrossRef]
13. Samanta, A.; Chakraborty, H.; Bhattacharya, M.; Ghosh, J.; Sreemany, M.; Bysakh, S.; Rane, R.; Joseph, A.; Jhala, G.; Mukherjee, S.; et al. Nanotribological Response of a Plasma Nitrided Bio-Steel. *J. Mech. Behav. Biomed. Mater.* **2017**, *65*, 584–599. [CrossRef] [PubMed]
14. de Sousa, R.R.M.; de Araújo, F.O.; da Costa, J.A.P.; Brandim, A.D.S.; de Brito, R.A.; Alves, C. Cathodic Cage Plasma Nitriding: An Innovative Technique. *J. Metall.* **2012**, *2012*, 1–6. [CrossRef]
15. de Sousa, R.R.M.; de Araújo, F.O.; Gontijo, L.C.; da Costa, J.A.P.; Nascimento, I.O.; Alves, C., Jr. Cathodic Cage Plasma Nitriding of Austenitic Stainless Steel (AISI 316): Influence of the Working Pressure on the Nitrided Layers Properties. *Mater. Res.* **2014**, *17*, 427–433. [CrossRef]
16. Manova, D.; Mändl, S.; Neumann, H.; Rauschenbach, B. Formation of Metastable Diffusion Layers in Cr-Containing Iron, Cobalt and Nickel Alloys after Nitrogen Insertion. *Surf. Coat. Technol.* **2017**, *312*, 81–90. [CrossRef]
17. Souza, G.; Foerster, C.; Silva, S.; Lipienski, C. Nanomechanical Properties of Rough Surfaces. *Mater. Res.* **2006**, *9*, 159–163. [CrossRef]
18. Oliveira, W.; Kurelo, B.C.E.S.; Ditzel, D.G.; Foerster, C.E.; Serbena, F.C.; de Souza, G.B. On the S-Phase Formation and the Balanced Plasma Nitriding of Austenitic-Ferritic Super Duplex Stainless Steel. *Appl. Surf. Sci.* **2018**, *434*, 1161–1174. [CrossRef]
19. Pastukh, I.M. Energy Model of Glow Discharge Nitriding. *Tech. Phys.* **2016**, *61*, 76–83. [CrossRef]
20. Kovács, D.; Quintana, I.; Dobránszky, J. Effects of Different Variants of Plasma Nitriding on the Properties of the Nitrided Layer. *J. Mater. Eng. Perform.* **2019**, *28*, 5485–5493. [CrossRef]
21. Tessier, F.; Navrotsky, A.; Niewa, R.; Leineweber, A.; Jacobs, H.; Kikkawa, S.; Takahashi, M.; Kanamaru, F.; DiSalvo, F.J. Energetics of Binary Iron Nitrides. *Solid State Sci.* **2000**, *2*, 457–462. [CrossRef]
22. Nascimento, F.C.; Lepienski, C.M.; Foerster, C.E.; Assmann, A.; Da Silva, S.L.R.; Siqueira, C.J.D.M.; Chinelatto, A. Structural, Mechanical, and Tribological Properties of AISI 304 and AISI 316L Steels Submitted to Nitrogen-Carbon Glow Discharge. *J. Mater. Sci.* **2009**, *44*, 1045–1053. [CrossRef]
23. Lepienski, C.; Nascimento, F.; Foerster, C.; da Silva, S.; Siqueira, C.D.M.; Alves, C. Glow Discharge Nitriding in AISI 304 at Different Nitrogen-Hydrogen Atmospheres: Structural, Mechanical and Tribological Properties. *Mater. Sci. Eng. A* **2008**, *489*, 201–206. [CrossRef]
24. Chiu, L.H.; Su, Y.Y.; Chen, F.S.; Chang, H. Microstructure and Properties of Active Screen Plasma Nitrided Duplex Stainless Steel. *Mater. Manuf. Process.* **2010**, *25*, 316–323. [CrossRef]

25. Oliver, W.C.; Pharr, G.M. An Improved Technique for Determining Hardness and Elastic Modulus Using Load and Displacement Sensing Indentation Experiments. *J. Mater. Res.* **1992**, *7*, 1564–1583. [CrossRef]
26. Borowski, T.; Adamczyk-Cieślak, B.; Brojanowska, A.; Kulikowski, K.; Wierzchoń, T. Surface Modification of Austenitic Steel by Various Glow-Discharge Nitriding Methods. *Mater. Sci.* **2015**, *21*, 376–381. [CrossRef]
27. Sumiya, K.; Tokuyama, S.; Nishimoto, A.; Fukui, J.; Nishiyama, A. Application of Active-Screen Plasma Nitriding to an Austenitic Stainless Steel Small-Diameter Thin Pipe. *Metals* **2021**, *11*, 366. [CrossRef]
28. Saha, R.; Nix, W.D. Effects of the Substrate on the Determination of Thin Film Mechanical Properties by Nanoindentation. *Acta Mater.* **2002**, *50*, 23–38. [CrossRef]
29. Fraczek, T.; Prusak, R.; Ogórek, M.; Skuza, Z. Nitriding of 316L Steel in a Glow Discharge Plasma. *Materials* **2022**, *15*, 3081. [CrossRef]
30. Borowski, T. Enhancing the Corrosion Resistance of Austenitic Steel Using Active Screen Plasma Nitriding and Nitrocarburising. *Materials* **2021**, *14*, 3320. [CrossRef]
31. Corujeira Gallo, S.; Dong, H. Study of Active Screen Plasma Processing Conditions for Carburising and Nitriding Austenitic Stainless Steel. *Surf. Coat. Technol.* **2009**, *203*, 3669–3675. [CrossRef]

Article

Rapid Alloy Surface Engineering through Closed-Vessel Reagent Pyrolysis

Cyprian Illing [†], Zhe Ren [†], Anna Agaponova[†], Arthur Heuer [†] and Frank Ernst *,[†]

Department of Materials Science and Engineering, Case Western Reserve University, 10900 Euclid Avenue, Cleveland, OH 44106, USA; cai7@case.edu (C.I.); zxr45@case.edu (Z.R.); ava132@case.edu (A.A.); ahh@case.edu (A.H.)
* Correspondence: fxe5@case.edu; Tel.: +1-216-368-0611
† These authors contributed equally to this work.

Abstract: For rapid surface engineering of Cr-containing alloys by low-temperature nitrocarburization, we introduce a process based on pyrolysis of solid reagents, e.g., urea, performed in an evacuated closed vessel. Upon heating to temperatures high enough for rapid diffusion of interstitial solute, but low enough to avoid second-phase precipitation, the reagent is pyrolyzed to a gas atmosphere containing molecules that (i) activate the alloy surface by stripping away the passivating Cr_2O_3-rich surface film (diffusion barrier) and (ii) rapidly infuse carbon and nitrogen into the alloy. We demonstrate quantitatively that this method can generate a subsurface zone with concentrated carbon and nitrogen comparable to what can be accomplished by established (e.g., gas-phase- or plasma-based) methods, but with significantly reduced processing time. As another important difference to established gas-phase processing, the interaction of gas molecules with the alloy surface can have auto-catalytic effects by altering the gas composition in a way that accelerates solute infusion by providing a high activity of HNCO. The new method lends itself to rapid experimentation with a minimum of laboratory equipment.

Keywords: alloy surface engineering; colossal supersaturation; nitrocarburization; reagent pyrolysis; auto-catalytic effect

Citation: Illing, C.; Ren, Z.; Agaponova, A.; Heuer, A.; Ernst, F. Rapid Alloy Surface Engineering through Closed-Vessel Reagent Pyrolysis. *Metals* **2021**, *11*, 1764. https://doi.org/10.3390/met11111764

Academic Editors: Shinichiro Adachi, Thomas Lindner and Francesca Borgioli

Received: 23 September 2021
Accepted: 27 October 2021
Published: 2 November 2021

Publisher's Note: MDPI stays neutral with regard to jurisdictional claims in published maps and institutional affiliations.

1. Introduction

The mechanical behavior and the corrosion resistance of a broad spectrum of Cr-containing structural alloys can be substantially improved by infusing concentrated interstitially dissolved carbon or nitrogen through the alloy surface at *low* temperature [1–9]. "Low temperature," in this context, refers to temperatures at which metal atom diffusion is effectively frozen over the processing time, such that, despite typically low equilibrium solubility limits of carbon and nitrogen, the infused solute atoms cannot precipitate as metal nitrides or carbides. On the other hand, since the diffusivity of interstitially dissolved atoms is comparatively high at low temperatures, the processing temperature can still be high enough to enable the solute to diffuse into considerable depth z and form a subsurface zone ("case") of concentrated solute level within a technically feasible processing time. The result is a graded subsurface diffusion profile with maximum nitrogen- or carbon concentrations that can reach $\approx 10^5$ times the equilibrium solubility limits at room temperature, corresponding to a "colossal supersaturation" with interstitial solute.

From a technical point of view, it is desirable to (i) maximize the fraction X of interstitial solute at each depth z below the surface (while avoiding precipitation), (ii) minimize the processing time τ_p, and (iii) maximize the mean depth $\bar{z}$ of the interstitial solute below the alloy surface. In the literature, the latter is often loosely characterized as "case depth" or "thickness" ζ of the infused layer, which is physically incorrect because a diffusion profile does not have a sharp depth. To design a process with high efficiency, from a fundamental point of view, there are three distinct components to consider:

1. The chemical potential of solute at the alloy surface.
2. The "transparency" of the surface for the interstitial solute.
3. The mobility of the solute in the alloy.

To provide a driving force for infusion, the process needs to deliver solute into the alloy surface with a chemical potential higher than that directly below the surface. Controlling the chemical potential at the surface is probably the most complex component and can be influenced by the largest variety of processing parameters. For example, solute atoms can be provided by a gas atmosphere, a salt bath, a solid–solid interface, or a plasma. In addition to primary processing parameters, catalytic or reactive responses of the particular alloy surface may influence the gas, salt, or plasma composition at the alloy surface—and thus the chemical potential of the solute and the driving force for infusion.

The transparency of the surface to solute atoms may be limited by microscopic mechanisms required for acquiring solute atoms from the environment, a surface layer of passivating oxide, or a "Beilby" layer [10,11], i.e., a layer of high defect density as generated by surface machining. The process of increasing surface transparency by removing such obstacles is known as "surface activation." This can occur in various ways, e.g., by chemical- or plasma etching, and it can occur just at the beginning or continuously throughout the entire infusion process [7–9,12–14].

The mobility of the solute in the alloy generally depends on the local solute concentration. For a given mobility, the transport rate of interstitial solute increases with the local gradient of the chemical potential. For the usual description of the flux density as the product of the concentration gradient with a diffusion coefficient (Fick's First "Law"), previous experimental work has revealed that the diffusion coefficients D_C, D_N of carbon and nitrogen in austenite strongly increase with the local atom fractions X_N, X_C, respectively [15–17]. For a given alloy composition, the main parameter through which $D[X]$ can be controlled is the specimen temperature T_s.

For technical applications, it is most practical to provide the solute from a gas-phase at atmospheric pressure (0.1 MPa). Building on this approach, Christiansen and Somers [18,19] have reported an elegant method where the alloy work piece is placed in the stream of a gas mixture consisting of a carrier gas and gaseous products originating from pyrolysis of a solid reagent while heating from room temperature T_0 to a maximum temperature of e.g., $T_{max} = 710\,\text{K}$ ($440\,^\circ\text{C}$) within a processing time of e.g., 2.7 ks (45 min). Upon reaching T_{max}, the article is cooled to room temperature in Ar within e.g., 0.6 ks (10 min) [18,19]. It was found that the gas mixture resulting from pyrolysis can effectively both activate the alloy surface and provide carbon and nitrogen to diffuse into the alloy. Although this "open-vessel" process described by Christiansen and Somers produced good results, the exact conditions at the specimen surface are difficult to infer and depend on many parameters. Especially since the reservoir of reagent is changing in time, the composition of the gas and the activities of its components at the specimen surface are not easy to control. Neither is the streaming velocity field at the specimen surface. Moreover, with that method, it is possible that the reagent and the specimen are at different temperatures, i.e., $T_r \neq T_s$. Therefore, the results obtained with a specific setup may be difficult to reproduce with another setup and scientific studies may be compromised by uncontrolled parameters.

The primary goal of the work reported here was to establish a new method for performing laboratory-scale low-temperature carburization, nitridation, or nitrocarburization by pyrolysis of solid reagents under simple and flexible, but, at the same time, well-defined and easily reproducible experimental conditions. Building on our earlier work on nitridation of Ti-base alloys [20,21] and nitrocarburization of Co–Cr–Mo alloys [22,23], our approach, illustrated in Figure 1, consists of *encapsulating* the reagent and the alloy workpiece in an inert container—specifically, a fused-silica ampoule. With this new "closed-vessel" approach, there is no net gas flow and all components are at the same temperature, which we denote as the processing temperature $T_p = T_r = T_s$. Another important difference is that this process can operate over a broad range of well-defined and adjustable pressure, which can be controlled via the effective reagent concentration c_r in the ampoule, e.g., expressed as number ν of moles per free ampoule volume V, Figure 1.

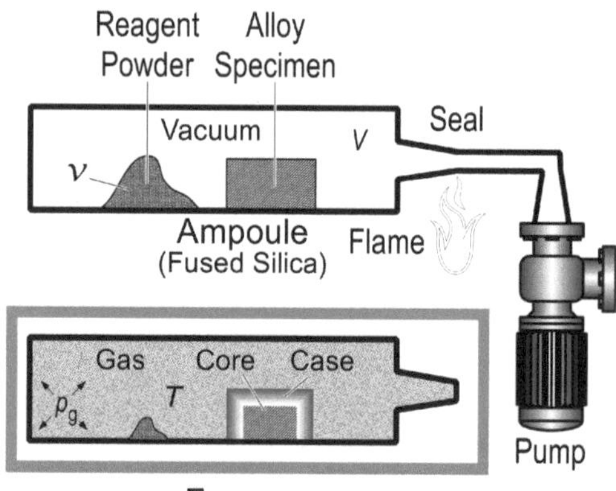

Figure 1. Concept of low-temperature nitrocarburization of alloys by closed-vessel reagent pyrolysis, the "Encapsulation Method" [22].

Another goal of our work was to analyze the pyrolysis products of the solid reagent (urea) and to understand which molecular species are key for nitrocarburization.

For brevity, in the following, we denote the approach of low-temperature nitrocarburization of alloys by reagent pyrolysis in a closed vessel as the "Encapsulation Method". In this article, we demonstrate the efficacy and potential of this method for exploiting and exploring pyrolysis of solid reagents for the purpose of alloy surface engineering by low-temperature infusion of carbon and nitrogen.

2. Material and Experimental Methods

The specimens for this work were as-machined Swagelok® ferrules. Figure 2 depicts the shape and size of a ferrule. The ferrules we used have an outer diameter $\varnothing \approx 10\,\mathrm{mm}$. These parts are used in gas tubing to enable gas-tight tube connectors ("fittings"). They are made from AISI-316L (austenitic Fe–Cr–Ni stainless steel containing Mo), but with a somewhat (1.05 times) higher Cr atom fraction and a significantly (1.2 times) higher Ni atom fraction than standard AISI-316L—within the allowed compositional range of AISI-316L. More precisely, the specification requires mass fractions $M_{Cr} \geq 0.170$ and $M_{Ni} \geq 0.120$ for Cr and Ni, respectively, while typical values for standard AISI-316L are $M_{Cr} \geq 0.165$ and $M_{Ni} \geq 0.105$. Table 1 shows the composition of the Ni- and Cr-rich AISI-316L by *atom* fractions. These specimens were chosen as they are produced under extremely well controlled conditions. This means that property changes observed after nitrocarburization can uniquely be attributed to the latter, rather than e.g., be caused by unknown changes in the microstructure between different samples from a less-defined material.

Table 1. Composition of the Ni- and Cr-Rich AISI-316L by Atom Fractions X.

Fe	Cr	Ni	Mn	Si
0.643	0.186	0.118	0.020	$\leq$0.015

Mo	N	C	P	S
0.012	$\leq$0.0039	$\leq$0.0014	$\leq$0.0008	$\leq$0.0005

Figure 2. Ferrule, made of a Ni-rich and Cr-rich austenitic stainless steel AISI-316L.

To establish the novel method of low-temperature infusion of interstitial solute, the first goal of our work, the ferrule specimens—after cleaning by ethanol and air-drying—were placed into a half-open fused-silica ampoule together with $CO(NH_2)_2$ (urea) powder as reagent. As urea is hygroscopic, it was prepared by baking at 370 K (97 °C) for 3.6 ks (1 h) and stored in a desiccator prior to usage. The open side of the tube was connected to a rotary pump via a PVC tube and sealed using an acetylene–oxygen torch while being evacuated by the pump to a residual gas pressure within (1..2) Pa ("..." denotes a continuous range). After sealing the ampoule, infusion of interstitial solute into the alloy was accomplished by heat-treating the ampoule in a tube furnace. Several heat-treating schemes were tested, all of them with keeping the ampoule ("a"), the reagent ("r"), the gas ("g"), and the specimen ("s") at the same temperature $T = T_a = T_r = T_g = T_s$ (Figure 1).

Initial experiments were carried out at a single specimen/reagent temperature $T_p = 720$ K for a processing time $t_p = 7.2$ ks (2 h). Then, we discovered that better results are obtained by a two-step process with $T_{p_1} = 620$ K (350 °C) for $t_{p_1} = 3.6$ ks (1.0 h) followed by $T_{p_2} = 720$ K (450 °C) for $t_{p_2} = 7.2$ ks (2.0 h). The results presented in this work refer to this particular two-step process.

The amount of urea was chosen such that—for the given volume of the ampoule—the net pressure p_r of decomposition products was 0.5 MPa (5 atm). Typically, in a fused-silica ampoule with a radius of 5.5 mm and a length of ≈ 200 mm, this requires 0.13 g of urea, corresponding to 2.2 mmol.

The pressure that develops under these conditions was *measured*. Specifically for this purpose, we designed an experimental procedure that involves heating an equivalent amount of urea in a cylindrical metal container closed by an initially planar, ≈ 0.5 μm thick sheet of an Al-alloy, held by a flange. The gas pressure that builds up during urea pyrolysis can then be determined from the plastic bulging of this sheet [24].

At the end of the two-step process, the ampoule was cooled in air and broken to extract the specimens. The effective diffusion depth of interstitial solute was experimentally determined and compared in three different ways from polished cross-sections: (i) indirectly by observing the *apparent* case depth ζ_{LOM} in light-optical metallographs after etching with a specific reagent, (ii) indirectly as the *apparent* case depth ζ_{HV} apparent in Vickers hardness–depth profiles, recorded at 3 different locations under a load of 25 g with a dwell time of 10 s, and (iii) directly from cross-sectional concentration-depth profiles obtained by AES (Auger electron spectrometry) performed by SAM (scanning Auger microprobe). The results were then calibrated by predetermined relative sensitivity factors for each element to quantify element fractions. In addition to these characterization techniques, we employed XRD (X-ray diffractometry) using a Bruker Discover D8 equipped with a Co-K$_\alpha$ source (wavelength $\lambda = 0.1789$ nm) in Bragg–Brentano setting.

To obtain further insight into the chemical reactions that take place during an ampoule process, we conducted complementary STA (simultaneous thermal analysis) measurements. These were performed in a Netzsch STA 449 F3 Jupiter paired with a Perseus system, consisting of a heated transfer pipe to a Bruker Alpha FTIR unit with a gas cell heated to 473 K (200 °C), located directly above the furnace.

The urea reagent powder, 8.2 mg, was contained in crucibles with *venting* lids. To study the pyrolysis products absent an alloy specimen, we employed crucibles of inert material—Al_2O_3—with a volume of 87 µL. To study the pyrolysis products in the presence of a AISI-316L specimen and to investigate the nitrocarburization it accomplished by the pyrolysis products, we pyrolyzed the urea, 8.7 mg, in crucibles of AISI-316L stainless steel with a volume of 27 µL. In other words, the AISI-316L crucibles not only served to contain the reagent powder, but also as an *alloy specimen* to be treated by nitrocarburization.

The lids of both crucibles had central holes to allow gas escape. Both measurements followed a 83 mK/s heating rate during which TGA (thermogravimetric analysis), DSC (differential scanning calirometry), and GP-FTIR (gas-phase Fourier-transform infrared spectrometry) measurements were acquired. The instrument furnace was evacuated and purged three times before the start of each sample run. Dry N_2 gas with a total flow rate of 1.2 mL/s was used to purge the system throughout the testing process. The pressure in the furnace was 0.1 MPa during testing.

The GP-FTIR spectra were extracted at temperatures showing the highest absorbance for each thermal event. Traces following wavenumbers for both NH_3 (at 965 cm^{-1}) and HNCO (at 2283 cm^{-1}) were also extracted for comparison with the TGA and DSC measurements.

3. Results

Figure 3 shows a microhardness–depth profile $h[z]$ of a nitrocarburized AISI-316L ferrule. Hardness values of "case" and "core" regions were evaluated and related to the indentations on a polished cross-section to bring out the differences. Two regions are revealed on the etched indented cross-section, the metallographic image of which shows an etch-resistant bright "case" layer and the microstructure of the "core". On the right side of $h[z]$, at a depth $h \geq 35$ µm, to which no significant amounts of interstitial solute could diffuse within the processing time, the diagram displays a hardness range of $h = (350 \pm 50)$ HV0.025. This level of hardness agrees well with the gray data points shown at the bottom left of the diagram, which were obtained from a *non-treated* (as-received) ferrule. The gray data points do not indicate any significant slope of $h[z]$, especially no increased hardness near the surface (as it might be expected from machining). Accordingly, the increased hardness after nitrocarburization (red data points) is the sole result of the treatment, not influenced by any (e.g., machining-induced) increased near-surface hardness prior to the treatment.

Compared to the base hardness $h = (350 \pm 50)$ HV0.025 exhibited by the non-treated ferrule and in the core of the nitrocarburized specimen, the hardness near the surface of the nitrocarburized specimen, at $z \approx 5$ µm, was measured to be (900 ± 20) HV0.025. With increasing depth z, i.e., from the surface deeper into the nitrocarburized specimen, the hardness gradually decreases, as expected from the graded solute-fraction–depth profile shown in Figure 5. Extrapolating the graph displayed within the z interval $(5..10)$ µm to the region directly below the surface (where h cannot be reliably measured) yields $h[0]$ within (1100 ± 100) HV0.025. Extrapolating to $h = h_b$ yields an apparent case thickness of $\zeta_{HV} \approx 12$ µm. Compared to the symmetrical pyramid indents at the "core" region, the indentations around the "case" region all exhibit asymmetry with regard to the vertical indent diagonal plane. This asymmetry can be explained from the gradation of the solute profile and the related gradation of the hardness profile, implying reduced hardness on the right side of the indents.

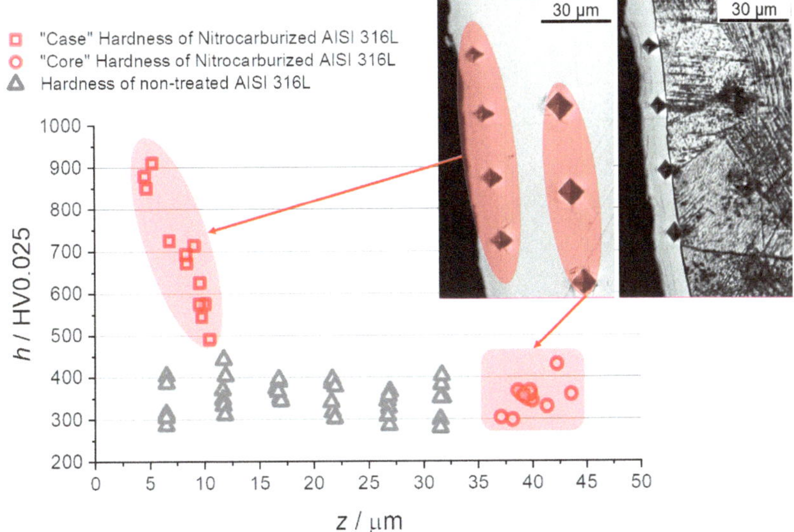

Figure 3. Microhardness–depth profile $h[z]$ recorded in the region of the solute-rich "case" and the region of the non-infused "core." Hardness values (red data points) in the two regions of the nitrocarburized AISI-316L are highlighted and related to the indents seen on a light-optical micrograph of the polished cross-section of the alloy specimen. The etched cross-section at the same location displays the "case" and "core" regions. The gray data points at the bottom left were obtained from a non-treated (as-received) AISI-316L ferrule.

Figure 4 shows a light-optical micrograph from part of a color-etched cross-section of a nitrocarburized AISI-316L ferrule. Parallel to the surface, the image features a bright conformal layer. This layer is the "case", i.e., the solute-rich zone under the alloy surface generated by the infusion process. More precisely, this is the subsurface zone in which nitrocarburization has introduced levels of carbon or nitrogen that made the material resistant to the specific etchant and etching conditions: As diffusion produces *graded* composition–depth profiles, the apparently sharp boundary to the alloy core does not correspond to the end of the diffusion profile. Rather, the *apparent* case thickness ζ_{LOM} depends on details of the metallographic etching. For example, work by Sun [25] indicates that a carbon fraction $X_C > 0.015$ is necessary to see the benefits of interstitially dissolved carbon on corrosion resistance in aqueous NaCl solution. In Figure 4, $\zeta_{\mathrm{LOM}} = (10.9 \pm 0.4)\ \mu\mathrm{m}$.

Figure 5 presents fraction–depth profiles obtained by SAM from a nitrocarburized, polished, and cross-sectioned AISI-316L specimen. The solid curves are Bézier curves intended to estimate the true nitrogen and carbon fractions in solid solution, accounting for nitride formation (see below) and excessive noise in the SAM data. (Owing to differentiation of the original—noisy—signal, noise in SAM data mainly increases background and scatter in regions of *low* signal [26]). The dashed curve displays their sum, i.e., the total fraction of interstitial solute as a function of depth.

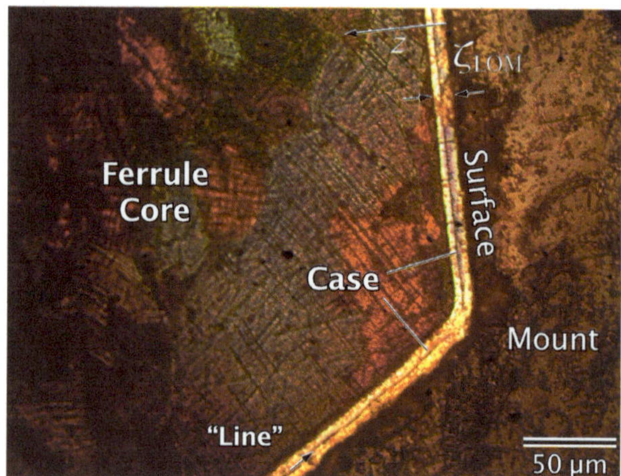

Figure 4. Light-optical micrograph of a color-etched cross-section of a nitrocarburized AISI-316L ferrule.

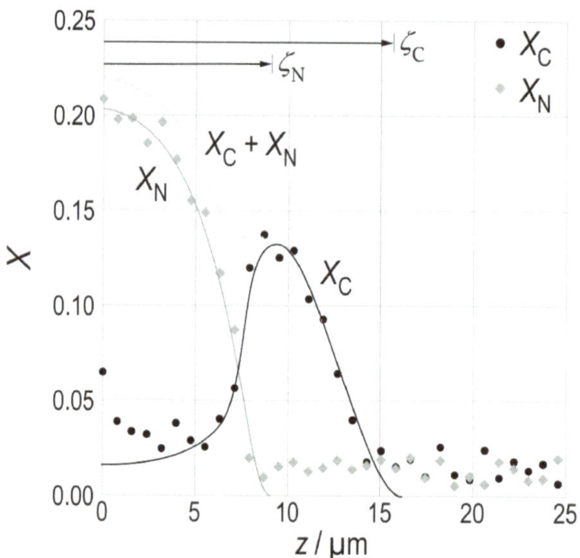

Figure 5. Fraction–depth profiles of nitrogen and carbon in an AISI-316L ferrule, obtained by SAM. The solid curves are Bézier curves intended to estimate the true fractions of carbon and nitrogen in solid solution, accounting for nitride formation and artifacts of SAM. The dashed curve displays their sum, i.e., the total fraction of interstitial solute as a function of depth.

The SAM data reveal a non-trivial depth-distribution of carbon and nitrogen with the following features:

1. A nitrogen-rich outer layer (directly below the surface) with a local thickness $\zeta_{\text{SAM-N}} = 8\,\mu\text{m}$. In this region, the carbon fraction X_C exhibits a positive slope, $dX_C/dz > 0$, corresponding to "uphill" diffusion.
2. A carbon-rich inner layer (directly below the outer layer) with a local thickness $\zeta_{\text{SAM-C}} = 8\,\mu\text{m}$. The bottom of this layer corresponds to the bottom of the case, i.e., $\zeta_{\text{SAM}} = \zeta_{\text{SAM-C}}$.

3. Evaluating $\zeta_i := \int_0^\infty X_i[z]\mathrm{d}z$, $i = \mathrm{C}, \mathrm{N}$ for the Bézier curves X_i in Figure 5 yields a ratio $X_\mathrm{N}/X_\mathrm{C} \approx 1.4$, indicating that the specimen assimilated correspondingly more nitrogen than carbon.

4. Although high X_N apparently limits the level of X_C, the total interstitial solute fraction $X[z] = X_\mathrm{C}[z] + X_\mathrm{N}[z]$ (dashed line in Figure 5) decreases monotonously with z—although the graph exhibits a small plateau in the transition region where the X_C and X_N cross.

The border between the outer, nitrogen-rich case and the inner, carbon-rich material manifests itself as a dark "line" in the LOM (light-optical microscopy) micrograph of Figure 4 (arrowed). One possible explanation for this observation is that the dark line is a groove caused by preferential attack of the metallographic etchant. As total solute fraction ($X_\mathrm{C} + X_\mathrm{N}$) in this region is *higher* than in the carbon-rich region below, this would imply that etch resistance requires either X_C or X_N—not just their sum—to exceed certain thresholds X_C^* and X_N^*, respectively. This conclusion would also explain why $\zeta_\mathrm{LOM} < \zeta_\mathrm{SAM}$. Comparing ζ_SAM in Figure 5 with ζ_LOM in Figure 4 suggests $X_\mathrm{C}^* > 0.05$. Another potential explanation for the observed dark line between the nitrogen-rich and the carbon-rich region could be grooving or step formation by creation of a local galvanic element.

The case depths ζ we observed in a multitude of experiments exhibits significant variation (e.g., compared a currently used industrial low-temperature carburization process [2]). For ζ_SAM, in particular, we observed a (sample) standard variation $s/\sqrt{Z} = 0.2\,\mu\mathrm{m}$ when measuring various locations in several samples.

In any of its forms (ζ_HV, ζ_LOM, ζ_SAM), the "case depth" sensitively depends on the specific shape of the fraction–depth profile at its tail (where $X_\mathrm{N}, X_\mathrm{C} \to 0$. This means that the exact value for the case depth is determined by the spatial distribution of only a small fraction of solute atoms. A more robust measure is the *mean solute depth*

$$\bar{z} := \frac{\int_0^\infty z X[z]\,\mathrm{d}z}{\int_0^\infty X[z]\,\mathrm{d}z}. \tag{1}$$

Corresponding evaluation of the Bézier curves in Figure 5 yields the following mean depths of carbon, nitrogen, and both interstitial solutes combined:

$$\bar{z}_\mathrm{C} = 9.2\,\mu\mathrm{m}, \tag{2}$$
$$\bar{z}_\mathrm{N} = 3.4\,\mu\mathrm{m}, \tag{3}$$
$$\bar{z} = 5.8\,\mu\mathrm{m}. \tag{4}$$

Figure 6 shows an X-ray diffractogram recorded from a nitrocarburized AISI-316L specimen in Bragg–Brentano setting. For comparison, the plot also shows a corresponding diffractogram of *non*-treated AISI-316L. In the $2\theta_\mathrm{B}$ of Figure 6, the A1 (FCC, face-centered-cubic) structure of the austenite generates two peaks, labeled as $(111)\,\gamma$ and $(200)\,\gamma$. For the Co-K$_\alpha$ radiation employed in this work, the X-ray penetration depth is within $(6..16)\,\mu\mathrm{m}$ for absorption within the range $(80..99)\,\%$. This implies that the diffractogram of the nitrocarburized specimen displays significant information from depths z within the entire range $[0, \zeta_\mathrm{SAM}]$. Regions closer to the surface contribute higher intensity than regions deeper below the surface. Based on the SAM fraction–depth profile of the nitrocarburized material in Figure 5, the features of this diffractogram can be interpreted as follows:

1. The two peaks of highest intensity, labeled $(111)\,\gamma_\mathrm{N}$ and $(200)\,\gamma_\mathrm{N}$, originate from the outer case, which is rich in nitrogen. Compared to the reference peaks from non-treated material, these peaks exhibit a pronounced shift to smaller $2\theta_\mathrm{B}$. This indicates a corresponding increase of the lattice parameter, caused by interstitially dissolved nitrogen expanding the interatomic spacings between the metal atoms.

2. The two peaks of third and fourth highest intensity, labeled $(111)\,\gamma_\mathrm{C}$ and $(200)\,\gamma_\mathrm{C}$, originate from the inner case, which is rich in carbon. Their intensities are lower than those of the corresponding peaks from the outer case because the inner case is deeper

below the surface. The peaks from the inner case also exhibit a shift to lower $2\theta_B$ versus the corresponding values for non-treated material. However, the peak shift is smaller because (i) the maximum carbon fraction in the inner case is smaller than the maximum nitrogen fraction in the outer case and (ii) interstitially dissolved carbon is less effective than nitrogen in increasing interatomic spacings between the metals atoms [27,28].

3. The peaks $(111)\,\gamma_N$, $(200)\,\gamma_N$, $(111)\,\gamma_C$, and $(200)\,\gamma_C$ are not mirror-symmetric. They exhibit shoulders on the right, i.e., towards higher $2\theta_B$, because in both the outer and the inner case the regions of highest interstitial solute fraction are closest to the surface, i.e., contribute higher diffracted intensity than regions of lower solute fraction in greater depth z.

4. The diffractogram from the nitrocarburized material also exhibits peaks with low intensity at the positions of the $(111)\,\gamma$ overlapped with the corresponding peaks from the non-treated material. These peaks originate from the non-infused material at depths $z > \zeta_{SAM}$, proving that diffractogram samples over the entire depth range $[0, \zeta_{SAM}]$ and beyond.

5. No additional peaks are observed, indicating that second phases—if formed at all—have a negligibly small volume fraction (<0.05).

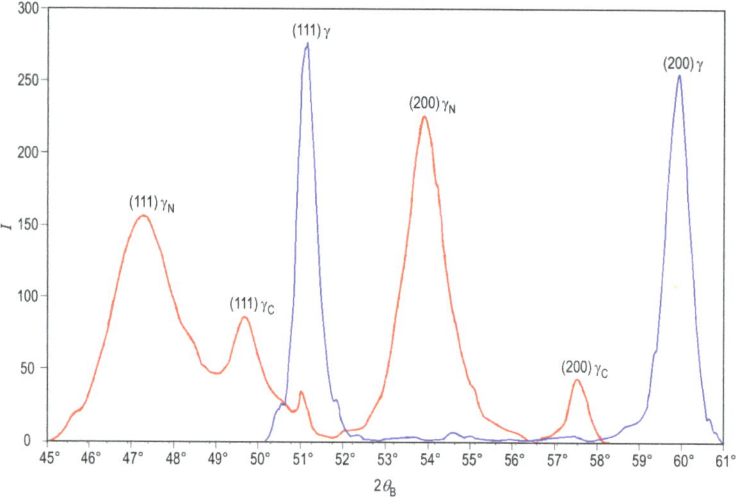

Figure 6. X-ray diffractograms of nitrocarburized (red graph) and non-treated (blue graph) AISI-316L ferrules, recorded with Co-K$_\alpha$ radiation in Bragg–Brentano setting. I: Relative intensity, θ_B: Bragg angle.

To address the second goal of our work, identifying the pyrolysis products of the reagent and their role in nitrocarburizing AISI-316L, we performed GP-FTIR under N_2 at ambient pressure. These experiments were carried out in two different forms: (i) Pyrolyzing urea in an inert (Al_2O_3) crucible. (ii) Pyrolyzing urea in a crucible of AISI-316L. This setup allowed us to study how the resulting solute–depth profiles $X[z]$ correlate with specific processing parameters, but also how the presence of the alloy surface impacts the spectrum of molecular species in the gas atmosphere.

For the two different crucibles, Al_2O_3 and AISI-316L, Figure 7 displays GP-FTIR data for NH_3 (ammonia) and HNCO (cyanic acid), respectively, as well as complementary TGA mass-change and DSC heat-flux data. Regions I, II, III, and IV are similar to those noted by Schaber et al. [29], but with adjustments to encompass the important thermal events observed. Region I, $T_I = (406..490)\,K$, is above the melting point and includes initial decomposition of $CO(NH_2)_2$ and the formation of $C_2H_5N_3O_2$ (biuret). Region II, $T_{II} = (490..523)\,K$, includes the decomposition of $C_2H_5N_3O_2$ and rapid formation of various aromatic compounds. Region III, $T_{III} = (523..633)$ K, indicates the beginning of primary $(CNOH)_3$ (cyanuric acid) decomposition. Region IV, $T_{IV} > 633$ K, is the final decomposition of other aromatic compounds formed earlier in small quantities.

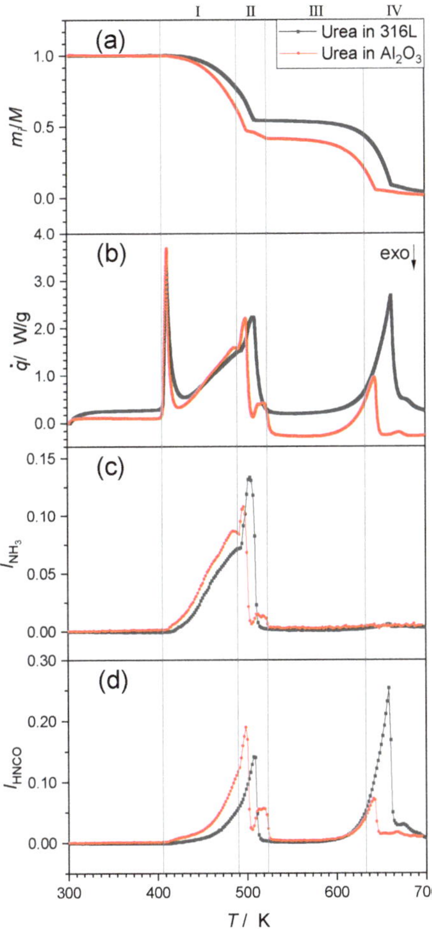

Figure 7. Urea pyrolysis in AISI-316L and Al_2O_3 crucibles heated at $5\,\mathrm{K/min}$ under N_2. (**a**) TGA results. (**b**) DSC results; $\dot{q}$ denotes specific heat flow. (**c**) GP-FTIR data; I_{NH_3} is the measured intensity of NH_3 at $965\,\mathrm{cm}^{-1}$. (**d**) GP-FTIR data; I_{HNCO} is the measured intensity of HNCO at $2283\,\mathrm{cm}^{-1}$.

The TGA data reveal that the molecular composition of the gas obtained by urea pyrolysis in the AISI-316L crucible as a function of temperature differs significantly from what is obtained in the Al_2O_3 crucible. The urea in the AISI-316L crucibles lost less mass in Regions I and II than the urea in the Al_2O_3 crucible. In the AISI-316L crucible, the urea also does not have the small, independent mass loss within Region II that accounts for some of the different mass loss in these regions in the Al_2O_3 crucible. Furthermore, the AISI-316L causes the urea to lose more mass in the decomposition observed in Region III/IV.

Like the TGA data, the DSC heat-flux data corroborate these differences. Specifically, the DSC peak for the small mass loss in Region II is not present for the urea sample in AISI-316L crucible and, likewise, the Region III/IV peak (corresponding to the larger mass loss with this sample) is larger and longer in duration in the AISI-316L crucible.

The GP-FTIR data for NH_3 and HNCO clarify the differences between these samples in different crucibles. Both samples have similar trends in the GP-FTIR traces for NH_3 and HNCO in Region I. However, in Region II, the one in AISI-316L crucible shows a larger NH_3 release and no subsequent emission like the small mass loss seen in that in Al_2O_3crucible. The Region II in AISI-316L crucible also has a smaller HNCO emission and likewise does not have an emission for the missing mass-loss event. Region III/IV has minimal signal intensity for NH_3 in both samples. However, the urea sample in AISI-316L crucible in Regions III and IV shows a higher intensity of HNCO and longer duration than the same trace for the Al_2O_3 crucible.

4. Discussion

4.1. Efficacy of the Encapsulation Method

For the low rate at which the temperature was changed to obtain the data of Figure 7, it can be assumed that the gas atmosphere is equilibrated at each temperature, i.e., there is no retardation. The results of Figure 7 suggest that already in Step 1 of the new two-step closed-vessel low-temperature nitrocarburization process, the reagent (urea) is pyrolyzed into gas molecules that activate the alloy surface. However, the temperature $T_{p_1} = 620\,\mathrm{K}$ is too low for significant diffusion of interstitial solute within t_{p_1}. Significant diffusion can only occur at $T_{p_2} = 720\,\mathrm{K}$.

Carbon- and nitrogen-fraction–depth profiles $X_C[z], X_N[z]$ with the characteristic features of those in Figure 5 have already been observed in earlier work on nitrocarburization [18,19,30]. As seen in Figure 5, the presence of nitrogen in the austenite reduces the carbon fraction near the surface. This has been explained by nitrogen locally elevating the activity coefficient of carbon [22,31]. Therefore, although the chemical potential of carbon globally fulfills the requirement $\partial_z \mu_C < 0$ for inward diffusion, $(\partial_z X_C > 0)$ below the nitrogen-rich near-surface zone, corresponding to an apparent "uphill" diffusion of carbon. Furthermore, as discussed in earlier work [16], the concave shape of $X_C[z], X_N[z]$ reflects a positive concentration dependence of the respective diffusion coefficients: $\partial_{X_C} D_C > 0, \partial_{X_N} D_N > 0$.

To investigate the efficacy of the new two-step closed-vessel low-temperature nitrocarburization process, we compare the mean depths (2)–(4) of interstitial solute to the mean depth of carbon accomplished by an industrial low-temperature gas-phase carburization process [16].

Figure 8 presents a carbon-fraction–depth profile of AISI-316L generated by that industrial process, measured by SAM. The processing temperature of $T_p = 726\,\mathrm{K}$ was practically the same as $T_{p_2} = 720\,\mathrm{K}$ in the second step of the new two-step closed-vessel nitrocarburization process introduced in this article. The processing time, however, was much longer: $t_p = 137\,\mathrm{ks}$ (38 h). The continuous line marked "SIM" in Figure 8 is a corresponding *simulated* profile, based on a concentration-dependent carbon diffusion coefficient obtained by Boltzmann–Matano analysis [16].

Based on the simulation shown in Figure 8 and the intermediate carbon-fraction–depth profiles $X_c[t_i]$ at intermediate time steps (not shown), Figure 9 shows how the mean carbon depth $\bar{z}_C$ evolves during the industrial process (data marked by hollow circles and connected by a continuous line). The solid symbols in Figure 9 indicate the corresponding data (2)–(4) for the Encapsulation Method, using the two-step process. The mean depth $\bar{z}_N$ of nitrogen is comparable to the mean depth $\bar{z}_C$ accomplished by the industrial process in the same amount of time (7.2 ks). However, the mean depth $\bar{z}_C$ of carbon is about 4 times the corresponding value accomplished by the industrial process. Conversely, the industrial process takes about 15 times longer to accomplish the same mean depth of carbon (dashed marker lines in Figure 9). The mean depth $\bar{z}_I$ of both interstitial solutes combined is more than 2 times larger than $\bar{z}_C$ after the industrial process, in which carbon reaches this mean depth only after 5 times longer processing.

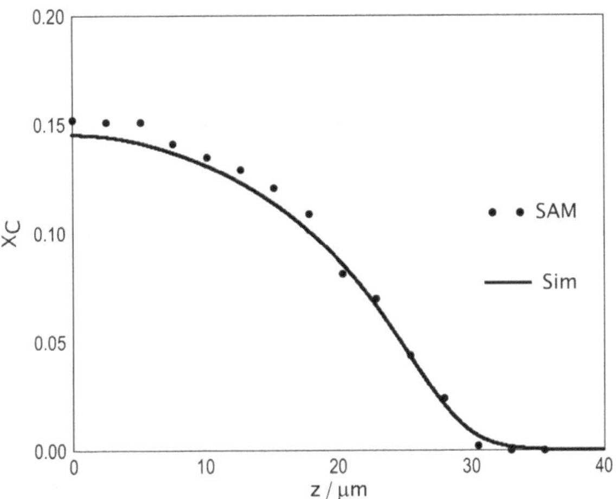

Figure 8. Carbon-fraction–depth profile of an industrial low-temperature gas-phase carburization process, measured by SAM, and corresponding simulated profile, based on concentration-dependent diffusion [16].

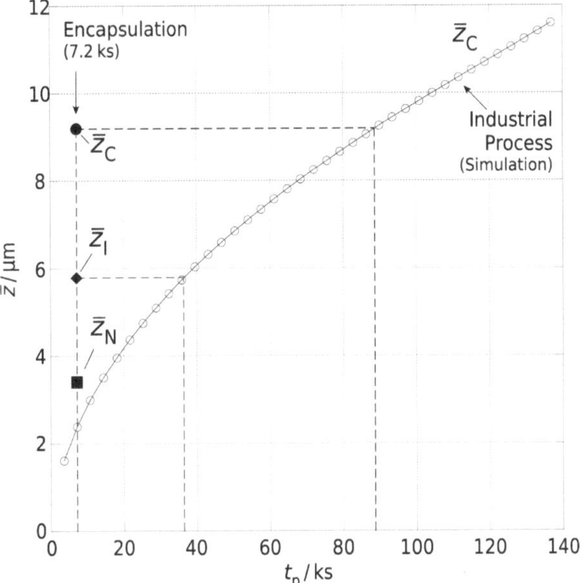

Figure 9. Mean depth of interstitial solute versus processing time t_p. The solid data points marked by $\bar{z}_\mathrm{N}$, $\bar{z}_\mathrm{C}$, and $\bar{z}_\mathrm{I}$ refer to the Encapsulation Method, indicating the mean depth of nitrogen, carbon, and both interstitial solutes (nitrogen and carbon combined), respectively. The hollow circular data points, connected by a continuous line, show *simulated* $\bar{z}_\mathrm{C}$ data for the industrial process [16].

The definition (1) of the mean solute depth does not include information about the actual level (fraction, concentration) of interstitial solute. As a figure of merit that equivalently reflects solute level as well as solute depth, we define the "penetration"

$$P := \int_0^\infty z\, X[z]\, \mathrm{d}z.$$

(5)

Figure 10 shows P as a function of processing time t_p. Data points referring to the Encapsulation Method are marked by solid symbols. P_N, P_C, and P_I indicate P for nitrogen, carbon, and both interstitial solutes (nitrogen and carbon combined), respectively, after the processing time of 7.2 ks (2 h, Step 2 of the process). (No significant interstitial solute diffusion will occur during Step 1 owing to insufficient temperature). Data points marked by hollow circles, connected by a continuous line, show corresponding P_C data for the established industrial low-temperature gas-phase carburization process [16]. These data (shown only up to 80 ks) were obtained by simulation, evaluating (5) for the intermediate steps that lead to the simulated carbon-fraction–depth profile labeled as "Sim" in Figure 8. Comparing the data in Figure 10 confirms that the Encapsulation Method is *very* effective. The industrial process needs 4 times longer (28 ks) to accomplish the same penetration of carbon—and 6 times longer to accomplish the same total integration of interstitial solute (dashed marker lines in Figure 10).

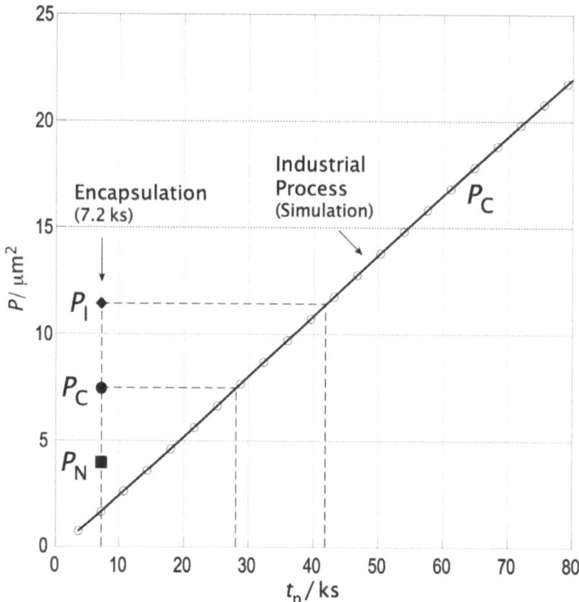

Figure 10. Interstitial solute penetration P, defined in (5), as a function of processing time t_p. The solid data points marked by P_N, P_C, and P_I refer to the Encapsulation Method, indicating P for nitrogen, carbon, and both interstitial solutes (nitrogen and carbon combined), respectively. The hollow circular data points, connected by a continuous line, show *simulated* P_C data for the industrial process [16].

4.2. Auto-Catalytic Effect of the Alloy Surface

Earlier work of our group and others demonstrated the efficacy of NH_3 for both surface activation and infusion of nitrogen [20,21,32–34]. In addition, surface infusion of nitrogen and carbon is provided by decomposition of $HNCO$, $(CNOH)_3$, $C_3H_4N_4O_2$ (ammelide), or $C_3H_5N_5O$ (ammeline). We expect small shifts in decomposition and evaporation temperatures as well as changes in gas composition because the Encapsulation Method operates at 0.5 MPa, while STA testing occurs at 0.1 MPa. As labeled in Figure 7, the pyrolysis of urea occurs in four major reaction regions, which have been adjusted here for clarity based on previous work [29]. Chemical equations for the following reactions can be found in Appendix A. Region I encompasses urea pyrolysis into NH_3 and HNCO according to (A1), $C_2H_5N_3O_2$ formation according to (A2), and small amounts of aromatic compound formation according to (A3)–(A6). In Region II, $C_2H_5N_3O_2$ decomposes by the reverse of (A2) into HNCO and urea (which decomposes as above or further reacts). Reactions producing the aromatic compounds occur at higher rates in this region. Furthermore, NH_3, HNCO and H_2O evolve according to (A3)–(A12). A smaller subsequent thermal event follows with the evolution of HNCO and NH_3 from the solid residue matrix. Only $(CNOH)_3$, $C_3H_4N_4O_2$ (ammelide), and $C_3H_5N_5O$ are noted as solids present at the start of reaction Region III. As these aromatic compounds (primarily $(CNOH)_3$) begin to pyrolyze, HNCO forms according to (A13). Region IV covers the conclusion of $(CNOH)_3$ decomposition and later melt with decomposition of $C_3H_4N_4O_2$ and $C_3H_5N_5O$ with the emission of HNCO and small amounts of NH_3, CO_2 (carbon dioxide), and H_2O. The presently reported results of urea pyrolysis in Al_2O_3 track closely with these reaction schemes from [29].

However, our data reveal significant differences between urea pyrolyzed in an *inert* Al_2O_3- and in a AISI-316L crucibles. The AISI-316L presence caused a more complete conversion of urea to $(CNOH)_3$ as demonstrated by the AISI-316L sample exhibiting the following features:

- A lower total mass loss in Regions I and II.
- Absence of a small mass-loss event and corresponding DSC peak in Region II.
- A larger mass loss in Region III, continuing into Region IV.
- A higher emission of NH_3 in Region II before the missing event.
- Lower HNCO emission in Region II.
- A larger HNCO emission in Regions III and IV.

Through catalytic or direct reaction with the alloy surface, more HNCO was converted to (primarily) $(CNOH)_3$ and less HNCO mass was lost in Regions I and II. The missing mass-loss emission event in Region II would normally be from NH_3 and HNCO leaving the solid residue. The absence of this event suggests that there was significantly less HNCO trapped in the solid residue or the reactions that would have released it did not occur in this range. The larger mass-loss event of Regions III/IV for the AISI-316L sample as well as the related HNCO emission during this stage suggests that the larger mass was primarily the $(CNOH)_3$ that formed in larger quantity because of the alloy surface effects.

The differences in the STA and GP-FTIR data for urea pyrolyzed in AISI-316L vs in Al_2O_3 indicate an auto-catalytic effect of the alloy surface on the spectrum pyrolysis products—or even a direct participation of surface atoms in chemical reactions. These differences in the condensed and gas phases caused by the presence of an alloy surface with the pyrolyzing reagent may provide substantial opportunities for low-temperature alloy surface engineering by solid-reagent pyrolysis. It is not yet known whether liquid phases on the alloy surface or molecular species in the gas phase contributes more to surface activation and infusion of carbon and nitrogen. In an open system, the delayed evolution of active gas species (NH_3 and HNCO) may be beneficial for both the overall treatment effectiveness and efficient reagent use because these species would be present at higher temperatures necessary for treatment instead of exhausting too early to be effective. However, the ampoule process we introduce here and analogous constant volume, closed-vessel processes largely bypass the effect of reagent decomposing too early because all pyrolysis products remain in close proximity to the alloy surface.

4.3. Model of Interstitial Infusion via the Encapsulation Method

Figure 11 shows a model of the micromechanisms underlying the rapid surface engineering of Cr-bearing alloys through closed-vessel reagent pyrolysis. Under the well-defined process parameters within the vessel, particularly a uniform temperature distribution, a controlled reagent quantity per volume of vessel, and a controlled reagent quantity per alloy surface area, ramping up the temperature initiates urea pyrolysis and the evolution of gaseous HNCO and NH_3. The pressure of in the vessel gradually increases as these species evolve. For a sufficiently slow temperature ramp, it can be assumed that the partial pressure of each species corresponds to its equilibrium value for the given temperature and solid-reagent ingot. By decomposing or modifying the passivating Cr-rich oxide film on the alloy surface, accompanied by the release of CO_2 and/or H_2O, HNCO and NH_3 activate the alloy surface (AISI-316L) for infusion of interstitial solute (carbon and nitrogen). Even though we separate the surface activation as a prerequisite step of the whole treatment in this model, the reactions may occur throughout the remainder of the process if chromium oxide should reform, keeping the surface transparent to the interstitial solute. Interaction of the pyrolysis products with the active alloy surface catalyze the formation of $(CNOH)_3$ and other aromatic species according to (A3) and (A4). These species remain present until they decompose at higher temperature. Depending on reagent content and composition, an organic film may form on the alloy surface [13]. Future work needs to show whether this film can directly provide carbon and nitrogen to diffuse into the alloy surface or first needs to decompose to gas species (e.g., HNCO), which then feed the surface with carbon and nitrogen. At the treatment temperature ($T_{P_2} = 720\,\mathrm{K} = 450\,^\circ\mathrm{C}$), the HNCO and NH_3 produced by the pyrolysis reactions may adsorb at the surface and further decompose to produce atomic carbon and nitrogen that diffuse into the surface. Breakdown byproducts include CO_2, H_2O, and H_2. This model can be generalized and applied to a wide array of reagents (e.g., guanidine, GuHCl and biguanide HCl) and alloys (e.g., Al6XN, Inconel 625, Hastelloy C-22).

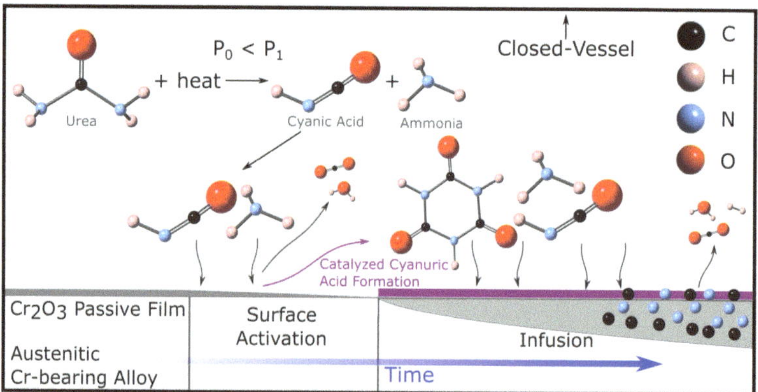

Figure 11. Model for surface engineering of Cr-bearing alloys by solid-reagent pyrolysis in a closed-vessel. (For clarity, minor reactions and species not shown).

5. Conclusions

The new "Encapsulation Method" introduced in this article, i.e., low-temperature nitrocarburization of alloys by closed-vessel reagent pyrolysis, is highly effective for surface engineering of structural alloys by infusing high concentrations of interstitial solute. The method allows for rapid experimentation at low cost and with a minimum of required laboratory equipment. Furthermore, it has the conceptual advantage of working with a closed-vessel system (ampoule) under well-defined conditions, which, in principle, are easy to control. This should facilitate the reproduction of experimental results. If variations in apparent case depth are observed, these may originate from the fact that the apparent case depth measured by different methods depends on rather insignificant details of the "tail" of the solute-fraction–depth profile. More robust figures of merit have been introduced in this paper, which also provide more meaningful quality parameters than "the" case depth for the success of surface engineering by infusion of interstitial solute.

For laboratory-scale experimentation, a great conceptual advantage of a closed system is the superior control of the gas composition (which can be analyzed as a function of numerous parameters in a suitable open system). Using sophisticated experimental methods we developed for this purpose, we succeeded in measuring (i) the total pressure that the products of pyrolysis build up inside the ampoule and (ii) the major constituents of the gas mixture that forms by pyrolysis of urea. The experimental results indicate that interaction of gas molecules with the alloy surface can have an auto-catalytic effect by altering the surrounding gas composition and condensed species in a way that accelerates solute infusion, especially by providing a higher activity of HNCO at high temperature.

The Encapsulation Method is substantially more effective in driving interstitial solute into the alloy than other low-temperature gas-phase diffusion processes currently used in the industry. As the latter are not limited by the availability of carbon from the gas atmosphere, the higher efficacy of the Encapsulation Method must result from better surface activation, the effect of nitrogen on the chemical potential of carbon, or enhanced mobility of the solute, particularly carbon.

Author Contributions: Conceptualization, F.E.; Data curation, C.I.; Formal analysis, C.I. and F.E.; Funding acquisition, F.E.; Investigation, C.I., Z.R. and A.A.; Methodology, C.I. and F.E.; Project administration, F.E.; Resources, F.E.; Software, F.E.; Supervision, A.H. and F.E.; Visualization, C.I. and Z.R.; Writing—original draft, C.I., Z.R., A.A. and F.E.; Writing—review & editing, C.I., Z.R. and F.E. All authors have read and agreed to the published version of the manuscript.

Funding: We acknowledge financial support from Swagelok, DOE project NEUP-3451 under contract 00128081, and NSF project DMR-1208812.

Data Availability Statement: The data we used for this article are contained within the article.

Acknowledgments: We would like to thank P Williams and S Marx of Swagelok for helpful discussions and for providing ferrule specimens and Vickers hardness data. We also thank S Collins at CWRU for many fruitful discussions. We especially thank Mike Purcel of CDS Analytical for performing pyrolysis tests. We acknowledge SCSAM (Swagelok Center for Surface Analysis of Materials) in the Case School of Engineering at Case Western Reserve University for the usage of instruments and the support SCSAM engineers have provided with the collection and analysis of experimental data.

Conflicts of Interest: The authors declare no conflict of interest. The funders had no role in the design of the study, nor in the analysis or interpretation of data, nor in the writing of the manuscript, nor in the decision to publish the results. As a service, the funders provided hardness measurements on specimens processed by us.

Appendix A

$$CO(NH_2)_2 \text{ (m)} \rightarrow HNCO \text{ (g)} + NH_3 \text{ (g)}, \tag{A1}$$

$$CO(NH_2)_2 \text{ (m)} + HNCO \text{ (g)} \rightarrow C_2H_5N_3O_2 \text{ (g)}, \tag{A2}$$

$$C_2H_5N_3O_2 \text{ (m)} + HNCO \text{ (g)} \rightarrow (CNOH)_3 \text{ (s)} + NH_3 \text{ (g)}, \tag{A3}$$

$$3\,HNCO \text{ (g)} \rightarrow (CNOH)_3 \text{ (s)}, \tag{A4}$$

$$CO(NH_2)_2 \text{ (m)} + 2\,HNCO \text{ (g)} \rightarrow C_3H_4N_4O_2 \text{ (s)} + 2\,H_2O \text{ (g)}, \tag{A5}$$

$$C_2H_5N_3O_2 \text{ (m)} + HNCO \text{ (g)} \rightarrow C_3H_4N_4O_2 \text{ (s)} + H_2O, \tag{A6}$$

$$2\,C_2H_5N_3O_2 \text{ (m)} \rightarrow (CNOH)_3 \text{ (s)} + HNCO \text{ (g)} + 2\,NH_3 \text{ (g)}, \tag{A7}$$

$$2\,C_2H_5N_3O_2 \text{ (m)} \rightarrow C_3H_4N_4O_2 \text{ (s)} + HNCO \text{ (g)} + 2\,NH_3 \text{ (g)} + H_2O \text{ (g)}, \tag{A8}$$

$$C_2H_5N_3O_2 \text{ (m)} + HCNO \text{ (g)} \rightarrow C_3H_5N_5O \text{ (s)} + H_2O, \tag{A9}$$

$$C_3H_4N_4O_2 \text{ (s)} + NH_3 \text{ (g)} \rightarrow C_3H_5N_5O \text{ (s)} + H_2O \text{ (g)}, \tag{A10}$$

$$2\,HNCO \text{ (g)} + CO(NH_2)_2 \text{ (m)} \rightarrow C_3H_5N_5O \text{ (s)} + 2\,H_2O \text{ (g)}, \tag{A11}$$

$$HNCO \text{ (g)} + C_2H_5N_3O_2 \text{ (m)} \rightarrow C_3H_5N_5O \text{ (s)} + 2\,H_2O \text{ (g)}, \tag{A12}$$

$$(CNOH)_3 \text{ (s)} \rightarrow 3\,HNCO \text{ (g)}. \tag{A13}$$

References

1. Dong, H. S-phase surface engineering of Fe—Cr, Co—Cr and Ni—Cr alloys. *Int. Mater. Rev.* **2010**, *55*, 65–98. [CrossRef]
2. Collins, S.R.; Williams, P.C.; Marx, S.V.; Heuer, A.H.; Ernst, F.; Kahn, H. Low-Temperature Carburization of Austenitic Stainless Steels. In *ASM Handbook: Vol. 4D, Heat Treating of Irons and Steels*; Dossett, J., Totten, G., Eds.; ASM International: Materials Park, OH, USA, 2014; Volume 4, pp. 451–460.
3. Ren, Z.; Heuer, A.H.; Ernst, F. Ultrahigh-strength AISI-316 austenitic stainless steel foils through concentrated interstitial carbon. *Acta Mater.* **2019**, *167*, 231–240. [CrossRef]
4. Ren, Z.; Ernst, F. Electronic Impact of Concentrated Interstitial Carbon on Physical Properties of AISI-316 Austenitic Stainless Steel. *Acta Mater.* **2019**, *173*, 96–105. [CrossRef]
5. Christiansen, T.L.; Somers, M.A.J. Low temperature gaseous surface hardening of stainless steel. *HTM J. Heat Treat. Mater.* **2011**, *66*, 109–115. [CrossRef]
6. Christiansen, T.L.; Ståhl, K.; Brink, B.K.; Somers, M.A.J. On the Carbon Solubility in Expanded Austenite and Formation of Hägg Carbide in AISI 316 Stainless Steel. *Steel Res. Int.* **2016**, *87*, 1395–1405. [CrossRef]
7. Borgioli, F. From Austenitic Stainless Steel to Expanded Austenite-S Phase: Formation, Characteristics and Properties of an Elusive Metastable Phase. *Metals* **2020**, *10*, 187. [CrossRef]
8. Collins, S.; Williams, P. Low-Temperature Colossal Supersaturation. *Adv. Mater. Process.* **2006**, *164*, 32–33.
9. Somers, M.; Christiansen, T. Gaseous processes for low temperature surface hardening of stainless steel. In *Thermochemical Surface Engineering of Steels*; Elsevier: Amsterdam, The Netherlands, 2015; pp. 581–614. [CrossRef]
10. Ge, Y.; Ernst, F.; Kahn, H.; Heuer, A.H. The Effect of Surface Finish on Low-Temperature + Acetylene-Based Carburization of 316L Austenitic Stainless Steel. *Metall. Mater. Trans. A* **2014**, *45*, 2338–2345. [CrossRef]
11. Ge, Y. Low-Temperature Acetylene-Based Carburization and Nitrocarburizing of 316L Austenitic Stainless Steel. Ph.D. Thesis, Case Western Reserve University, Cleveland, OH, USA, 2013.
12. Christiansen, T.; Drouet, M.; Martinavičius, A.; Somers, M. Isotope exchange investigation of nitrogen redistribution in expanded austenite. *Scr. Mater.* **2013**, *69*, 582–585. [CrossRef]
13. Illing, C.A.W. Chemical Mechanisms and Microstructural Modification of Alloy Surface Activation for Low-Temperature Carburization. Master's Thesis, Case Western Reserve University School of Graduate Studies, Cleveland, OH, USA, 2018.
14. Li, Z.; Illing, C.; Heuer, A.H.; Ernst, F. Low-Temperature Carburization of AL-6XN Enabled by Provisional Passivation. *Metals* **2018**, *8*, 997. [CrossRef]
15. Christiansen, T.; Somers, M.A. Determination of concentration dependent diffusion coefficient of nitrogen in expanded austenite. *Int. J. Mater. Res.* **2008**, *24*, 159–167. [CrossRef]
16. Ernst, F.; Avishai, A.; Kahn, H.; Gu, X.; Michal, G.M.; Heuer, A.H. Enhanced Carbon Diffusion in Austenitic Stainless Steel Carburized at Low Temperature. *Metall. Mater. Trans. A* **2009**, *40*, 1768–1780. [CrossRef]
17. Gu, X.; Michal, G.M.; Ernst, F.; Kahn, H.; Heuer, A.H. Concentration-Dependent Carbon Diffusivity in Austenite. *Metall. Mater. Trans. A* **2014**, *45*, 3790–3799. [CrossRef]
18. Christiansen, T.L.; Hummelshoj, T.S.; Somers, M.A.J. A Method of Activating an Article of Passive Ferrous or Non-Ferrous Metal Prior to Carburizing, Nitriding and/or Nitrocarburizing. Patent EP2278038-A1, 2011.
19. Christiansen, T.L.; Hummelshoj, T.S.; Somers, M.A.J. Method of Activating an Article of Passive Ferrous or Non-Ferrous Metal Prior to Carburising, Nitriding and for Nitrocarburising. Patent US8845823-B2, 2014.
20. Liu, L.; Ernst, F.; Michal, G.M.; Heuer, A.H. Surface Hardening of Ti Alloys by Gas-Phase Nitridation: Kinetic Control of the Nitrogen Surface Activity. *Metall. Mater. Trans.* **2005**, *36*, 2429–2434. [CrossRef]
21. Ernst, F.; Michal, G.M.; Oba, F.; Liu, L.; Blush, J.; Heuer, A.H. Gas-Phase Surface Alloying under "Kinetic Control", A Novel Approach to Improving the Surface Properties of Titanium Alloys. *Int. J. Mater. Res.* **2006**, *97*, 597–606. [CrossRef]
22. Ren, Z.; Eppell, S.; Collins, S.; Ernst, F. Co–Cr–Mo alloys: Improved wear resistance through low-temperature gas-phase nitro-carburization. *Surf. Coat. Technol.* **2019**, *378*, 124943. [CrossRef]
23. Ren, Z. Intrinsic Properties of "Case" and Potential Biomedical Applications. Ph.D. Thesis, Case Western Reserve University, Cleveland, OH, USA, 2019.
24. Agaponova, A. Encapsulation Method for Surface Engineering of Corrosion-Resistant Alloys by Low-Temperature Nitro-Carburization. Master's Thesis, Case Western Reserve University, Cleveland, OH, USA, 2015.
25. Sun, Y. Corrosion behaviour of low temperature plasma carburised 316L stainless steel in chloride containing solutions. *Corros. Sci.* **2010**, *52*, 2661–2670. [CrossRef]
26. Ernst, F. Background of SAM Atom-Fraction Profiles. *Mater. Charact.* **2017**, *125*, 142–151. [CrossRef]
27. Wriedt, H.A.; Murray, J.L. The N-Ti (Nitrogen-Titanium) System. In *Binary Alloy Phase Diagrams*; Massalski, T., Murray, J., Bennet, L.H., Baker, H., Eds.; American Society for Metals: Metals Park, OH, USA, 1987; pp. 176–186.
28. Ridley, N.; Stuart, H. Partial Molar Volumes from High-Temperature Lattice Parameters of Iron–Carbon Austenites. *Met. Sci. J.* **1970**, *4*, 219–222. [CrossRef]
29. Schaber, P.M.; Colson, J.; Higgins, S.; Thielen, D.; Anspach, B.; Brauer, J. Thermal Decomposition (Pyrolysis) of Urea in an Open Reaction Vessel. *Thermochim. Acta* **2004**, *424*, 131–142. [CrossRef]
30. Wu, D.; Ge, Y.; Kahn, H.; Ernst, F.; Heuer, A.H. Diffusion Profiles After Nitrocarburizing Austenitic Stainless Steel. *Surf. Coat. Technol.* **2016**, *279*, 180–185. [CrossRef]

31. Gu, X.; Michal, G.M.; Ernst, F.; Kahn, H.; Heuer, A.H. Numerical Simulations of Carbon and Nitrogen Composition-Depth Profiles in Nitrocarburized Austenitic Stainless Steels. *Metall. Mater. Trans. A* **2014**, *45*, 4268–4279. [CrossRef]
32. Christiansen, T.; Somers, M.A.J. Controlled dissolution of colossal quantities of nitrogen in stainless steel. *Metall. Mater. Trans. A (Phys. Metall. Mater. Sci.)* **2006**, *37A*, 675–682. [CrossRef]
33. Christiansen, T.; Somers, M.A.J. Low temperature gaseous nitriding and carburising of stainless steel. *Surf. Eng.* **2005**, *21*, 445–455. [CrossRef]
34. Somers, M.A.J.; Christiansen, T.; MØLLER, P. Case-Hardening of Stainless Steel. Patent WO2004007789A2, 2004.

Article

Effect of Deformation Structure of AISI 316L in Low-Temperature Vacuum Carburizing

Hyunseok Cheon [1,2], Kyu-Sik Kim [3], Sunkwang Kim [1], Sung-Bo Heo [1], Jae-Hun Lim [4], Jun-Ho Kim [1,*] and Seog-Young Yoon [2,*]

[1] Dongnam Division, Korea Institute of Industrial Technology (KITECH), Jungang-ro 33-1, Yangsan 50623, Korea; chs1006@kitech.re.kr (H.C.); skkim82@kitech.re.kr (S.K.); hsb85@kitech.re.kr (S.-B.H.)
[2] School of Materials Science and Engineering, Pusan National University, Busandaehak-ro 63beon-gil, Busan 46241, Korea
[3] Department of Materials Science and Engineering, Inha University, Inha-ro 100, Incheon 22212, Korea; kskim87@inha.edu
[4] BMT Co., Ltd., Sanmakgongdannam11-gil 35, Yangsan 50568, Korea; prime619@superlok.com
* Correspondence: jhkim81@kitech.re.kr (J.-H.K.); syy3@pusan.ac.kr (S.-Y.Y.)

Abstract: The effect of plastic deformation applied to AISI 316L in low-temperature vacuum carburizing without surface activation was investigated. To create a difference in the deformation states of each specimen, solution and stress-relieving heat treatment were performed using plastically deformed AISI 316L, and the deformation structure and the carburized layer were observed with EBSD and OM. The change in lattice parameter was confirmed with XRD, and the natural oxide layers were analyzed through TEM and XPS. In this study, the carburized layer on the deformed AISI 316L was the thinnest and the dissolved carbon content of the layer was the lowest. The thickness and composition of the natural oxide layer on the surface were changed due to the deformed structure. The natural oxide layer on the deformed AISI 316L was the thickest, and the layer was formed with a bi-layer structure consisting of an upper Cr-rich layer and a lower Fe-rich layer. The thick and Cr-rich oxide layer was difficult to decompose due to the requirement for lower oxygen partial pressure. In conclusion, the oxide layer is the most influential factor, and its thickness and composition may determine carburizing efficiency in low-temperature vacuum carburizing without surface activation.

Keywords: plastic deformation; deformed structure; low-temperature vacuum carburizing; expanded austenite; natural oxide layer

Citation: Cheon, H.; Kim, K.-S.; Kim, S.; Heo, S.-B.; Lim, J.-H.; Kim, J.-H.; Yoon, S.-Y. Effect of Deformation Structure of AISI 316L in Low-Temperature Vacuum Carburizing. *Metals* **2021**, *11*, 1762. https://doi.org/10.3390/met11111762

Academic Editor: Shinichiro Adachi

Received: 10 September 2021
Accepted: 30 October 2021
Published: 2 November 2021

Publisher's Note: MDPI stays neutral with regard to jurisdictional claims in published maps and institutional affiliations.

1. Introduction

Austenitic stainless steels (ASSs) are the materials used in various industries due to their formability and weldability based on excellent corrosion resistance. However, ASSs have the disadvantage of a lack of durability due to their relatively low mechanical properties (hardness, strength, and fatigue resistance) in manufacturing parts, so a thermochemical process is required [1,2]. While conventional thermochemical treatment methods such as carburizing and nitriding have been applied, the inherent corrosion resistance of ASSs is impaired due to precipitates such as carbide and nitride [3,4]. To overcome this problem, a low-temperature hardening process, which is carried out below precipitate formation temperature, has been developed [5]. The low-temperature surface hardening process is a method in which carbon or nitrogen is supersaturated in the interstitial site of the FCC at a relatively low temperature [6–8]. The interstitial atoms can be supersaturated due to the influence of alloying elements, thereby increasing surface hardness, wear resistance, and fatigue resistance, because those induced the compressive stress at the surface as the lattice expands due to supersaturation. For this reason, the low-temperature hardening layer, which is supersaturated by the interstitial atoms, is called expanded austenite (or the supersaturated layer), and the inherent corrosion resistance of ASSs can be also maintained.

However, there is one important thing to assess before carrying out the thermochemical surface hardening process that can improve the mechanical properties and maintain the inherent corrosion resistance of ASSs. It is precisely to consider the effect of plastic working imparted to the material before the thermochemical surface hardening process. Plastic working causes metallographic changes in ASSs such that the grain size is reduced, strain-induced martensite is formed, or the formation of Cr precipitates is accelerated, and these changes affect the result of the thermochemical surface hardening process. It has been reported [9–11] that the carburizing and nitriding performed in a low-temperature region are also affected by whether plastic deformation of the material is applied or not. They reported that on equiaxed tensile strained AISI 304, strain-induced martensite was formed thereby increasing the formation rate of the low-temperature nitride layer and CrN precipitates. Other research results have been published that show that the formation rate of the low-temperature nitride layer was increased when deformation was applied to the surface through the Surface Mechanical Attrition Treatment (SMAT) process [12,13]. On the other hand, in the case of AISI 316L, in which martensite was not easily formed by plastic working, the depth of the low-temperature hardening layer was not measurably affected by whether to generate plastic deformation or not [14].

However, the results of the above research, although desirable, may tend to be inconsistent when the process method is changed. This is because the researchers removed the natural oxide layer and Beilby layer of ASSs with a surface activation step using electro-polishing or halogen gases (NF_3, HCl, etc.) before low-temperature surface hardening [15,16]. If these layers were present, it would have been difficult to confirm the clarity of the study because the natural oxide layer and Beilby layer interfere during the diffusion and solid solution of nitrogen. The removal process of these layers (surface activation) is considered an essential step in the low-temperature surface hardening of ASSs using gas atmosphere, but it is not required in low-temperature vacuum carburizing, which is being recently researched [17–20]. This is because low-temperature vacuum carburizing is the technology that simultaneously performs the removal of the oxide layer (surface activation) using the strong active characteristics of acetylene, and also because the natural oxide layer is spontaneously decomposed due to the low partial pressure of oxygen generated by the reaction of the process gases in a vacuum atmosphere [21]. Therefore, the state of the natural oxide layer can be a major factor in the carburizing efficiency of the low-temperature vacuum carburizing process. This is because the deformed structure and phase formed by plastic working not only affect the diffusion of carbon atoms but also the formation of a natural oxide film. Research on the correlation between a plastically deformed structure and carburizing efficiency in low-temperature vacuum carburizing has been reported in [18], but it focuses mainly on the deformed structure and diffusion. It was reported that the deformed structure was formed as nano-sized grains on the outermost surface, thus causing a low carburizing efficiency because diffusion was interrupted by the changed grains. In addition, it also showed that the plastically worked material was inefficient for carburizing because the oxide layer was formed thickly by the deformed structure (nano-sized grain) and that it can be removed with a suitable range of oxygen partial pressure by controlling the ratio of acetylene and hydrogen. However, research data are insufficient to apply low-temperature vacuum carburizing quickly and easily to the industries. Therefore, it is necessary to improve the low-temperature vacuum carburizing efficiency by studying the natural oxide film changes because the deformed structure including the diffusion interference factors has already been studied.

In this study, the step of low-temperature vacuum carburizing without surface activation was performed on AISI 316L which had a plastically deformed structure and a natural oxide layer. In order to control the accumulated strain and microstructure, solution and stress-relieving heat treatment were performed on the plastic-worked AISI 316L, and the formation rate and characteristics of the low-temperature carburized layer according to the decreasingly deformed structure by heat treatment were investigated. Based on the experimental results, the changes in the natural oxide layer and deformed structure of ASSs due to plastic deformation and recovery are confirmed, and its effect on the low-temperature vacuum carburizing process is discussed.

2. Materials and Methods

2.1. Material and Heat Treatment

The specimens of cold-worked AISI 316L (ASTM A193-B8M CL2) used in this study were manufactured by drawing and were made in the form of disks (diameter (D) and thickness (T) of 32 and 7 mm, respectively). The specimen showed a tensile strength of 758 MPa, yield strength of 675 MPa, elongation of 28%, reduction in area of 68%, and average hardness of 342 $HV_{0.3}$. The stress-free specimen was prepared via solution heat treatment at 1353 K for 3 h using the cold-worked specimen. To obtain stress-relieving specimens, the cold-worked specimens were subjected to heat treatment in an argon atmosphere for 1 h at 973 and 623 K, which are the stress-relieving temperatures of general ASSs. After heat treatment, the Vickers hardness of the material was measured as 150 $HV_{0.3}$, 250 $HV_{0.3}$, and 280 $HV_{0.3}$, respectively, in the order of the heat-treatment temperature. After each treatment, the specimens were ground with silicon carbide papers from 180 to 2000 grit. Mechanical polishing with a 1 μm diamond suspension was conducted to achieve a mirror-like surface finish. The polished specimens were exposed to an atmosphere of 50 °C and 50% humidity for 3 h in the environmental test chamber in order to make a uniform passive layer. Immediately after the exposure, the low-temperature vacuum carburizing process was performed.

2.2. Low-Temperature Vacuum Carburizing

Low-temperature vacuum carburizing was performed in a commercial vacuum carburizing furnace (VH556-10, Rübig, Austria). High purity acetylene (C_2H_2, 99.80%) and hydrogen (H_2, 99.999%) were used as the process gases. Low-temperature vacuum carburizing was carried out for 10 and 14 h with a carburizing potential (Kc) of 1.41 at a working pressure of 800 Pa and a temperature of 723 K. After carburizing, the specimens were cooled to a temperature below 473 K in 60 s under a pressure of 5 bars of nitrogen gas to control any unexpected irregular phase transformation. At the end of this process, the specimens attained the ambient temperature in 10 min.

2.3. Characterization

The specimens after carburizing were etched in aqua regia for 3 min and observed using optical microscopy (OM; Nikon Eclipse LV 150NL, Nikon, Tokyo, Japan) The changes in the average grain size and grain boundary dependent on the heat treatment were analyzed by electron backscattered diffraction (EBSD; SU6600, Hitachi, Tokyo, Japan). The EBSD data were analyzed using orientation imaging microscopy software (OIM Analysis, EDAX, Draper, UT, USA). (X-ray diffraction (XRD) analysis was conducted for phase identification at room temperature using the conventional symmetric Bragg–Brentano geometry by an X-ray generated by monochromatic Cu-Kα radiation (Ultima IV, Rigaku, Tokyo, Japan). The XRD measurements were conducted at 30 to 90 degrees and used radiation operating at 40 kV and 30 mA with a step size of 0.02. The detector used was a dual-type ultra-silicon strip (D/TeX) with an 8 mm first slit and 0.3 mm second slit, and the radius of the horizontal goniometer combined with D/Tex was 185 mm. The $(111)_{\gamma}$ peak of austenite was defined with ICDD pdf #00-033-0397. An X-ray photoelectron spectroscopy (Sigma probe XPS system, Thermo fisher Sci., Waltham, MA, USA) depth profile was obtained with a beam size of 400 μm to analyze the thickness and composition of the natural oxide layer in the initial and heat-treated specimens. Al-Kα X-rays (hν = 1486.74 eV) were used as the primary radiation source. Analysis parameters were 1 eV for the survey scan and 0.1 eV for the high resolution scan energy step size, in CAE scan mode with 37° of take-off angle (TOA). The ion beam sputter raster size was 2 mm, and the etched rate was 0.18 nm/s. The XPS spectra were recorded in the depth direction from the surface every 10 s for a total of 300 s. The XPS spectra were recorded for a 2p3/2 level of Ni, Cr, and Fe and for a 1 s level of O and C. Quantitative elemental analyses were performed using Avantage software (Thermo Fisher Sci., Waltham, MA, USA). The quantification was carried out using relative sensitivity factors (RSF) provided in the database of the Avantage software. In addition, to analyze the natural oxide layer of the specimens, specimens were prepared in a cross-section using a focused ion beam and observed under 200 kV using transmission electron microscopy (JEM 2010, JEOL, Tokyo, Japan). The TEM specimen preparation process was as follows: a pure Pt coating was applied to the polished material to prevent further oxidation and contamination. After that, specimens were collected using FIB and TEM images were measured.

3. Results and Discussions

The heat treatment was respectively conducted to compare the state of the plastic deformation and the gradually relieving stress, because diffusion, the driving force behind low-temperature vacuum carburizing, is affected by the state of the material. These AISI 316L specimens subjected to either plastic deformation, solution heat-treatment, or stress-relieving treatment (623 K, 973 K) were measured by Electron Backscatter Diffraction (EBSD) to observe the microstructure and strain distribution.

Figure 1 shows EBSD results images and values of the low-angle grain boundary (LAGB), high-angle grain boundary (HAGB), and kernel average misorientation (KAM) for each specimen. The grain boundary, which is the main diffusion pathway, showed differences according to the deformation states of the specimens. The plastically deformed (Deformed AISI 316L) specimen was found to have the highest amount of LAGB and HAGB. Stress-relieving heat-treated (SR AISI 316L) specimens were shown to have a lower amount of HAGB and LAGB relative to the deformed AISI 316L due to grain growth and recovery. In the case of the solution heat-treated (Solution AISI 316L) specimen, HAGB and LAGB were significantly reduced as recrystallization and grain growth had occurred. The change in misorientation (KAM value) was confirmed since the misorientation of the local grain is also a factor that can affect the diffusion of atoms. From the KAM map in Figure 1e–h, it was confirmed that the Deformed AISI 316L had many deformed structures due to plastic working, and these were uniformly distributed throughout the material. On the other hand, in the SR AISI 316L (623 K, 973 K) specimens of Figure 1f,g, it was confirmed that the misorientation within the grain structure was reduced as the heat treatment temperature increased. In case of the Solution AISI 316L specimen, the deformed structure disappeared due to recrystallization and the misorientation was only observed in a few grain boundaries. The plastic equivalent strain was confirmed in order to represent at a glance each specimen in a metallurgically different state by plastic deformation and recovery heat treatment. The plastic equivalent strain rate of specimens in this study was quantified and compared using the following Equation (1) revealed in the study of Githinji et al. [22]. KAM_a and ε are average KAM values and equivalent strain, respectively. The equivalent strain of each specimen was derived as 0.74 (Deformed AISI 316L), 0.66 (SR at 623 K), 0.58 (SR at 973 K), and 0.02 (Solution at 1323 K), showing a similar trend to the EBSD analysis result.

$$KAM_a = 3.314 - \frac{3.014}{\left[1 + \left(\frac{\varepsilon}{0.770}\right)^{1.178}\right]} \tag{1}$$

The EBSD results showed that the Deformed AISI 316L had a lot of grain boundaries which are the main path of diffusion and the misorientations which indicate geometrically necessary dislocation (GND). When the deformation was relieved through heat treatment, the grain boundary and the misorientation tended to decrease as the heat treatment temperature increased. The plastic equivalent strain and KAM value did not decrease significantly up to 973 K, but it was confirmed that the grain gradually recovered and the misorientation inside the grain progressively decreased with an increasing heat treatment temperature. Therefore, it was predicted that the Deformed AISI 316L would have the highest carburizing efficiency when low-temperature vacuum carburizing was carried out.

This was because the diffusion path was the greatest in the Deformed AISI 316L among the plastically deformed specimens. In order to confirm whether the carburizing result according to this prediction occurred, specimens with different amounts of deformation were subjected to low-temperature vacuum carburizing.

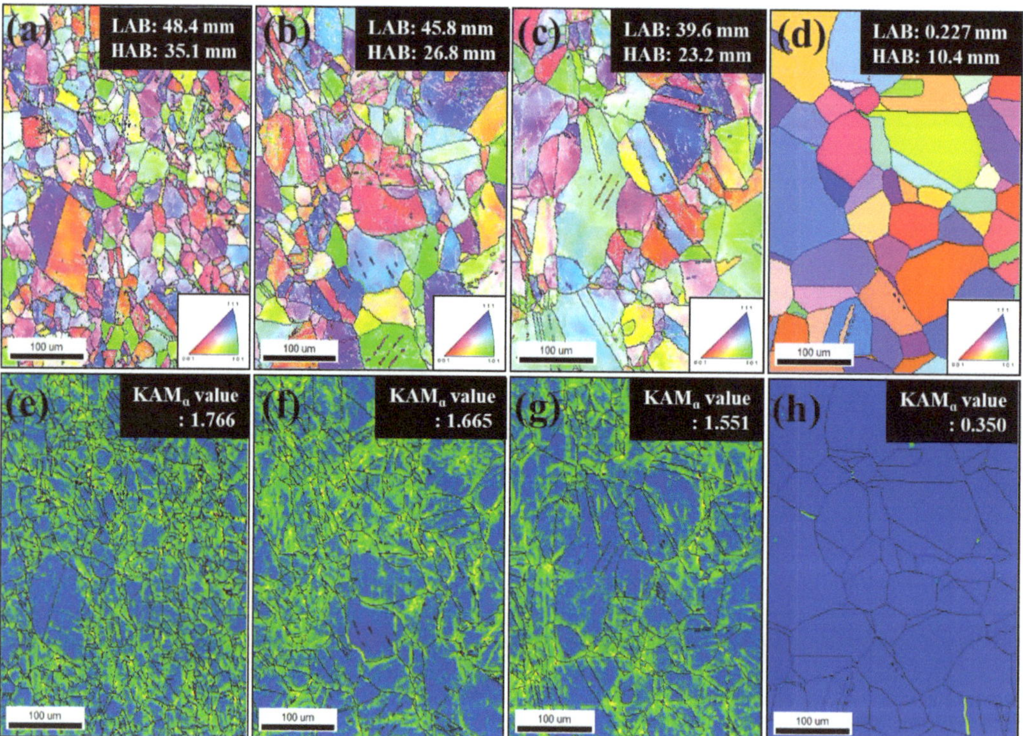

Figure 1. Inverse pole figure and KAM map of AISI 316L with different deformed states; (**a**,**e**) plastically deformed AISI 316L, (**b**,**f**) 623 K stress-relieving treated, (**c**,**g**) 973 K stress-relieving treated, and (**d**,**h**) 1353 K solution-treated AISI 316L.

Deformed, Solution, and SR (623 K, 973 K) AISI 316L specimens were subjected to the low-temperature vacuum carburizing process unrequired for the surface activation step for 10 h. The formation rate of the carburized layer was different depending on the strain rate within the specimen, and the low-temperature carburizing layer was shown in Figure 2. In the Deformed AISI 316L specimen, the carburizing layer was not formed uniformly (Figure 2a) and only an island-shaped carburizing structure at a depth of approximately 9.2 μm was observed. On the contrary, a uniform carburized layer of about 14.8 μm was formed in the Solution AISI 316L specimen (Figure 2d). It can be seen that the SR AISI 316L specimens with an intermediate deformation between Deformed AISI 316L and Solution AISI 316L were non-uniformly formed from 3.8 to 13.5 μm in terms of carburizing layer thicknesses. In order to confirm whether a similar pattern was observed when the carburizing time was increased, low-temperature vacuum carburizing was carried out for 14 h and the results were shown in Figure 2's inset images. After carburizing for 14 h, the thicknesses of the carburized layers that formed uniformly on all specimens increased to 14.5~18.7 μm. However, grain boundaries and deformed structures originating from the base material were observed within the carburized layer, except for the Solution AISI 316L specimen. In general, the defects (e.g., twins and grain boundary) could not be observed within the carburized layer due to expansion of the lattice by the interstitial solid solution of carbon atoms after low-temperature carburizing. The situation in this study (the situation in which the defects were observed) that the grain boundaries were observed within the carburized layer may have appeared because the lattice expansion did not sufficiently occur due to low carbon content dissolved into the layer. Therefore, it was considered that the difference in dissolved carbon content occurred according to the deformed state of the specimen before carburizing. To confirm in more detail the point predicted from the OM result, the carburized layer was analyzed through XRD.

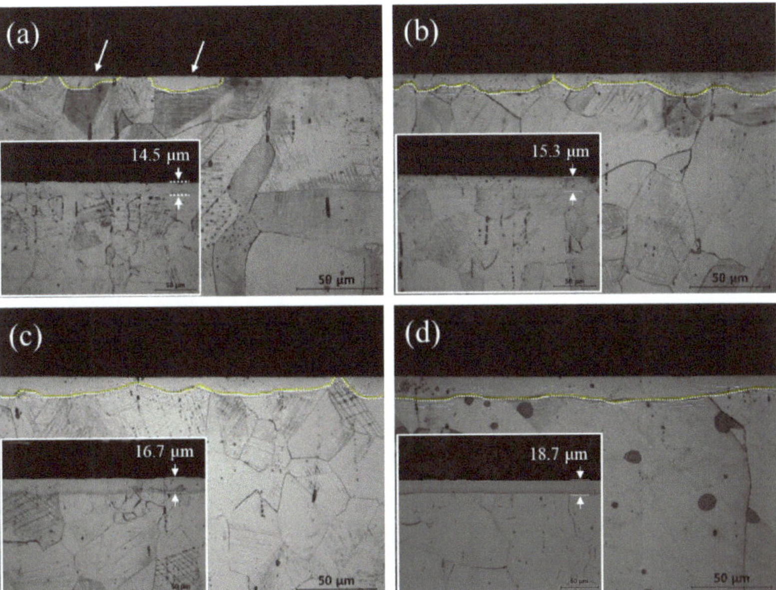

Figure 2. Cross-section of microstructure of austenitic stainless steel after low-temperature vacuum carburizing process for 10 h and 14 h (inset images); (**a**) plastically deformed AISI316L, (**b**) 623 K stress-relieving treated, (**c**) 973 K stress-relieving treated, and (**d**) 1353 K solution-treated AISI 316L.

The information from the ICDD pdf file 00-033-0397 exhibited the common name as AISI 304. Its chemical composition is different from AISI 316L. In the case of 316L with a higher Mo content, the (111) $_\gamma$ peak occurs at a lower angle because the lattice constant must be larger due to the influence of Mo. However, in the XRD results of cold-worked AISI 316L and heat-treated AISI 316L before carburizing in this study, (111) $_\gamma$ peaks were confirmed in the range of 43.60°~43.65°. This opposite peak shift in the base material was presumed to be the effect of plastic deformation. However, the XRD results were used for the purpose of comparing after carburizing only according to the degree of deformation of the material. Thus, the values of Figure 3b were calculated based on two theta of (111) $_\gamma$ of the ICDD pdf 00-033-0397 (43.60°). The (111) $_\gamma$ peaks of austenite after the low-temperature carburizing process were shifted more toward the lower two theta than the peak of the base material because lattice expansion occurred due to a supersaturation of carbon without precipitation. Furthermore, the peaks were shifted to the lower two theta when a higher amount of carbon was dissolved. Figure 3 shows the results of X-ray diffraction analysis, confirming the lattice expansion of the specimen before and after carburizing. In the diffraction pattern of the (111) $_\gamma$ plane measured on the specimen subjected to carburizing treatment for 14 h, it was confirmed that low-temperature carburizing was performed well because all peaks of the carburized specimens were shifted to a lower two theta compared to the base material before carburizing (Figure 3a). It was also shown that the peaks of the (111) $_\gamma$ plane tended to be narrower and shifted to a lower angle in accordance with a reduced deformed structure by heat treatment (from 42.85° to 42.65°). The lattice expansion rate was calculated by d-spacing obtained using Bragg's law and shown in Figure 3b. The lattice of the Deformed AISI 316L specimen was expanded by 1.67%. As the temperature of the heat treatment increased, the lattice expansion rate increased by 1.78%, 2.01%, and 2.12%, respectively. Figure 3b also showed the predicted carbon content derived by Equation (2) based on the lattice constants. The predicted carbon content in Figure 3b was derived using the research results of Hummelshøj et al. [23]. Equation (2) is shown below (a is the lattice constants, X_c is the weight percent of carbon):

$$a = 0.35965 \,[\text{nm}] + \left(0.0028 \pm 9.07 \times 10^{-5}\right) x_c (\text{wt.\%C}) \tag{2}$$

The carbon content in the Deformed AISI 316L specimen was predicted to be 2.05 wt.%C, and the carbon content in the SR AISI 316L specimens increased when the stress-relieving treatment temperature increased. In the case of the Solution AISI 316L specimen in which the deformed structure was almost removed, the highest carbon content among the specimens was shown—2.64 wt.%C. From the above results, it was confirmed that the lattice expansion and the dissolved carbon content were increased when the decreased deformed structure was made by plastic deformation. Similar to the OM analysis result, the XRD result also confirmed that the lattice expansion of the Solution AISI 316L specimen was the greatest, and the carbon content of the carburized layer was relatively uniform through the small FWHM.

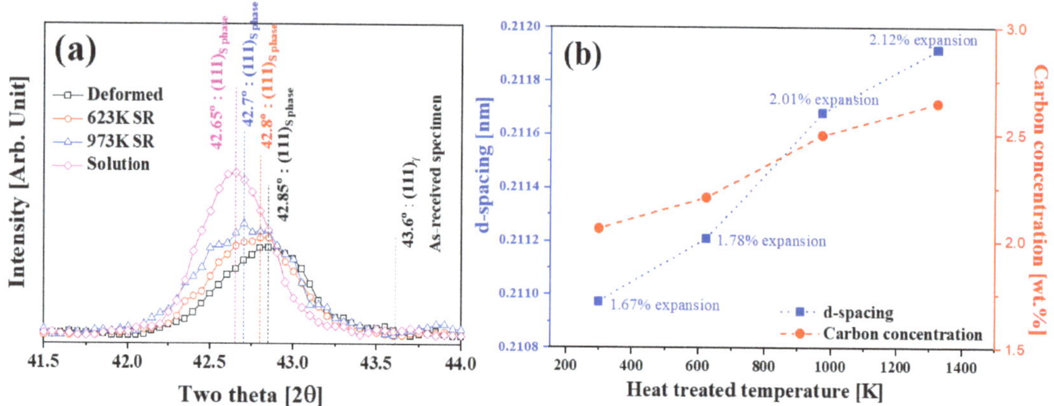

Figure 3. The results of phase analysis of AISI316L carburized for 14 h; (**a**) peak shift and broadening of (111)$_\gamma$ plane, (**b**) d-spacing of (111)$_\gamma$ plane and estimated carbon concentrations.

According to other studies [9,10,24], it was found that the increase in dislocation density and the formation of strain-induced martensite in accordance with plastic working did not affect the formation of the carburized layer or rather increased the carburizing rate. In this study, however, it was confirmed that the carbon content and thickness of carburized layer were decreased after low-temperature carburizing when the many deformed structures remained. Since these conflicting results, it may be surmised that other reasons be at play due to differences in the process. By comparison with the results of other studies, it was inferred that there were other reasons why the formation of the low-temperature carburized layer was disturbed by the deformed structure. As in the results of Li et al. [25], the nano-grained Beilby layer generated by plastic working may interrupt the diffusion of carbon. Alternatively, it may be because the natural oxide layer was formed differently from the conventional state on the nano-sized deformed layer generated with plastic working, as in the research of Ge [18]. As a result of this study, carried out without the surface activation step, it was confirmed that the carburized structure (expanded austenite) was formed only in some regions in the Deformed AISI 316L specimen during the 10 h carburization, and in the case of the Solution AISI 316L specimen, the shape of the layer was formed in the same process. When the carburizing was performed for 14 h, although the shape of the carburized layers was almost formed in all specimens whether the deformed structure remained or not, the thicknesses were different for each specimen. In carburizing for 14 h, the difference in the thickness of the carburized layer meant that the diffusion of the interstitial atoms was disturbed by the deformed structure in each specimen. In addition, as previously mentioned in [18], there was another reason for the difference in carburizing efficiency, because the natural oxide layer formed differently from the conventional state on the nano-sized deformed layer generated with the plastic working. It was regarded that the required time for decomposition of the natural oxide layer during the carburizing process was different, because all specimens in this study stabilized the natural oxide layer in the same environment after polishing. Therefore, the natural oxide layer of specimens before carburizing, in which the deformation state was changed by plastic working and heat treatment, was concretely analyzed.

The natural oxide layer formed on the surface of all carburized specimens was analyzed in the depth direction by X-ray photoelectron spectroscopy and the result is shown in Figure 4a. Iron (Fe) content on the outmost surface was lower than 20 at.% but gradually increased below a depth of 7 nm. In addition, it was shown in the surface region that Fe content tended to be relatively higher as the heat treatment temperature was increased. Chromium (Cr) content was less than 7 at.% at the outermost surface, but it was immediately increased in an inward direction and stabilized at 12~14 at.% below a depth of 10 nm. However, it was shown that Cr content was increased rapidly to about 15 at.% at a depth of 2 nm for the Deformed and SR 623 K AISI 316L specimen, and then decreased and stabilized below a depth of 7 nm. In the other two specimens, also, Cr content increased from 7 at.% to 10~12 at.% up to a depth of 2 nm; however, there was no significant change after 2 nm. Oxygen content was measured about 45–53 at.% at the outermost surface in all specimens, and rapidly decreased in a depth direction and stabilized below a depth of 5 nm. A large amount of the oxygen content was measured in the order of the quantity of retained deformed structure, and the thickness of the natural oxide layer (*1/2 point of maximum oxygen content (please note: this follows the SEMI F60 standard for indirect verification)) was also 2.1 nm (Solution), 2.8 nm (SR 973 K), 3.2 nm (SR 623 K), and 3.9 nm (Deformed). In other words, the oxygen content and the thickness of natural oxide layer increased as the amount of deformed structure increased. This tendency was confirmed more specifically with the images taken by transmission electron microscopy shown in Figure 4. In the SR 623 K AISI 316L specimen, the deformed structure was mostly maintained and a natural oxide layer with a thickness of about 2.7 nm was observed (Figure 4b, black arrows). Nevertheless, the existence of oxide clusters over 5 nm was observed in some areas (Figure 4b, white arrow) On the other hand, in the case of the Solution AISI 316L specimen, the deformed structure was not observed, and a uniform oxide layer of 1.2 nm was formed without oxide clusters (Figure 4c).

As mentioned in the previous results, the thickness and composition of the natural oxide layer were changed according to the difference and presence in the deformed structure formed by plastic working and post-heat treatment. The natural oxide layer of the Deformed AISI 316L specimen was the thickest when the deformation structure was reduced due to an increase in heat-treatment temperature, and the thickness of the natural oxide layer decreased linearly. The composition of the natural oxide layer showed a tendency divided into particular differences with a heat-treatment temperature based on about 623 K. In terms of diffusion, the formation of a thick oxide layer in the Deformed AISI 316L specimen may be supposed to be due to the many diffusion paths generated by plastic working. Additionally, it was considered that the main diffusion path was the high-angle grain boundary. Although the LAGB significantly decreased only in the Solution AISI 316L specimen, the thickness and composition of the natural oxide layer were gradually changed according to the heat-treatment temperature. This means that pipe diffusion of oxygen atoms through the dislocation of strain-hardened austenitic stainless steel did not significantly affect the formation of the natural oxide layer. Therefore, it was considered that diffusion of oxygen atoms and oxygen-friendly metal-based atoms through the high-angle grain boundary had a greater effect on the formation of the natural oxide layer [26]. In this aspect, the results of composition analysis, that the Cr_xO_y was formed more dominantly than Fe_xO_y until a depth of 5 nm in the natural oxide layer on the Deformed AISI 316L specimen, can be considered as follows: Cr atoms that have higher oxygen affinity than other metals were diffused relatively quickly along the high-angle grain boundary, and when more high-angle grain boundaries generated by plastic working remained, a relatively high Cr concentration was induced on the surface region of the natural oxide layer. This was because Cr (and substitutional metal) atoms were fixed in the grain boundaries and the dislocations at the non-equilibrium state generated by plastic working, and then the Cr atoms rapidly diffused outward through the path when the surface was exposed to oxygen from the outside [27,28]. As a result, a relatively thick oxide layer was formed and the Cr-rich oxide layer was more predominantly formed on the outermost surface because there were many grain boundaries from the fine grains by plastic working. This result was also found by Gui et al. [29], who reported that a natural oxide layer having a bi-layer structure composed of an upper Cr-rich oxide film and a lower Fe-rich oxide film was formed on the fine grain in the surface region by plastic working.

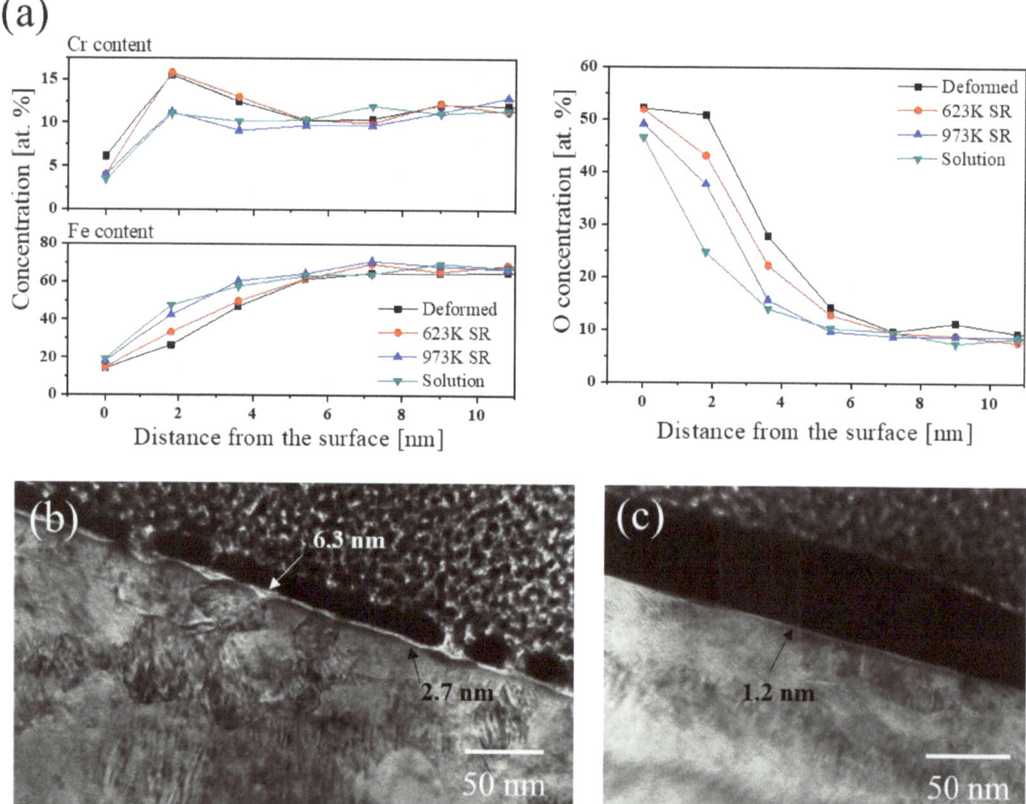

Figure 4. (**a**) The results of depth profiling of Fe, Cr, and O by XPS analysis of specimens heat treated in various temperature conditions before low-temperature vacuum carburizing (non-treatment, 623 K, 973 K, and 1323 K), and microstructures of natural oxide layer by TEM after heat treatment at (**b**) 623 K and (**c**) 1323 K before low-temperature vacuum carburizing.

The natural oxide layers, with different thicknesses and compositions, also affected the carburizing efficiency. For the same reason as a natural oxide layer, the Deformed AISI 316L specimen with the most dislocations and grain boundaries should have higher carburized efficiency or not negative significant influence since carbon diffusion is faster than stress-relived specimens. However, the opposite result was obtained; it is confirmed from this result that the decomposition of the natural oxide layer affects the carburizing efficiency more in the absence of the surface activation process. In low-temperature vacuum carburizing only using acetylene and hydrogen, there were typically two known decomposition methods of the oxide layer (surface activation). These were influenced by the thickness and composition of the oxide layer. In the first method, as mentioned in our previous result [20], the free radicals, which were generated during the carburizing reaction, removed the natural oxide layer [17,18]. The second method was to keep the oxygen partial pressure low so that the oxide layer spontaneously decomposed [18]. When the free radicals generated during the reaction of unsaturated hydrocarbon gas decomposed the thick and/or dense natural oxide layer, it took long time to remove it. Additionally, the oxygen partial pressure for spontaneous decomposition of the Cr-rich oxide layer was about 10^{-10} bar lower than that of the Fe-rich oxide layer. Therefore, the thicker the Cr-rich oxide film, the more difficult it is to remove. Thus, it was considered that the removal of a thick Cr-rich oxide layer formed by plastic working was difficult during the low-temperature vacuum carburizing process. As a result, the carburizing rate is inevitably slowed.

In this study, it was conjectured that the carburizing efficiency was relatively low because of the thick Cr-rich natural oxide layer and non-uniform natural oxide cluster that formed in some regions of the Deformed AISI 316L specimen. In addition, since the removal time of the natural oxide layer and the clusters increased relatively, the diffusion time was lacking compared to the total process time so that the carbon content was relatively reduced. In order to confirm this conjecture, the natural oxide layer was removed using a commonly known mixture of nitric acid and hydrofluoric acid (nitric–hydrofluoric acid). After that, low-temperature vacuum carburizing was performed to confirm carburizing efficiency (Figure 5). When the natural oxide layer was effectively removed, a carburizing layer of similar thickness was formed regardless of the deformed structure. In conclusion, the thickness and composition of the natural oxide layer were the most important factors for carburizing efficiency in the low-temperature vacuum carburizing process without a surface activation step.

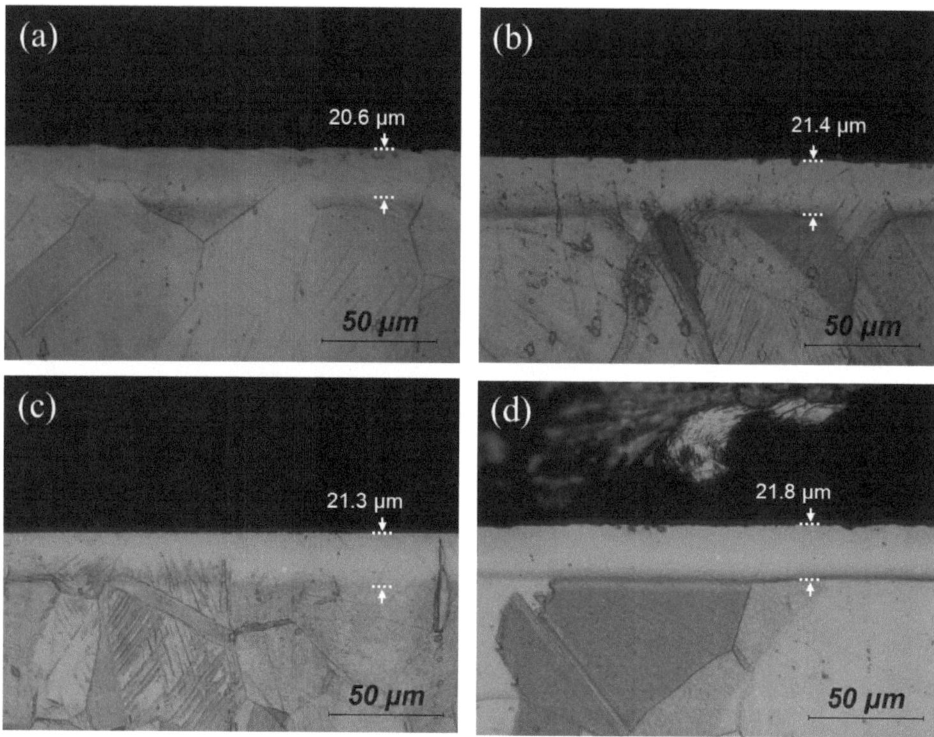

Figure 5. Cross-section of microstructure of austenitic stainless steel after 14 h low-temperature vacuum carburizing process with surface activation; (**a**) plastically deformed AISI316L, (**b**) 623 K stress-relieving treated, (**c**) 973 K stress-relieving treated, and (**d**) 1353 K solution-treated AISI 316L.

4. Conclusions

In this study, the effect of plastic deformation applied to AISI 316L in the low-temperature vacuum carburizing process without the surface activation step was investigated. In the case of AISI 316L, with many deformed structures, the natural oxide layer was formed more thickly, with a bi-layer structure consisting of an upper Cr-rich layer and a lower Fe-rich layer, and the Cr-rich oxide layer being more predominantly formed on the outermost surface. When the low-temperature vacuum carburizing was performed for a relatively short time, the carburized layer on the specimen with a deformed structure was thin, or only carburized clusters were formed, and even if the carburizing time was increased, the carburizing efficiency was not improved. However, when the deformed structure was reduced by heat treatment, the natural oxide layer was thinner and the Cr-rich oxide layer was reduced, thereby increasing carburizing efficiency. Consequentially, when the thick and Cr-rich oxide layer was formed by a deformed structure, it was considered that the efficiency of surface activation by free radicals was further reduced and the lower oxygen pressure required for spontaneous decomposition of oxide was demanded, thereby slowing the carburizing reaction. Therefore, it was found that to obtain a uniform carburized layer on the plastic worked AISI 316L, a longer process time was required in the low-temperature vacuum carburizing. In addition, it was confirmed that when the plastic worked AISI 316L was subjected to heat treatment to recover the deformed structure, the low-temperature vacuum carburizing efficiency could be improved. In conclusion, a deformed structure by plastic working is the most important factor for carburizing efficiency in the low-temperature vacuum carburizing process without a surface activation step because it changes the composition and the thickness of the natural oxide layer.

Author Contributions: Conceptualization, H.C. and K.-S.K.; methodology, S.K.; validation, K.-S.K., J.-H.K. and J.-H.L.; formal analysis, H.C. and S.-B.H.; resources, J.-H.L.; writing—original draft preparation, H.C. and K.-S.K.; writing—review and editing, S.K., S.-Y.Y. and J.-H.K.; project administration, J.-H.K.; supervision, S.-Y.Y. All authors have read and agreed to the published version of the manuscript.

Funding: This work was supported by the Renewable Energy Program of the Korea Institute of Energy Technology Evaluation and Planning (KETEP) grant funded by the Korea government Ministry of Trade, Industry and Energy (MOTIE) (Grant No. 20203030040060, Development of ultra-high pressure hydrogen piping valve (105MPa), dispenser, heat exchanger, and monitoring system with hydrogen embrittlement suppression effect and leak detection function).

Data Availability Statement: Not Applicable.

Conflicts of Interest: The authors declare no conflict of interest.

References

1. Michal, G.M.; Ernst, F.; Heuer, A.H. Carbon paraequilibrium in austenitic stainless steel. *Metall. Mater. Trans. A Phys. Metall. Mater. Sci.* **2006**, *37*, 1819–1824. [CrossRef]
2. Bell, T. Surface engineering of austenitic stainless steel. *Surf. Eng.* **2002**, *18*, 415–422. [CrossRef]
3. Collins, S.R.; Williams, P.C.; Marx, S.V.; Heuer, A.H.; Ernst, F.; Kahn, H. Low-Temperature Carburization of Austenitic Stainless Steels. In *ASM Handbook, Volume 4D: Heat Treating of Irons and Steels*; Dossett, J., Totten, G., Eds.; ASM International: Materials Park, OH, USA, 2014; Volume 4, pp. 451–460. ISBN 978-1-62708-066-8.
4. Tsujikawa, M.; Yoshida, D.; Yamauchi, N.; Ueda, N.; Sone, T.; Tanaka, S. Surface material design of 316 stainless steel by combination of low temperature carburizing and nitriding. *Surf. Coat. Technol.* **2005**, *200*, 507–511. [CrossRef]
5. Ichii, K. Structure of the ion-nitrided layer of 18-8 stainless steel. *Technol. Rep. Kansai Univ.* **1986**, *27*, 135.
6. Cao, Y.; Ernst, F.; Michal, G.M. Colossal carbon supersaturation in austenitic stainless steels carburized at low temperature. *Acta Mater.* **2003**, *51*, 4171–4181. [CrossRef]
7. Ernst, F.; Avishai, A.; Kahn, H.; Gu, X.; Michal, G.M.; Heuer, A.H. Enhanced carbon diffusion in austenitic stainless steel carburized at low temperature. *Metall. Mater. Trans. A Phys. Metall. Mater. Sci.* **2009**, *40*, 1768–1780. [CrossRef]
8. Martin, F.J.; Natishan, P.M.; Lemieux, E.J.; Newbauer, T.M.; Rayne, R.J.; Bayles, R.A.; Kahn, H.; Michal, G.M.; Ernst, F.; Heuer, A.H. Enhanced corrosion resistance of stainless steel carburized at low temperature. *Metall. Mater. Trans. A Phys. Metall. Mater. Sci.* **2009**, *40*, 1805–1810. [CrossRef]
9. Bottoli, F.; Winther, G.; Christiansen, T.L.; Somers, M.A.J. Influence of Plastic Deformation on Low-Temperature Surface Hardening of Austenitic Stainless Steel by Gaseous Nitriding. *Metall. Mater. Trans. A Phys. Metall. Mater. Sci.* **2015**, *46*, 2579–2590. [CrossRef]
10. Bottoli, F.; Winther, G.; Christiansen, T.L.; Somers, M.A.J. Influence of Plastic Deformation on Low Temperature Surface Hardening of Austenitic and Precipitation Hardening Stainless Steels by Gaseous Nitriding. *HTM J. Heat Treat. Mater.* **2015**, *70*, 228–238. [CrossRef]
11. Peng, Y.; Gong, J.; Jiang, Y.; Fu, M.; Rong, D. The effect of plastic pre-strain on low-temperature surface carburization of AISI 304 austenitic stainless steel. *Surf. Coat. Technol.* **2016**, *304*, 16–22. [CrossRef]
12. Chemkhi, M.; Retraint, D.; Roos, A.; Garnier, C.; Waltz, L.; Demangel, C.; Proust, G. The effect of surface mechanical attrition treatment on low temperature plasma nitriding of an austenitic stainless steel. *Surf. Coat. Technol.* **2013**, *221*, 191–195. [CrossRef]
13. Balusamy, T.; Sankara Narayanan, T.S.N.; Ravichandran, K.; Park, I.S.; Lee, M.H. Influence of surface mechanical attrition treatment (SMAT) on the corrosion behaviour of AISI 304 stainless steel. *Corros. Sci.* **2013**, *74*, 332–344. [CrossRef]
14. Peng, Y.W.; Gong, J.M.; Jiang, Y.; Fu, M.H.; Rong, D.S. Influence of Plastic Pre-Strain on Low-Temperature Gas Carburization of 316L Austenitic Stainless Steel. *Appl. Mech. Mater.* **2016**, *853*, 178–183. [CrossRef]
15. Collins, S.R.; Schiroky, G.H.; Marx, S.V.; Williams, P.C. Concurrent flow of activating gas in low temperature carburization. U.S. Patent 10,246,766 B2, 2 April 2019.
16. Somers, M.A.J.; Christiansen, T.L. Gaseous processes for low temperature surface hardening of stainless steel. In *Thermochemical Surface Engineering of Steels: Improving Materials Performance*; Mittemeijer, E.J., Somers, M.A.J., Eds.; Woodhead Publishing Series in Metals and Surface Engineering; Woodhead Publishing: Sawston, UK, 2014; pp. 581–614. ISBN 978-0-85709-592-3.
17. Christiansen, T.L.; Hummelshøj, T.S.; Somers, M.A.J. Gaseous carburising of self-passivating Fe–Cr–Ni alloys in acetylene–hydrogen mixtures. *Surf. Eng.* **2011**, *27*, 602–608. [CrossRef]
18. Ge, Y. Low-Temperature Acetylene-Based Carburization and Nitrocarburizing of 316L Austenitic Stainless Steel. Ph.D. Thesis, CASE Western Reserve University, Cleveland, OH, USA, 2013.
19. Ge, Y.; Ernst, F.; Kahn, H.; Heuer, A.H. The Effect of Surface Finish on Low-Temperature Acetylene-Based Carburization of 316L Austenitic Stainless Steel. *Metall. Mater. Trans. B Process Metall. Mater. Process. Sci.* **2014**, *45*, 2338–2345. [CrossRef]
20. Song, Y.; Kim, J.H.; Kim, K.S.; Kim, S.; Song, P.K. Effect of C2H2/H2 Gas mixture ratio in direct low-temperature vacuum carburization. *Metals* **2018**, *8*, 493. [CrossRef]

21. Hsieh, M.-C.; Ge, Y.; Kahn, H.; Michal, G.M.; Ernst, F.; Heuer, A.H. Volatility Diagrams for the Cr-O and Cr-Cl Systems: Application to Removal of Cr2O3-Rich Passive Films on Stainless Steel. *Metall. Mater. Trans. B* **2012**, *43*, 1187–1201. [CrossRef]
22. Githinji, D.N.; Northover, S.M.; Bouchard, P.J.; Rist, M.A. An EBSD Study of the Deformation of Service-Aged 316 Austenitic Steel. *Metall. Mater. Trans. A* **2013**, *44*, 4150–4167. [CrossRef]
23. Hummelshøj, T.S.; Christiansen, T.L.; Somers, M.A.J. Lattice expansion of carbon-stabilized expanded austenite. *Scr. Mater.* **2010**, *63*, 761–763. [CrossRef]
24. Peng, Y.; Chen, C.; Li, X.; Gong, J.; Jiang, Y.; Liu, Z. Effect of low-temperature surface carburization on stress corrosion cracking of AISI 304 austenitic stainless steel. *Surf. Coat. Technol.* **2017**, *328*, 420–427. [CrossRef]
25. Li, Z.; Illing, C.; Heuer, A.; Ernst, F. Low-Temperature Carburization of AL-6XN Enabled by Provisional Passivation. *Metals* **2018**, *8*, 997. [CrossRef]
26. Gulbransen, E.A. Kinetic and Structural Factors Involved in Oxidation of Metals. *Ind. Eng. Chem.* **1949**, *41*, 1385–1391. [CrossRef]
27. Benafia, S.; Retraint, D.; Yapi Brou, S.; Panicaud, B.; Grosseau Poussard, J.L. Influence of Surface Mechanical Attrition Treatment on the oxidation behaviour of 316L stainless steel. *Corros. Sci.* **2018**, *136*, 188–200. [CrossRef]
28. Gupta, R.K.; Birbilis, N. The influence of nanocrystalline structure and processing route on corrosion of stainless steel: A review. *Corros. Sci.* **2015**, *92*, 1–15. [CrossRef]
29. Gui, Y.; Zheng, Z.J.; Gao, Y. The bi-layer structure and the higher compactness of a passive film on nanocrystalline 304 stainless steel. *Thin Solid Films* **2016**, *599*, 64–71. [CrossRef]

Article

Identification of Expanded Austenite in Nitrogen-Implanted Ferritic Steel through In Situ Synchrotron X-ray Diffraction Analyses

Bruna C. E. Schibicheski Kurelo [1,*], Carlos M. Lepienski [1], Willian R. de Oliveira [2], Gelson B. de Souza [2], Francisco C. Serbena [2], Rodrigo P. Cardoso [3], Julio C. K. das Neves [1] and Paulo C. Borges [1]

1   Postgraduate Program in Mechanical and Materials Engineering, Federal Technological University of Paraná, Campus Ecoville, Curitiba 81280-340, Brazil; cmlepienski@utfpr.edu.br (C.M.L.); jkneves@utfpr.edu.br (J.C.K.d.N.); pborges@utfpr.edu.br (P.C.B.)
2   Laboratory of Mechanical Properties and Surfaces, Department of Physics, State University of Ponta Grossa, Campus Uvaranas, Ponta Grossa 84030-900, Brazil; oliveira.willianrafael@gmail.com (W.R.d.O.); gelsonbs@uepg.br (G.B.d.S.); fserbena@uepg.br (F.C.S.)
3   LabMat, Department of Mechanical Engineering, Federal University of Santa Catarina, Campus Trindade, Florianópolis 88040-900, Brazil; rodrigo.perito@labmat.ufsc.br
*   Correspondence: brunakurelo@utfpr.edu.br or bruna_schibicheski@hotmail.com

Abstract: The existence and formation of expanded austenite in ferritic stainless steels remains a subject of debate. This research article aims to provide comprehensive insights into the formation and decomposition of expanded austenite through in situ structure analyses during thermal treatments of ferritic steels. To achieve this objective, we employed the Plasma Immersion Ion Implantation (PIII) technique for nitriding in conjunction with in situ synchrotron X-ray diffraction (ISS-XRD) for microstructural analyses during the thermal treatment of the samples. The PIII was carried out at a low temperature (300–400 °C) to promote the formation of metastable phases. The ISS-XRD analyses were carried out at 450 °C, which is in the working temperature range of the ferritic steel UNS S44400, which has applications, for instance, in the coating of petroleum distillation towers. Nitrogen-expanded ferrite (α_N) and nitrogen-expanded austenite (γ_N) metastable phases were formed by nitriding in the modified layers. The production of the α_N or γ_N phase in a ferritic matrix during nitriding has a direct relationship with the nitrogen concentration attained on the treated surfaces, which depends on the ion fluence imposed during the PIII treatment. During the thermal evolution of crystallographic phase analyses by ISS-XRD, after nitriding, structure evolution occurs mainly by nitrogen diffusion. In the nitrided samples prepared under the highest ion fluences—longer treatment times and frequencies (PIII 300 °C 6 h and PIII 400 °C 3 h) containing a significant amount of γ_N—a transition from the γ_N phase to the α and CrN phases and the formation of oxides occurred.

Keywords: UNS S44400 alloy; nitriding at low temperature; α_N phase; γ_N phase; in situ thermal evolution of crystallographic phases; synchrotron XRD

Citation: Schibicheski Kurelo, B.C.E.; Lepienski, C.M.; de Oliveira, W.R.; de Souza, G.B.; Serbena, F.C.; Cardoso, R.P.; das Neves, J.C.K.; Borges, P.C. Identification of Expanded Austenite in Nitrogen-Implanted Ferritic Steel through In Situ Synchrotron X-ray Diffraction Analyses. *Metals* 2023, 13, 1744. https://doi.org/10.3390/met13101744

Academic Editor: Andrea Di Schino

Received: 30 June 2023
Revised: 30 September 2023
Accepted: 10 October 2023
Published: 14 October 2023

1. Introduction

The formation of the nitrogen-expanded ferrite (α_N) phase, a metastable nitrogen-rich in the solid solution phase, predominates in ferritic stainless steel surfaces modified at low temperatures by conventional plasma nitriding (PN), while at high temperatures (above 400 °C), iron ($\varepsilon - Fe_{2-x}N$ ($0 \leq x \leq 1$) and $\gamma'-Fe_4N$) and chromium nitrides (CrN and Cr_2N) are the predominant phases [1–6]. The products of PN are highly influenced by the treatment temperature and chemical composition of the steel [7]. Plasma Immersion Ion Implantation (PIII) is a process in which ions accelerated by high voltages in the plasma sheath are physically inserted in the sample surface. Hence, in PIII, the mean ion energy in the process sums with other important factors from PN (temperature and ion availability on the surfaces) to attain higher levels of nitrogen in the lattice in a solid solution. It enables the necessary thermodynamic conditions to produce new phases [8]. High ion doses can be attained by controlling the PIII parameters, such as voltage, duty cycle, and plasma ionization fractions. Most of the literature reports the formation of ε and γ' nitrides and the α_N phase in the modified layers of low-temperature PIII-nitrided ferritic steels at temperatures ranging from 300 °C to 400 °C [8–12].

A less common phase produced in ferritic stainless steels is the nitrogen-expanded austenite, γ_N, which is typically produced in austenitic stainless steels [7,13,14]. In those materials, studies using X-ray diffraction (XRD) and electron diffraction by Transmission Electron Microscopy (TEM) reported the γ_N phase as a cubic structure with a high density of dislocations and stacking faults, where nitrogen preferentially occupies the octahedral sites and tends to the γ'-Fe_4N (BCC) ordering [14–18].

The γ_N phase, which has a nitrogen content much larger than that in the α_N phase, is often identified on the ferrite grains of the duplex stainless steels after PIII and PN processes, indicating that an α_N to γ_N phase transition occurs [19–21]. In only ferritic alloys, Jiraskova et al. [22] identified the γ_N phase in a PIII nitrided X10CrAl18 ferritic stainless steel without Ni in its composition. In addition, Piekoszewski et al. identified the formation of expanded austenite in pure iron after a high-intensity nitrogen pulsed-plasma treatment [23]. In addition, the nitriding at high temperature can also cause an α to y transition in the surface microstructure depending on the incorporation of nitrogen in the process [24]. The colossal interstitial nitrogen supersaturation in the ferrite (α) phase, similar to that observed for the γ_N phase, was also predicted by theoretical works and identified in nitriding works in duplex steels that indicate that both phases are formed by means of a spinodal-like decomposition of ferrite into Cr-rich and Fe-rich ferrite phases [7,25,26]. However, in those experimental works that feature this mechanism, the matrix remains with its BCC crystal structure, and nitrides are formed. In the present study, conversely, the XRD patterns show a clear tendency towards an FCC crystal structure.

The analyses presented in this work indicate that necessary conditions to produce γ_N depend on the nitrogen concentration increase in the modified layers. In addition, the nitrogen concentration in the γ_N phase is related to the widening of the peaks observed in the X-ray diffractograms [27,28]. Therefore, the γ_N phase formation was investigated by XRD by varying ion fluences in PIII treatments of the ferritic UNS S44400 stainless steel alloy. The thermal evolution of crystalline phases using ISS-XRD allowed the clear separation of α_N and γ_N from the nitrided phases and also allowed the obtainment of concrete evidence of the formation of a nitrogen-expanded austenite (γ_N) phase in the ferritic stainless steel. A similar protocol was successfully employed by [29] to analyze the γ_N thermal decomposition in austenitic stainless steel.

PIII at a low temperature, which produces predominantly expanded phases in modified surface layers, can be used in stainless steels in order to increase the duration of parts and equipment for application in the most diverse areas of the industry, improving the wear resistance and improving or not changing the corrosion resistance [6,7,11,12,30]. In certain applications, such as petroleum refining, in furnaces and gas turbines of the chemical and food industries and automotive exhaust systems, these parts will be subjected to high temperatures. Thus, to apply these materials in these environments, studies on the stability and thermal evolution of the expanded phases are necessary, emphasizing the importance of this work.

2. Materials and Methods

The UNS S44400 (AISI 444) ferritic stainless steel (nominal composition in wt%: 19.5 Cr, 0.2 Ni, 2.5 Mo, 0.14 Ti, 0.17 Nb, 0.5 Mn, 0.5 Si, 0.04 P, 0.02, C, 0.03 N, and Fe—balance) samples were shaped as discs with 12 mm diameter and 2 mm thickness and were metallographically prepared to obtain a mirror-polished finishing and were cleaned in ultrasonic baths. Prior to nitriding, the samples surfaces were sputter-cleaned by a 50%Ar–50%H_2 (vol.) gas mixture for 20 min at 300 °C. After the end of the process, the voltage was turned off. The samples were then nitrided by PIII at 9 kV voltage, 30 μs pulse width, and different frequencies (Table 1) to obtain temperatures of 300 and 400 °C and different ion doses on the surface [27]. The heating time to reach the treatment temperature was approximately 30 min for all the conditions. The treatments were carried out for 3 and 6 h, at 3 Pa in the chamber, with a 60% N_2–40% H_2 (vol.) gas mixture. The sample names begin with the name of the nitriding technique, PIII, followed by the treatment temperature and time. The heat-treated samples were named with the suffix HT.

Table 1. Average parameters of the Plasma Immersion Ion Implantation treatments.

Sample	Nitriding Parameters				
	Temperature (°C)	Time (h)	Frequency—f (Hz)	Current Density—j (mA/cm²)	Ion Fluence—Γ (ions/cm²)
Untreated	---	---	---	---	---
PIII 300 °C 3 h	300 °C	3	728	7.8 ± 0.2	2.5×10^{18}
PIII 300 °C 6 h	300 °C	6	728	7.7 ± 0.2	5.0×10^{18}
PIII 400 °C 3 h	400 °C	3	981	13.1 ± 0.1	5.7×10^{18}

The ISS-XRD analyses (Figure 1a) were carried out in a furnace with a Kapton window enabling X-ray diffraction with synchrotron radiation in the XRD2 beamline of the Brazilian Synchrotron Light Laboratory (LNLS). The Kapton window is not suitable for vacuum pressure operations; for this reason, the process was conducted under He flow, which reduced but did not eliminate the oxidation of the surfaces. In these analyses, the X-ray beam energy was adjusted to 7 keV (λ = 1.773 Å) to prevent Fe fluorescence, and a Mythen linear detector was employed. The size of the X-ray beam under the sample was 3 mm × 0.5 mm. The heat treatment consisted of the following steps: (i) heating period from room temperature to 450 °C; (ii) isothermal period at 450 °C; and (iii) cooling period. After step (ii), the furnace was turned off and the sample was naturally cooled until reaching room temperature. These periods can be seen on the heating ramp (Figure 1b).

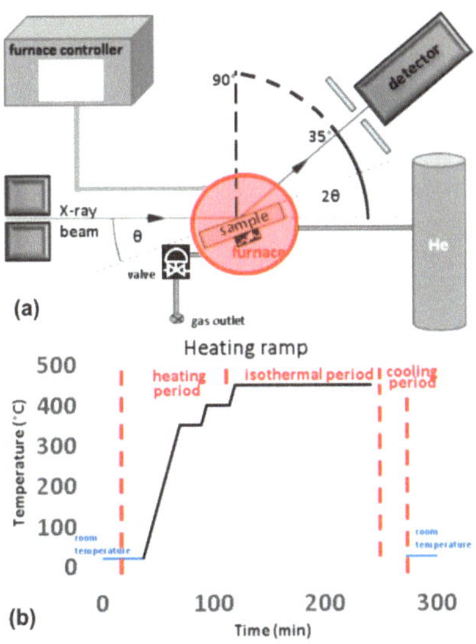

Figure 1. (**a**) Schematic diagram of ISS-XRD experiment, where the temperature varied according to (**b**) heating ramp.

The thermal cycle was chosen after several tests using other types of stainless steels: austenitic, ferritic, and duplex. In the literature, the threshold temperature for the decomposition of the α_N phase is 300 °C [14]. The tests indicated the temperature of 350 °C as the beginning of variation in the lattice parameter in expanded phases, while the 450 °C temperature indicated an ideal parameter to evaluate the evolution of the phases. In addition, this temperature is approximately the maximum temperature for the application of UNS S44400 steel in petroleum distillation towers. A period of 5 h was adequate for the entire heating, isothermal, and cooling cycles. Higher temperatures led to very abrupt variations in the expanded phases, leading to their decomposition as well as the formation of a thicker oxide film, while temperatures below 450 °C did not provide a measurable variation in the phases in the period available for each sample to be tested.

In the heating period, a 10 °C/min heating rate was applied, and XRD spectra were acquired at 350 °C and 400 °C; in both cases, the acquisition started after 1 min of temperature stabilization. The isothermal period, performed at 450 °C, was 124 min long, and XRD data were collected in intervals of 18 min. In this time interval, the scans were performed for the 2θ range from 35° to 90°, and the acquisition time was 2.5 s for each step scan of 0.2°, totaling 11.5 min in each scan. During the heat treatments, the grazing incidence angle of the X-ray beam was fixed at $\theta = 10°$, which allowed a penetration depth up to about 4 μm on the surface of the samples. ISS-XRD tests were also performed at room temperature before the heating period and after the cooling period; XRD patterns were collected with grazing angles of $\theta = 10°$ and $\theta = 2°$ observed, with the latter reducing penetration to ~0.8 μm for near-surface analyses. During all XRD tests, the position of the samples remained unchanged inside the furnace.

The intensity of the XRD counts was normalized by the value of the average electron beam current in the synchrotron ring during the test. The lattice parameters and interplanar distances at each temperature were calculated by considering thermal expansion coefficients of the phases [31–35] and crystallographic PDF cards: 33–397 (γ-austenite); 6–696 (α-ferrite); 86–231 (γ'-Fe$_4$N); 49–1664 and 72–2126 (ε-Fe$_{2\text{-}3}$N); 11–65 (CrN); 33–664 (Fe$_2$O$_3$); and 38–1479 (Cr$_2$O$_3$). These PDF cards were selected according to the literature about the structures present in stainless steels after plasma nitriding. The Fityk open-source software [36] was used for nonlinear curve fitting of the XRD patterns. The variation in the lattice parameters of the expanded phases was calculated using the method described in Oliveira et. al. [37] from a weighted mean using the integrated peak area as weight and the peak center positions.

3. Results and Discussions

The cross-section FEG-SEM and OM images for the nitrogen-implanted super ferritic steel before and after thermal treatment (HT), etched with Murakami's reagent, are presented in Figure 2. With the increase in treatment time from 3 h to 6 h, the 300 °C nitrided samples, as expected, presented a small increase in the modified layer thickness. When the nitriding temperature increased to 400 °C, the increase in the modified layer thickness was much more significant. After heat treatments, the thicknesses grew due to the nitrogen diffusion into the substrate.

The averaged layer thicknesses of the nitrogen-implanted ferritic steel before and after thermal treatment (HT) measured in cross sections of the samples are presented in Table 2.

Nitrogen contents in all modified layers before and after heat treatment were measured by the Wavelength Dispersive X-ray Spectrometry (WDS/WDX) technique; the results are shown in Figure 3. The WDS spot under the analysis conditions is approximately 700 nm in diameter, so there is difficulty in measuring the nitrogen content in thin layers such as those obtained in the 300 °C treatments. The behavior of the nitrogen profiles agrees with the layer thicknesses measured in the cross sections of the samples and is presented in Table 2.

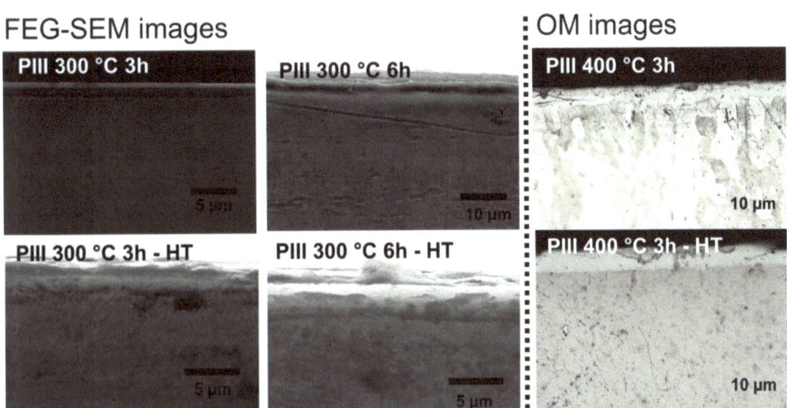

Figure 2. Cross-section FEG-SEM and OM images for the nitrogen-implanted super ferritic steel before and after thermal treatment (HT), etched with Murakami's reagent.

Table 2. Average layer thicknesses of the nitrogen-implanted super ferritic steel before and after thermal treatment (HT). The layer thicknesses were measured by SEM and Optical Microscopy on the etched cross-section samples.

Sample	Thickness (μm)
PIII 300 °C 3 h	1.1 ± 0.1
PIII 300 °C 3 h HT	2.1 ± 0.3
PIII 300 °C 6 h	1.3 ± 0.2
PIII 300 °C 6 h HT	3.2 ± 0.2
PIII 400 °C 3 h	9.3 ± 1.3
PIII 400 °C 3 h HT	13.3 ± 1.4

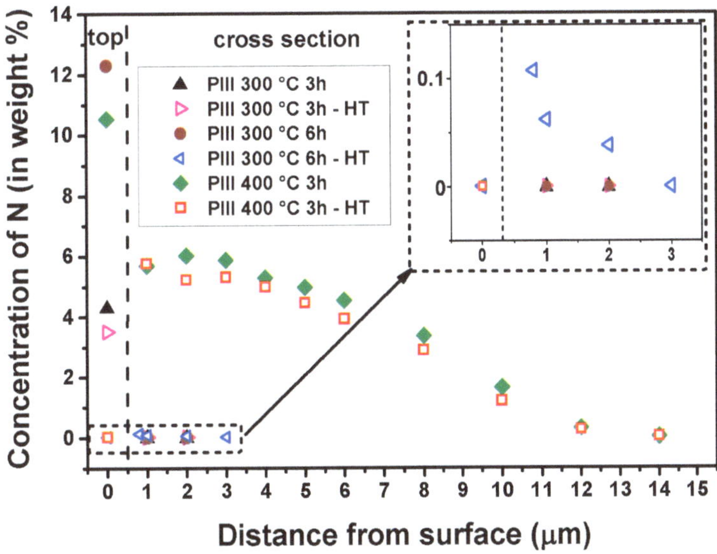

Figure 3. The nitrogen concentration, in wt.%, obtained by WDS point analysis measured on top and cross sections of nitrided samples before and after heat treatment.

The XRD of untreated (non-nitrided) UNS S44400 (AISI 444) steel samples showed only the expected peaks of the ferrite phase at room temperature, as shown in the lowermost XRD pattern in Figure 4a. The predominant phase produced by nitriding in the condition of 300 °C for 3 h was the α_N phase, which is identified by the presence of broad peaks in the XRD displaced to lower angles (greater interplanar distances) concerning the position of the corresponding α phase peaks [3,7,11]. This shift of the XRD peaks (interplanar distances) is due primarily to the presence of nitrogen in the solid solution, because the nitrogen concentration gradient in the nitrided surface layer causes an increase in the lattice parameter that is dependent on the nitrogen concentration in the crystal lattice [27,28,38]. The α_N phase peaks are broad due to the continuous decrease in the nitrogen concentration with depth, which results in the overlapping of several substoichiometries of the nitrogen in the α_N phase. Table 3 presents the calculated values for the average lattice parameters from the XRD patterns obtained at room temperature before and after the heat treatment (HT). The lattice parameters of the α phase, calculated from the peak positions at 51.6° (α (110)) and 75.9° (α (200)), were 2.88 Å. In the α_N phase, the lattice parameters calculated at 51.1° (α_N (110)) and 74.6° (α_N (200)) were 2.91 Å and 2.92 Å, respectively. After the heat treatment, the peak positions of the α_N phase peaks were displaced to 51.3° (α_N (110)) and 75.1° (α_N (200)), and the calculated lattice parameters changed to 2.90 Å and 2.91 Å. The average lattice parameters values, calculated from the interplanar distances determined from the peaks of the α_N phase, are shown in Table 3, and the deviation in the values was small, indicating that the values obtained by the XRD analysis were then consistent with each other. Figure 4a also presents diffractograms of the 300 °C 3 h nitrided sample obtained in situ during the heating period of the heat treatment. During the heating period, α phase peaks from the substrate were displaced to lower angles (higher interplanar distances) due to thermal expansion, as expected. In contrast, peaks related to the α_N phase shifted to lower interplanar distances, indicating that a reduction in nitrogen in the solid solution occurred, probably related to the inward diffusion of N atoms. These diffractograms presented little or no nitride contribution. Although iron nitrides were not identified, their position is indicated only to facilitate a comparison.

Table 3. Calculated values for average lattice parameters from XRD patterns obtained at room temperature before and after heat treatment (HT).

Sample	Average Lattice Parameter (Å)	
	α_N	γ_N
PIII 300 3 h	2.91 ± 0.01	---
PIII 300 3 h HT	2.90 ± 0.01	---
PIII 300 6 h	2.91 ±0.01	3.85 ± 0.03
PIII 300 6 h HT	2.91 ± 0.01	3.71 ± 0.06
PIII 400 3 h	2.90 ± 0.03	3.86 ± 0.04
PIII 400 3 h HT	2.90 ± 0.02	3.81± 0.04

The broad peak at 49° lost intensity, was displaced to smaller interplanar distances in the heating period, and disappeared after the isothermal period, indicating that it is related to an expanded phase containing nitrogen in the solid solution. This peak may be related to the α_N phase containing the highest N content, or it may be related to the beginning of an α_N to γ_N phase transition on the sample surface. Accordingly, the position of this peak can be related to the highest-intensity peak of the γ_N phase.

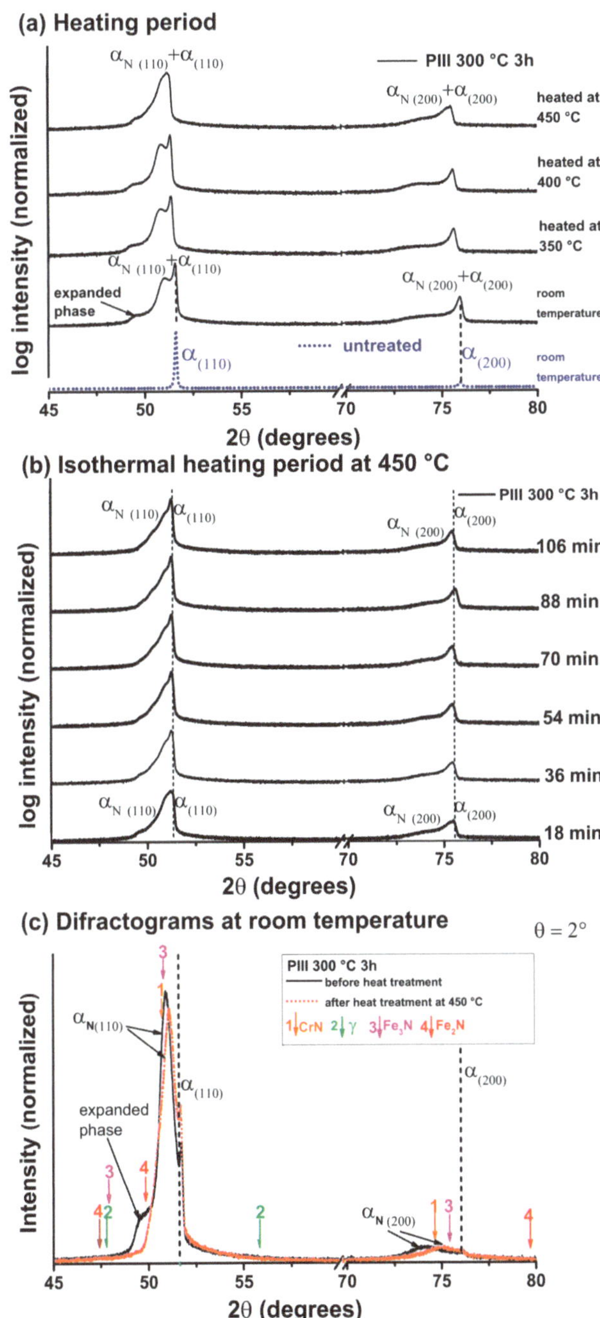

Figure 4. X-ray diffractograms of UNS S44400 steel samples nitrided by PIII at 300 °C for 3 h obtained in situ ($\theta = 10°$) during heat treatment: (**a**) in the heating period and (**b**) in the isothermal period at 450 °C for 2 h. The diffractograms of the nitrided samples with $\theta = 2°$, before and after the heat treatment, are presented in (**c**). A diffractogram of the untreated sample is presented for comparison.

During the isothermal period, shown in Figure 4b, interplanar distances of the α_N phase were further reduced, accompanied by a slight decrease in the intensities of the peaks, indicating the partial diffusion of nitrogen in the solid solution due to nitrogen diffusion under a constant temperature.

Figure 4c compares the diffractograms of the PIII 300 °C 3 h nitrided sample before and after the heat treatment, both obtained at room temperature. Neither Cr nitrides nor oxides could be identified in any of the cases. At the temperature and time of analysis used, the iron nitrides were stable, as identified by Brink (2014) in a nitrided austenitic stainless steel [31]. Despite an increase in the nitrides or their appearance after the heat treatment, the intensities decreased in regions of the diffractograms related to them and also to the α_N phase. This confirms the fact that iron nitrides in the 300 °C 3 h sample were absent or that only a few were present before and after the heating period. Thus, for the 300 °C 3 h nitrided sample, the thermal evolution of the modified layer resulted predominantly from nitrogen diffusion. Neither the formation of Cr nitrides nor oxides was observed. This corroborates the secure application of such modified surface in conditions with temperatures below 450 °C.

In agreement with the XRD results, the nitrogen content analyses in Figure 2 show a reduction in the nitrogen concentration measured at the top in the PIII 300 °C 3 h sample when compared to the PIII 300 °C 3 h sample after heat treatment (PIII 300 °C 3 h HT) that predominantly presents the α_N phase, indicating nitrogen diffusion from the surface to the bulk. It is also observed that the XRD beam with $\theta = 10°$ (Figure 4a,b) permeates the modified layer due to the beam penetration being ~4 um, while the thickness of the modified layer is smaller (Table 2). In this condition, the α phase was identified in the substrate. In the $\theta = 2°$ analyses condition (Figure 4c), the beam penetration was ~0.8 um, and the α phase was identified at a depth smaller than the thickness of the modified layer after thermal evolution. Thus, the expanded phases were partially converted into the α phase after the heat treatment in the modified layer.

The X-ray diffractogram of the UNS S44400 steel sample nitrided by PIII at 300 °C for 6 h is shown in Figure 5a. In addition to α_N, this sample presented the γ_N phase in the modified layer. Firstly, the treatment time in this condition was twice as high as the previous one, 300 °C for 3 h, leading to a significant increase in ion fluence (Table 1). Secondly, the large amount of Cr in the alloy (about 19.5% wt.) may have facilitated the N entrapment in Cr sites [7], similarly to the trapping–detrapping model developed for austenitic steels [14], promoting an enrichment of the α_N phase with N and, eventually, its transition to the γ_N phase [22,27,31,39], since nitrogen is an austenite stabilizer element. In the nitrogen trapping model by the Cr sites, first proposed by Parascandola [12], a weak bond in the order of -0.193 eV between Cr and N occurs in the γ_N phase [13]. After the nitrogen trapping or Cr–N weak bond, the expanded phase remains with a nitrogen limit up to 38% at., without forming any precipitates [2,17]. Higher contents of elements such as Cr, Mo, Ti, and Nb in the alloy also favor the formation of the γ_N phase by increasing the nitrogen solubility [13]. Thus, the formation of the γ_N phase is strongly influenced by the Cr content in the alloy. In addition, the γ_N phase has a mixture of lattice parameters, as this phase is formed with different stoichiometries. The grains might contain crystalline structures with different lattice parameters, as the nitriding process is dependent on the crystalline orientation [40,41]. The broadening of the XRD peaks is caused by the strains induced in the crystal lattice due to the introduction of nitrogen in the solid solution, while the peaks overlapping correspond to the various non-stoichiometric Fe-N concentrations in the austenite CFC structure [14,27,42]. In agreement with the XRD results, the nitrogen content analyses in Figure 2 show an increase in the nitrogen concentration measured at the top in the PIII 300 °C 6 h sample rich in the γ_N phase when compared to the PIII 300 °C 3 h sample that presents only the α_N phase. After the heat treatment, the nitrogen content measured at the top of the PIII 300 °C HT sample was approximately zero due to the oxide film formed on the surface of this sample, preventing the correct detection of nitrogen on the surface.

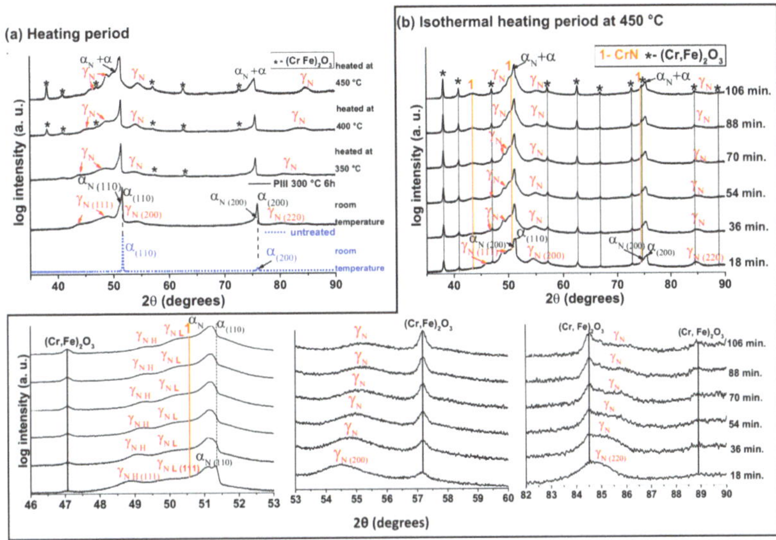

Figure 5. X-ray diffractograms of UNS S44400 steel samples nitrided by PIII at 300 °C for 6 h, obtained in situ (θ = 10°): (**a**) in the heating period and (**b**) in the isothermal period at 450 °C for 2 h with a detailed view of the different phases at different 2θ angular positions.

It is straightforward to consider the overlapping of γ'-Fe$_4$N and ε-Fe$_2$ + $_x$N nitride peaks (with x between 0 and 1) with α$_N$ and γ$_N$ peaks in the diffractogram of the 300 °C 6 h sample. Nitrides are generally formed after the supersaturation of nitrogen in solid solutions [43,44]. However, the results of the in situ XRD in the heating period until 450 °C, shown in Figure 5a, indicate that the overlapped peaks laying at lower angles (greater interplanar distances) from the α$_N$ phase peak at 50.6° were predominantly composed of the γ$_N$ phase peaks. During the heating period, peaks related to the γ$_N$ phase shifted to smaller interplanar distances due to that the partial diffusion of nitrogen in the solid solution was seen, in the same fashion as observed for the α$_N$ phase in the 300 °C 3 h sample (Figure 1a). No other peaks were left in the 47.64°–50.8° range after the decay of the expanded phases, which would be expected for iron nitrides, since they are thermodynamically stable in these conditions, as previously mentioned [31].

In the isothermal period at 450 °C, shown in Figure 5b, peaks related to the γ$_N$ phase were resolved into two contributions, the low-expansion (γ$_{N\,L}$) and the high-expansion γ$_N$ phase (γ$_{N\,H}$), similar behavior to that observed in austenitic steels [27]. The metastable phases γ$_N$ and α$_N$ were displaced even more to lower interplanar distances due to interstitial solid solution reduction as a consequence of the nitrogen diffusion in the lattice. At the end of the isothermal period, the integrated areas of the γ$_N$ peaks were observed to decrease by ~53%, whereas the CrN peaks increased in intensity. These findings corroborate a possible phase transition from γ$_N$ to α + CrN, as evidenced in the literature on other heat-treated stainless steels [22,27,29,31,39]. If the temperature in the isothermal period was increased, the γ$_N$ to α + CrN transition would occur in a more pronounced way, as observed by Tschiptschin et. al. [29].

During the heat treatment, (Cr, Fe)$_2$O$_3$ oxide peaks appeared in the diffractogram, which became more evident with the treatment time. Oxides form a film that grows externally to the modified layer, and the peaks related to these phases are not broad nor displaced. Their presence serves as a reference point to state peaks related to the change in the expanded phases regarding width and position as the treatment time increased. A hypothesis explaining the tendency of oxide formation in nitrided stainless steels considers the immobilization of chromium in the presence of nitrogen due to their chemical affinity:

N is trapped in Cr sites, producing a short-range ordering between Cr and N in the γ_N phase, as proposed by Cao and Norell [45]. As a result, the Cr activity reduces, making it difficult to form protective chromium oxides in the presence of oxygen (residual, in this case) and at high temperatures.

Figure 6a presents the diffractograms, obtained at room temperature with $\theta = 2°$, of the samples nitrided at 300 °C for 6 h before and after the heat treatment. After the heating period, despite the formation of oxides on the sample surface that reduced the beam penetration into the surface, the intensity of the peaks related to the α and CrN phases increased when compared to the diffractogram obtained before the heat treatment. This observation corroborates the hypothesis of the γ_N phase undergoing transformation into α + CrN. In the UNS S44400 steel, the γ_N phase cannot become γ due to the absence of Ni and other austenitizing elements in its chemical composition. The peak positions of iron nitrides are indicated in Figure 6a, and one could infer the possible contribution of these peaks, with very low intensity. However, their contributions to the diffractograms are unlikely since peak intensities in the iron nitride region decreased after the heat treatment. The most probable situation before the heat treatment was the modified layer being composed predominantly of γ_N and α_N phases. The approximate lattice parameters of the γ_N phase, calculated from the peak positions of the γ_N phase at 47.4° (γ_N (111)), 54.5° (γ_N (200)), and 81.3° (γ_N (220)), were 3.82 Å, 3.87 Å, and 3.85 Å, respectively. After the heat treatment, the γ_N phase peak positions were displaced to 49.3° (γ_N (111)), 55.75° (γ_N (200)), and 85.8° (γ_N (220)), and the calculated lattice parameters changed to 3.68 Å, 3.79 Å, and 3.68 Å. In the average lattice parameter value (shown in Table 3), calculated from the interplanar distances determined from the γ_N phase peaks, there was a small deviation in the values, indicating that the values obtained by the XRD analyses were then consistent with each other, corroborating the presence of this phase.

By comparing diffractograms obtained at room temperature with $\theta = 2°$ (Figure 6a) and with $\theta = 10°$ (Figure 6b), the γ_N phase was identified as to be concentrated closer to the surface in the modified layer (due to the relative peak intensities), which are those more enriched with nitrogen by the N implantation process. Additionally, the heat treatment gave rise to broad peaks in the 2θ diffraction angles of approximately 51.6° and 75.9°, located close to the α phase peaks. These peaks are related to the transition from γ_N to α + CrN predicted in the literature [22,27,31,39], resulting from the formation of microregions containing nanocrystalline α phase grains that can generate tensile stresses at the grain interfaces with regions rich in γ_N and α_N phases. After the transition, a nitrogen redistribution is possible, as is the formation of randomly oriented crystals of the α_N phase.

Similarly observed for the PIII 300 °C 3 h sample, the XRD analyses with $\theta = 10°$ (Figures 6 and 7b) carried out on the PIII 300 °C 6 h sample before and after the heat treatment were influenced by the bulk (see Table 2). Thus, the α phase was identified in these diffractograms. In the XRD analyses carried out with $\theta = 2°$ (Figure 7a), the α phase was also identified in the diffractograms after the heat treatment due to the transition of γ_N to α + CrN, discussed previously.

In Figure 7a, X-ray diffractograms of UNS S44400 steel samples nitrided at 400 °C for 3 h are shown. In this sample, where ion fluency (Table 1) increased due to the increase in frequency, the presence of the γ_N phase was more evident than in the 300 °C 6 h condition. In addition to the γ_N phase, diffractograms presented peaks relative to α_N and CrN phases. During the heating period, although it may seem that the γ_N phase peaks did not change position until 350 °C, it actually means that the effects of thermal expansion and N diffusion occurred simultaneously and were (fortuitously) equivalent, so that the angular positions remained almost unchanged. At 450 °C (Figure 7a), the reduction in interplanar distances instead of thermal expansion was evident in the peaks related to the expanded phases due to nitrogen diffusion in the heating period.

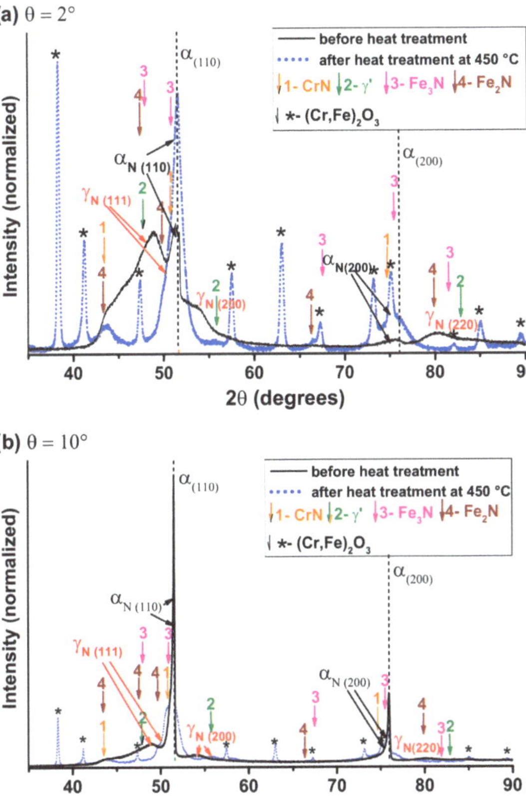

Figure 6. Diffractograms of samples nitrided at 300 °C for 6 h obtained at room temperature, before and after the heat treatment with different grazing incidence angle (θ): (**a**) $\theta = 2°$ and (**b**) $\theta = 10°$.

During the isothermal period, shown in Figure 7b, it became more evident that γ_N peaks separated into two contributions, low-expansion ($\gamma_{N\,L}$) and high-expansion ($\gamma_{N\,H}$) γ_N phases, the same as observed in the 300 °C 6 h sample (Figure 5b). All γ_N and α_N peaks shifted to lower interplanar distances at this period. Prior to the heating period, lattice parameters calculated for $\gamma_{N\,H}$ at 47.5° (γ_N (111)), 54.4° (γ_N (200)), and 80.2° (γ_N (220)) were 3.81 Å, 3.87 Å, and 3.89 Å, respectively. After the heat treatment, the peak positions of the γ_N phase displaced to 48.0° (γ_N (111)), 54.8° (γ_N (200)), and 82.2° (γ_N (220)), which corresponded to the lattice parameters 3.77 Å, 3.85 Å, and 3.81 Å. These values were similar to each other before and after the heating period, as also observed for the 300 °C 6 h condition. However, the change in the lattice parameter after the heat treatment was smaller in the 400 °C 3 h sample than in the 300 °C 6 h, perhaps because N at the outermost surface (probed by X-rays) would need a much longer time to undergo a significant thermal diffusion within the thicker layer. In the diffractograms collected during the isothermal period, shown in Figure 7b, the CrN phase contribution increased with time.

There was no influence of the substrate in any of the XRD analyses carried out on the sample at 400 °C for 3 h due to the modified layer being much thicker (see Table 2) than the penetration depth of the XRD beam (~4 μm). Therefore, the alpha phase was not identified in the XRD analyses of this sample before the heat treatment at room temperature. After heating, as observed for the sample PIII 300 °C 6 h, the results indicate a transition from γ_N to α + CrN.

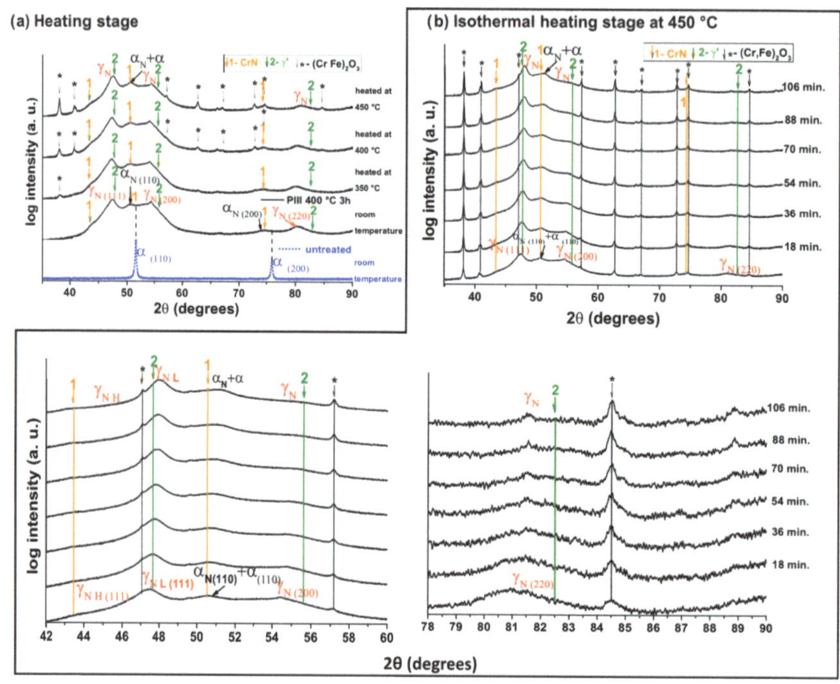

Figure 7. X-ray diffractograms of UNS S44400 steel samples nitrided by PIII at 400 °C for 3 h, obtained in situ with θ = 10°: (**a**) in the heating period; (**b**) in the isothermal period at 450 °C for 2 h with a detailed view of the different phases at different 2θ angular positions.

After the heat treatments, significant formations of iron oxides and chromium nitrides were observed only in the 300 °C 6 h and 400 °C 3 h samples, nitrided with the highest ionic fluences and both presenting the γ_N phase in the modified layer. As previously mentioned, the formation of oxides in these samples can also be related to the entrapment process of N by Cr [45]. The precipitation of chromium nitrides can be attributed to the high nitrogen content attained on the surfaces rich in the γ_N phase, which may hinder Cr mobility and decelerate nitrogen diffusion [27,39,46]. Furthermore, according to the literature, in the crystalline lattice, there is greater coherence between CrN and γ_N phases than between CrN and γ phases, favoring the formation of CrN under high-nitrogen-content conditions [39]. Peaks of the γ' and ε phases should have appeared in the diffractograms, but less significantly than the expanded phases γ_N and α_N; however, it was not possible to clearly index these phases due to the strong overlapping of peaks.

4. Conclusions

In the nitrided sample at 300 °C for 3 h, which is a typical treatment time in PIII, the α_N phase was the predominant phase produced. Augmented ion fluences, attained either by increasing the treatment time from 3 h to 6 h (300 °C 6 h sample) or by increasing the temperature (400 °C 3 h sample), allowed the formation of the γ_N phase in the modified UNS S44400 stainless steel surfaces.

During the heat treatment of the treated surfaces, the structure evolution was monitored in situ, and the transformation of metastable expanded phases was featured mainly by a reduction in the lattice parameter, related to nitrogen diffusion. The ε and γ' iron nitrides, if present on the nitrided surfaces, were in low amounts and not clearly detected in the XRD analyses.

During the isothermal period, the formation of CrN and contributions of the α phase peaks in the diffractograms were more evident, corroborating the possible transition from γ_N into α + CrN. A significant formation of oxides was observed after the heat treatments on the surfaces nitrided under the highest ion fluences and with the γ_N phase on the modified layer. Although it needs further investigation, in fact, this evidence points to a formation of the γ_N phase.

The α_N phase in the sample PIII 300 °C 3 h remained as the predominant phase after the heat treatment, with no identifiable decay into CrN and oxides. These findings are of interest for corrosion protection applications.

Author Contributions: Conceptualization, B.C.E.S.K. and C.M.L.; methodology, B.C.E.S.K., C.M.L., W.R.d.O. and G.B.d.S.; formal analysis, B.C.E.S.K., C.M.L., W.R.d.O., G.B.d.S., F.C.S., R.P.C., J.C.K.d.N. and P.C.B.; investigation, B.C.E.S.K., C.M.L., R.P.C., J.C.K.d.N. and P.C.B.; resources, C.M.L.; data curation, B.C.E.S.K.; writing—original draft preparation, B.C.E.S.K. and C.M.L.; writing—review and editing, B.C.E.S.K., C.M.L., W.R.d.O., G.B.d.S., F.C.S., R.P.C., J.C.K.d.N. and P.C.B. All authors have read and agreed to the published version of the manuscript.

Funding: This work was part of the NESAP project (PRONEX CNPq/Fundação Araucária 15/2017). Bruna Kurelo was supported by a postdoctoral research scholarship, PNPD, granted by the Coordination for the Improvement of Higher Education Personnel during the completion of this research.

Data Availability Statement: The data used in this research are available in the article.

Acknowledgments: The authors thank the Villares Metals Company for the donation of the SFSS, C-LABMU/UEPG, and CMCM/UTFPR for the use of characterization facilities and LNLS for the synchrotron XRD analyses (proposal 20190153).

Conflicts of Interest: The authors declare no conflict of interest.

References

1. Alphonsa, J.; Mukherjee, S.; Raja, V.S. Study of Plasma Nitriding and Nitrocarburising of AISI 430F Stainless Steel for High Hardness and Corrosion Resistance. *Corros. Eng. Sci. Technol.* **2018**, *53*, 51–58. [CrossRef]
2. De Araújo, E.; Bandeira, R.M.; Manfrinato, M.D.; Moreto, J.A.; Borges, R.; Vales, S.D.S.; Suzuki, P.A.; Rossino, L.S. Effect of Ionic Plasma Nitriding Process on the Corrosion and Micro-Abrasive Wear Behavior of AISI 316L Austenitic and AISI 470 Super-Ferritic Stainless Steels. *J. Mater. Res. Technol.* **2019**, *8*, 2180–2191. [CrossRef]
3. Gontijo, L.C.; Machado, R.; Casteletti, L.C.; Kuri, S.E.; Nascente, P.A.P. X-Ray Diffraction Characterisation of Expanded Austenite and Ferrite in Plasma Nitrided Stainless Steels. *Surf. Eng.* **2010**, *26*, 265–270. [CrossRef]
4. Li, L.; Liu, R.; Liu, Q.; Wu, Z.; Meng, X.; Fang, Y. Effects of Initial Microstructure on the Low-Temperature Plasma Nitriding of Ferritic Stainless Steel. *Coatings* **2022**, *12*, 1404. [CrossRef]
5. Tadepalli, L.D.; Gosala, A.M.; Kondamuru, L.; Bairi, S.C.; Subbiah, R.; Singh, S.K. A Review on Effects of Nitriding of AISI409 Ferritic Stainless Steel. *Mater. Today Proc.* **2020**, *26*, 1014–1020. [CrossRef]
6. Jimenez, L.B.V.; Umemura, M.T.; Calderón-Hernández, J.W.; Magnabosco, R.; Pinedo, C.E.; Tschiptschin, A.P. Plasma Nitriding of 410S Ferritic/Martensitic Stainless Steel: Microstructure, Wear and Corrosion Properties. *Tecnol. Metal. Mater. Mineração* **2023**, *20*, e2809. [CrossRef]
7. Borgioli, F. The "Expanded" Phases in the Low-Temperature Treated Stainless Steels: A Review. *Metals* **2022**, *12*, 331. [CrossRef]
8. Blawert, C.; Mordike, B.L.; Rensch, U.; Oettel, H. The Effect of HV in the Nitriding of Ferritic Steels by Plasma Immersion Ion Implantation. *Surf. Coatings Technol.* **2001**, *142–144*, 376–383. [CrossRef]
9. Schreiber, G.; Rensch, U.; Oettel, H.; Blawert, C.; Mordike, B.L. Thermal Stability of PI 3 Nitrided Surface Layers on Ferritic Steels. *Surf. Coat. Technol.* **2003**, *170*, 447–451. [CrossRef]
10. Manova, D.; Eichentopf, I.-M.; Hirsch, D.; Mandl, S.; Neumann, H.; Rauschenbach, B. Influence of Microstructure on Nitriding Properties of Stainless Steel. *IEEE Trans. Plasma Sci.* **2006**, *34*, 1136–1140. [CrossRef]
11. Schibicheski Kurelo, B.C.E.; de Souza, G.B.; Serbena, F.C.; Lepienski, C.M.; Borges, P.C. Mechanical Properties and Corrosion Resistance of AN-Rich Layers Produced by PIII on a Super Ferritic Stainless Steel. *Surf. Coatings Technol.* **2020**, *403*, 126388. [CrossRef]
12. Luiz, L.A.; Kurelo, B.C.E.S.; de Souza, G.B.; de Andrade, J.; Marino, C.E.B. Effect of Nitrogen Plasma Immersion Ion Implantation on the Corrosion Protection Mechanisms of Different Stainless Steels. *Mater. Today Commun.* **2021**, *201*, 8131–8135. [CrossRef]
13. Schibicheski Kurelo, B.C.E.; de Souza, G.B.; Serbena, F.C.; Lepienski, C.M.; Chuproski, R.F.; Borges, P.C. Improved Saline Corrosion and Hydrogen Embrittlement Resistances of Superaustenitic Stainless Steel by PIII Nitriding. *J. Mater. Res. Technol.* **2022**, *18*, 1717–1731. [CrossRef]

14. Borgioli, F. From Austenitic Stainless Steel to Expanded Austenite-s Phase: Formation, Characteristics and Properties of an Elusive Metastable Phase. *Metals* **2020**, *10*, 187. [CrossRef]

15. Blawert, C.; Mordike, B.L.; Jirásková, Y.; Schneeweiss, O. Structure and Composition of Expanded Austenite Produced by Nitrogen Plasma Immersion Ion Implantation of Stainless Steels X6CrNiTi1810 and X2CrNiMoN2253. *Surf. Coatings Technol.* **1999**, *116*, 189–198. [CrossRef]

16. Fernandes, F.A.P.; Casteletti, L.C.; Gallego, J. Microstructure of Nitrided and Nitrocarburized Layers Produced on a Super-austenitic Stainless Steel. *J. Mater. Res. Technol.* **2013**, *2*, 158–164. [CrossRef]

17. Stróż, D.; Psoda, M. TEM Studies of Plasma Nitrided Austenitic Stainless Steel. *J. Microsc.* **2010**, *237*, 227–231. [CrossRef] [PubMed]

18. Che, H.L.; Lei, M.K. Microstructure of Perfect Nitrogen-Expanded Austenite Formed by Unconstrained Nitriding. *Scr. Mater.* **2021**, *194*, 113705. [CrossRef]

19. Blawert, C.; Weisheit, A.; Mordike, B.L.; Knoop, R.M. Plasma Immersion Ion Implantation of Stainless Steel: Austenitic Stainless Steel in Comparison to Austenitic-Ferritic Stainless Steel. *Surf. Coatings Technol.* **1996**, *85*, 15–27. [CrossRef]

20. Alphonsa, J.; Raja, V.S.; Mukherjee, S. Study of Plasma Nitriding and Nitrocarburizing for Higher Corrosion Resistance and Hardness of 2205 Duplex Stainless Steel. *Corros. Sci.* **2015**, *100*, 121–132. [CrossRef]

21. Núñez de la Rosa, Y.E.; Palma Calabokis, O.; Borges, P.C.; Ballesteros Ballesteros, V. Effect of Low-Temperature Plasma Nitriding on Corrosion and Surface Properties of Duplex Stainless Steel UNS S32205. *J. Mater. Eng. Perform.* **2020**, *29*, 2612–2622. [CrossRef]

22. Jirásková, Y.; Blawert, C.; Schneeweiss, O. Thermal Stability of Stainless Steel Surfaces Nitrided by Plasma Immersion Ion Implantation. *Phys. Stat. Sol. (a)* **1999**, *175*, 537–548. [CrossRef]

23. Piekoszewski, J.; Sartowska, B.; Waliś, L.; Werner, Z.; Kopcewicz, M.; Prokert, F.; Stanisławski, J.; Kalinowska, J.; Szymczyk, W. Interaction of Nitrogen Atoms in Expanded Austenite Formed in Pure Iron by Intense Nitrogen Plasma Pulses. *Nukleonika* **2004**, *49*, 57–60.

24. Manne, V.; Singh, S.K.; Sateesh, N.; Ram, S. A Review on Influence of Nitriding on AISI430 Ferritic Stainless Steel. *Mater. Today Proc.* **2020**, *26*, 1010–1013. [CrossRef]

25. Sasidhar, K.N.; Meka, S.R. Thermodynamic Reasoning for Colossal N Supersaturation in Austenitic and Ferritic Stainless Steels during Low-Temperature Nitridation. *Sci. Rep.* **2019**, *9*, 7996. [CrossRef]

26. Wang, D.; Kahn, H.; Ernst, F.; Heuer, A.H. "Colossal" Interstitial Supersaturation in Delta Ferrite in Stainless Steels: (II) Low-Temperature Nitridation of the 17-7 PH Alloy. *Acta Mater.* **2017**, *124*, 237–246. [CrossRef]

27. Manova, D.; Mändl, S.; Neumann, H.; Rauschenbach, B. Formation of Metastable Diffusion Layers in Cr-Containing Iron, Cobalt and Nickel Alloys after Nitrogen Insertion. *Surf. Coatings Technol.* **2017**, *312*, 81–90. [CrossRef]

28. Manova, D.; Lutz, J.; Gerlach, J.W.; Neumann, H.; Mändl, S. Relation between Lattice Expansion and Nitrogen Content in Expanded Phase in Austenitic Stainless Steel and CoCr Alloys. *Surf. Coatings Technol.* **2011**, *205*, S290–S293. [CrossRef]

29. Tschiptschin, A.P.; Nishikawa, A.S.; Varela, L.B.; Pinedo, C.E. Thermal Stability of Expanded Austenite Formed on a DC Plasma Nitrided 316L Austenitic Stainless Steel. *Thin Solid Films* **2017**, *644*, 156–165. [CrossRef]

30. Borgioli, F. The Corrosion Behavior in Different Environments of Austenitic Stainless Steels Subjected to Thermochemical Surface Treatments at Low Temperatures: An Overview. *Metals* **2023**, *13*, 776. [CrossRef]

31. Brink, B.; Ståhl, K.; Christiansen, T.L.; Somers, M.A.J. Thermal Expansion and Phase Transformations of Nitrogen-Expanded Austenite Studied with in Situ Synchrotron X-Ray Diffraction. *J. Appl. Crystallogr.* **2014**, *47*, 819–826. [CrossRef]

32. Somers, M.A.J.; van der Pers, N.M.; Schalkoord, D.; Mittemeijer, E.J. Dependence of the Lattice Parameter of γ' Iron Nitride, Fe_4N, on Nitrogen Content; Accuracy of the Nitrogen Absorption Data. *Metall. Trans. A* **1989**, *20*, 1533–1539. [CrossRef]

33. Mayrhofer, P.H.; Tischler, G.; Mitterer, C. Microstructure and Mechanical/Thermal Properties of Cr-N Coatings Deposited by Reactive Unbalanced Magnetron Sputtering. *Surf. Coatings Technol.* **2001**, *142–144*, 78–84. [CrossRef]

34. Chen, H.Y.; Tsai, C.J.; Lu, F.H. The Young's Modulus of Chromium Nitride Films. *Surf. Coatings Technol.* **2004**, *184*, 69–73. [CrossRef]

35. Bian, L.; Chen, Z.; Wang, L.; Li, F.; Chou, K. Oxidation Resistance, Thermal Expansion and Area Specific Resistance of Fe-Cr Alloy Interconnector for Solid Oxide Fuel Cell. *J. Iron Steel Res. Int.* **2017**, *24*, 77–83. [CrossRef]

36. Wojdyr, M. Fityk: A General-Purpose Peak Fitting Program. *J. Appl. Crystallogr.* **2010**, *43*, 1126–1128. [CrossRef]

37. de Oliveira, W.R.; Chuproski, R.F.; Valadão, G.M.; Cintho, O.M.; de Souza, E.C.F.; Serbena, F.C.; de Souza, G.B. Symmetry between the Anisotropic N Behavior in the Lattice under High Pressures and the Formation of Expanded Austenite. *J. Alloys Compd.* **2021**, *871*, 159509. [CrossRef]

38. Fernandes, B.B.; Mändl, S.; Oliveira, R.M.; Ueda, M. Mechanical Properties of Nitrogen-Rich Surface Layers on SS304 Treated by Plasma Immersion Ion Implantation. *Appl. Surf. Sci.* **2014**, *310*, 278–283. [CrossRef]

39. Christiansen, T.; Somers, M.A.J. Decomposition Kinetics of Expanded Austenite with High Nitrogen Contents. *Int. J. Mater. Res.* **2006**, *97*, 79–88. [CrossRef]

40. Galdikas, A. The Anisotropic Stress-Induced Di Ff Usion and Trapping of Nitrogen in Austenitic Stainless Steel during Nitriding. *Metals* **2020**, *10*, 1319. [CrossRef]

41. Moskalioviene, T.; Galdikas, A. Crystallographic Orientation Dependence of Nitrogen Mass Transport in Austenitic Stainless Steel. *Metals* **2020**, *10*, 615. [CrossRef]

42. Christiansen, T.; Somers, M.A.J. On the Crystallographic Structure of S-Phase. *Scr. Mater.* **2004**, *50*, 35–37. [CrossRef]

43. Ye, S.; Cao, Y. Structure of Iron Nitrides under Different Nitridation Temperatures. In Proceedings of the AMITP 2016, Guilin, China, 24–25 September 2016; Zheng, Z., Zhuo, X., Eds.; Atlantis Press: Guilin, China, 2016; pp. 6–9.
44. Rohith, K.V.; Saravanan, P.; Sakar, M.; Balakumar, S. Insights into the Nitridation of Zero-Valent Iron Nanoparticles for the Facile Synthesis of Iron Nitride Nanoparticles. *RSC Adv.* **2016**, *6*, 45850–45857. [CrossRef]
45. Cao, Y.; Norell, M. Role of Nitrogen Uptake during the Oxidation of 304L and 904L Austenitic Stainless Steels. *Oxid. Met.* **2013**, *80*, 479–491. [CrossRef]
46. Manova, D.; Lotnyk, A.; Mändl, S.; Neumann, H.; Rauschenbach, B. CrN Precipitation and Elemental Segregation during the Decay of Expanded Austenite. *Mater. Res. Express* **2016**, *3*, 066502. [CrossRef]

Article

Assessment of the Pitting, Crevice Corrosion, and Mechanical Properties of Low-Temperature Plasma-Nitrided Inconel Alloy 718

Yamid Nuñez de la Rosa [1,2,*], Oriana Palma Calabokis [1,2], Vladimir Ballesteros-Ballesteros [1], Cristian Lozano Tafur [1] and Paulo C. Borges [2]

1 Faculty of Engineering and Basic Sciences, Fundación Universitaria Los Libertadores, Bogotá 111221, Colombia; calabokis@alunos.utfpr.edu.br (O.P.C.)
2 Department of Mechanical Engineering, Campus Ecoville, Grupo de Materiais, Tribologia e Superfícies (GrMatS), Universidade Tecnológica Federal do Paraná, Curitiba 81280-340, Brazil; pborges@utfpr.edu.br
* Correspondence: yenunezd@libertadores.edu.co

Abstract: A comparative study on the mechanical properties, scratch resistance, and localized corrosion (pitting and crevice) of plasma-nitrided Inconel alloy 718 (UNS NO7718: IN 718) was carried out. Thermochemical treatment was performed at low temperatures (400 and 450 °C) for 4 h. The treatment formed layers with thicknesses of 7.17 ± 0.89 µm (400 °C) and 7.96 ± 0.48 µm (450 °C). The XRD and nanohardness analyses indicated the formation of a hard layer composed of the expanded austenite phase (γ_N), CrN at 400 °C, and CrN + γ at 450 °C, with a maximum indentation hardness of 12 and 12.5 GPa, respectively, when compared to the 5 GPa substrate hardness. The scratching tests (2–8 N) showed that with increasing load, the nitrided surfaces had a transition from 100% microcutting to a combination of microplowing/cutting, with the presence of cracks. The critical load of the nitrided surfaces was 3 N for 400 °C and 4 N for 450 °C. The untreated condition maintained a crack-free combined mechanism regardless of the load. For the same load, the nitrided surfaces held lower coefficient of friction values and higher scratch resistance values, which were more pronounced at 450 °C. The linear polarization tests (3.56 wt.% NaCl) showed pitting corrosion in all samples, with the 450 °C condition being less resistant. Nitriding at 400 °C increased the crevice corrosion resistance of Inconel, while at 450 °C, it severely damaged it. Nitriding at 400 °C brought concomitant gains in hardness and scratch and crevice corrosion resistance when compared to the as-received IN 718.

Keywords: Inconel 718; plasma nitriding; nanohardness; scratch resistance; crevice corrosion

Citation: Nuñez de la Rosa, Y.; Palma Calabokis, O.; Ballesteros-Ballesteros, V.; Tafur, C.L.; Borges, P.C. Assessment of the Pitting, Crevice Corrosion, and Mechanical Properties of Low-Temperature Plasma-Nitrided Inconel Alloy 718. *Metals* **2023**, *13*, 1172. https://doi.org/10.3390/met13071172

Academic Editors: Francesca Borgioli, Shinichiro Adachi, Thomas Lindner and Frank Czerwinski

Received: 23 April 2023
Revised: 16 June 2023
Accepted: 20 June 2023
Published: 23 June 2023

1. Introduction

Inconel 718 (UNS NO7718: IN 718) is a nickel-chrome-based austenitic superalloy hardened via precipitation [1]. It is commonly used in the aeronautic, aero-spatial, petrol, and gas industries [2–4]. These Ni-based alloys are widely employed in adverse conditions involving high pressure, temperatures between −250 and 750 °C, and aggressive environments, showing good corrosion resistance [1,2,5,6]. However, it also exhibits poor wearing performance, limiting its use in several applications [7].

Plasma nitriding has been established for decades as a successful method in surface engineering [8]. It aims to modify metallic alloy surfaces by introducing nitrogen to improve their tribological performance. When treated at low temperatures, a hard supersaturated interstitial solid solution layer (known as nitrogen-expanded austenite γ_N or S phase) might form on the surface of stainless steels [8–13] and Ni-Cr alloys [1,3,6,7,14–16]. When compared to the number of studies on stainless steel nitriding, as reviewed, for example, in [8], the research on Inconel alloy nitriding is sparse. Surface modification via the nitriding of

IN 718 has been developed through various techniques. Regardless of the type of nitriding, when it is carried out at temperatures $\leq 450\ ^{\circ}C$, the formation of a γ_N phase has been confirmed [1,3,6,7,14–17]. However, nitriding at a higher temperature ($>450\ ^{\circ}C$) formed deleterious phases, identified as CrN, without the γ_N phase [4,6,7,14–16,18]. As for stainless steel nitriding, the growth in the layer thickness with increasing temperature or time was widely reported [7,14,18,19], following diffusion-controlled parabolic growth [6,19]. According to [18], time was more effective than temperature for increases in thickness. Besides, upwards of a two-fold hardness increase due to nitriding treatment is in agreement with all studies evaluating nitrided IN 718 Vickers microhardness, regardless of the layer thickness and composition [3,4,6,7,14–16,19]. Refs. [1,3–8,14–16,18] and Table 1 summarize the existing studies on the nitriding methods applied to IN 718 alloy, which evaluated wear and corrosion performance.

Table 1. Summary of Inconel 718 alloy nitriding research results in chronological order.

Ref.	Nitriding Process		Summary of Results
[19]	Plasma nitriding T; t: 550 to 750 $^{\circ}$C; 1 to 16 h P: 8 mbar/6 Torr %: N_2:H_2; 1:1, 1:3, 3:1, and 9:1	W	(Dry pin-on drum)—The nitride layer reduced the friction coefficient (CoF) by 3.7 times. Wear volume increase of up to 60% in nitrided samples, ascribed to the substrate fracture and separation.
[14]	Intensified plasma-assisted processing (IPAP). T; t: 450 and 490 $^{\circ}$C; 3 h P: 50 mTorr %: N_2:Ar; 4:1 V: -1000 V	W	(Dry pin-on disc) The average CoF for the unprocessed and nitrided alloy is comparable, but the latter showed a tremendous improvement in wear resistance.
		C	Open circuit potential (OCP): 0.1 N NaCl. The higher potential of nitrided IN 718 (+200 mV vs. -25 mV for untreated) suggests that IPAP nitriding at 450 $^{\circ}$C has a beneficial effect on corrosion resistance.
[18]	Plasma DC T; t: 400 to 600 $^{\circ}$C; 1 and 4 h. P: 500 Pa/3.75 Torr %: N_2:H_2; 1:1 V: 500 V	W	(Dry pin-on disc) The CoF also decreased with hardness growth. The untreated specimens exhibit the highest wear rate that decreased upon an increase in the nitriding time and temperature.
[6]	Salt Bath nitriding—molten salt (M_2CO_3: M: K, Na, etc.), ($CO(NH_2)_2$), others. T; t: 425 to 500 $^{\circ}$C; 4 to 16 h	W/C	(Erosion-corrosion: 3.5% NaCl + 5 v% H_2SO_4 + Al_2O_3)—mass loss reduction (up to 77%) in samples nitrided up to 475 $^{\circ}$C. Treatment at 500 $^{\circ}$C increased mass loss from 1.5 to 3.25 times.
[1]	Ultrasonic nanocrystal surface modification (UNSM) and Gas nitriding (GN) T; t: 460 $^{\circ}$C; 5 h P: 1 atm/760 Torr %: Ammonia	W	(Sliding dry wear: high-frequency reciprocating) Up to 98% reduction in the wear volume in both nitriding conditions. CoF remained around 0.8 in all conditions.
		C	Both treatments increased corrosion resistance in Tafel tests (3.5% NaCl).
[4]	Liquid nitriding: KCNO, NaCNO, KCl, NaCl, K_2CO_3, Na_2CO_3, and Li_2CO_3 T; t: 500 $^{\circ}$C; 16 h	W/C	(Dry Ball-on-disk: at 25, 100, and 200 $^{\circ}$C)—wear rate reduction of around 99.9% at all temperatures. (Erosion-corrosion: 3.5% NaCl + 5 HCl + SiO_2)—nitrided alloy exhibits an improvement of almost 80.3% in the erosion-corrosion resistance compared with its untreated counterpart.

Table 1. *Cont.*

Ref.	Nitriding Process		Summary of Results
[7]	Hot wall plasma nitriding (HWPN) T; t: 400, 450, and 500 °C; 6 h P: 4 mbar/3 Torr %: N_2:H_2; 3:1	W	(Pin-on disk)—The wear rates of samples nitrided at 400, 450, and 500 °C were about 3, 4, and 6 times lower compared to the untreated, and the CoF was reduced from 1 to 0.55.
		C	(Tafel 3.5% NaCl)—the 400 °C-samples showed the highest potential (E_{corr}) and lowest corrosion density (i_{corr}), followed by 450 °C. Higher i_{corr} and same E_{corr} for 500 °C-samples compared to the untreated ones.
[3]	Arc Enhanced Glow Discharge plasma nitriding T; t: 450 °C; 1 to 2 h P: 0.01 mbar/0.75 Torr %: N_2:Ar; H_2; 1:5.2:0.4 V: −250 V	W	(Dry pin-on disk)—nitriding increased the wear resistance significantly, so wear was not measurable.
[15]	Triode plasma nitriding (TPN) T: 400, 425, and 450 °C/700 °C t: 4 and 20 h/1, 2 and 4 h P: 0.4 Pa/0.03 Torr %: N_2:Ar; 7:3 V = −200 V	C	(Linear polarization 0.1N NaCl)—Samples nitrided at 400 and 450 °C for 20 h kept the electrochemical behavior of the untreated sample, while the one nitrided at 700 °C-1 h reduced passivation considerably and increased i_{corr} 11.5 times.

T: temperature, t: time, P: pressure, %: gas mixture composition, W: wear, and C: corrosion.

The results obtained (Table 1) revealed that several nitriding methods are effective for increasing IN 718 tribological performance and corrosion resistance when carried out at low temperatures ($\leq$450 °C). These improvements are associated with γ_N formation. Another finding was that, unlike stainless steel nitriding [8,9,13,20], the precipitation of Cr nitrides in the Inconel nitrided layer does not always hamper corrosion resistance [1,6,14–16]. The explanation for this fact has not been reported.

In the literature, few studies evaluate crevice corrosion in Ni-based alloys [21,22]. Among the reports found, the study developed by Mulford and Tromans [21] on IN 600 and IN 625 stands out, and their results showed that although both alloys were resistant to crevice corrosion (at 20, 55, and 80 °C), IN 625 was more resistant due to Mo addition in the alloy. Miller and Lillard [22] found out that the higher the potential applied to the IN 625 alloy was (E < 0.40 V vs. SCE), the shorter the time for the appearance of crevice corrosion. It seems relevant to emphasize that the Inconel alloy nitriding studies listed above (Table 1) did not present crevice corrosion analyses. In fact, crevice corrosion evaluation is also scarce in nitrided stainless steel [9,10,20] despite nitrogen's beneficial and proven effects on this type of corrosion [23–26]. Thus, evaluating the IN 718 nitriding effect on this type of corrosion is necessary.

The literature review showed that the evaluation of wear had been addressed through various tests listed in Table 1, among which the scratch test has yet to be reported. Furthermore, few investigations evaluated the corrosion resistance of nitrided IN 718. It was studied through open circuit potential, the Tafel curve, and linear polarization tests. This panorama demonstrates that there are still several gaps to be filled in the characterization of nitrided IN 718. This study investigates the viability of nitriding a Ni53/Cr19/Fe19 (IN 718) alloy using DC plasma nitriding in an N_2–H_2–Ar atmosphere. Based on the literature (Table 1), two treatment temperatures were selected: 400 °C and 450 °C. This paper evaluates the scratch wear and pitting and crevice corrosion resistance in nitrided samples and correlates them with the microstructural and topographical characterization of the nitrided and untreated surfaces.

2. Materials and Methods

2.1. Materials

The material studied was an IN 718 (UNS NO7718) alloy, as received by the manufacturer (solubilization at 1089 °C for 1 h, followed by aging at 788 °C for 7 h). The alloy chemical composition was determined using the positive material identification technique (PMI, X-ray fluorescence XRF: Olympus Vanta series M, (EVIDENT EUROPE GMBH, Hamburg, Germany) (Table 2). The IN 718 was received in bars 14.6 cm in diameter, from which the 5 mm-thick disks were obtained via wire electro-erosion. Then, $20 \times 20 \times 5 \ mm^3$ pieces were produced via waterjet cutting.

Table 2. Percentages (wt.%) of the IN 718 alloy main elements.

Ni	Cr	Fe	Nb	Mo	Ti	Al	Co
53.45	18.28	18.81	4.94	2.83	0.93	0.5	0.23

2.2. Low-Temperature Plasma Nitriding (LTPN)

The samples were sanded using SiC paper up to #1200 in grain size. Next, they were cleaned with pure acetone in an ultrasound bath and dried in warm air. The samples were then immediately placed inside a pulsed cold wall DC reactor similar to the one used in previous research [10–13]. The plasma nitriding treatment was carried out in two phases (Table 3). The first phase was sputtering, which aimed to remove the passive layer and clean the surface. The second phase was the nitriding treatment at 400 °C and 450 °C, named N400 and N450, respectively. Cross-sections of the nitrided samples were prepared metallographically and attacked with oxalic acid (10% m/v, $H_2C_2O_4$), applying a 5 V voltage for 10–15 s.

Table 3. Conditions of the low-temperature plasma-nitriding process.

Condition	Sputter-Cleaning	DC Plasma Nitriding
Gas mixture $N_2/H_2/Ar$ (%)	0/75/25	75/20/10
T_{on} (μs)	170–230	95–110 (400 °C), 160–200 (450 °C)
T_{off} (μs)	210–250	250
Treatment time (h)	0.5	4
Temperature (°C)	300	400 (N400), 450 (N450)
Peak voltage (V)	500	500
Pressure (Torr)	2.5	2.8
Total gas flow rate (SCCM)	200	200

2.3. X-ray Diffraction (XRD)

After the LTPN treatment, the phases formed in the nitrided layer were identified using grazing incidence X-ray diffraction (GIXRD: Shimadzu, model XRD-7000, Shimadzu Corporation, Tokyo, Japan) with Cu-Kα radiation (λ = 1.54060 Å; 30 kV; 30 mA) and a 3° incidence angle. The measurements were performed at the top of the sample's surface over the 2θ range from 30° to 80°. The phase constituents of the XRD spectra were identified by comparing them with crystallographic powder data based on JCPDS cards: for austenite (γ), 00-006-0696, and for chrome nitrides, 00-011-0065. The identification of the nitrogen-expanded ($γ_N$) peaks was based on the literature [1,3,6,7,14–16]. The penetration depths for the angles (2θ) of 43.5° and 50.58° corresponding to the peaks of γ(111) and γ(200), respectively, were approximately 5.51 μm and 6.15 μm. These values were calculated using the X-Ray attenuation length tool available at CXRO (The center for X-ray optics, Berkeley, CA, USA) [27], considering Cu-Kα radiation and 2θ incidence. The depths were determined by considering the four main elements (Ni, Fe, Cr, and Nb) for the stoichiometric fraction of the IN 718 alloy. Additionally, the depth of 2θ = 37.5° of the CrN compound was calculated, assuming a layer composed solely of CrN, resulting in 4.30 μm. These calculations were performed for 2θ incidence.

2.4. Microscopy Techniques

Optical microscopy (MO: Olympus BX51RF, Olympus Corporation Shinjuku Monolith, Tokyo, Japan) and scanning electron microscopy (SEM: Zeiss, model EVO MA 15, Carl Zeiss Ltd., Cambridge, UK) were employed to determine the layer thickness and to observe the microstructure and surfaces after the corrosion and scratch tests. Nitrogen profiles (weight percentage: wt.%) of the cross-sections were acquired using high-resolution field emission scanning electron microscopy FE-SEM (Tescan Lyra 3, TESCAN ORSAY HOLDING, Brno—Kohoutovice, Czech Republic) via energy-dispersive X-ray spectroscopy (EDX, Oxford XMAX 80, Aztec Software Oxford SP1 3.3, NanoAnalysis, Wiesbaden, Germany). Four linear EDX analyses and two mapping EDX analyses were performed for each nitriding condition on the cross-section of the samples at 15,000× magnification.

2.5. Indentation Hardness

Hardness measurements were carried out at the top of the samples using instrumented nano-indentation equipment (Zwick-Roell ZNH nanoindenter®, GmbH & Co. KG, Ulm, Germany). The quasi-continuous stiffness measurement (QCSM) method was employed to measure hardness as a function of the tip penetration during loading. The QCSM method increases the accuracy of measurements when compared to cyclic load-unload typical tests [28]. This technique is ideal for measuring the mechanical properties of thin films, which change with indentation depth, such as in surface modification thermochemical treatments [28]. A diamond tip with Berkovich geometry was used, and the load was gradually increased over 24 cycles up to the 1000 mN (100 g) maximum load. A matrix of 25 indentations separated by a 100 μm gap was delineated in the regions of interest.

2.6. Scratch Test

Linear scratch tests were performed using a tribometer (CETR-UMT-2MT s/n T1471 Bruker Nano Inc., Campbell, CA, USA), applying a 2–8 N constant load at a speed of 0.017 mm/min. Three scratches (5 mm in length) were produced per load in each load/sample condition combination. The conical stylus (diamond Rockwell C) has an apex angle of 120°, and the cone ends in a hemispherical tip 200 μm in radius. The friction coefficient (CoF) was recorded during scratching using a coupled load cell, which automatically measures tangential force and controls the normal force.

Scratch resistance was determined via the scratch hardness number (HS$_P$, in Pa) and abrasion factor (F$_{ab}$, nondimensional). The HS$_P$ was determined following ASTM G171, according to Equation (1) [29]. A total of 25 measurements for scratch width were determined for each load/sample condition using an optical microscope with an incident light polarization filter (OM: Olympus BX51RF, Olympus Corporation Shinjuku Monolith, Tokyo, Japan).

$$HS_P = \frac{8P}{\pi w^2} \tag{1}$$

where P = normal force (N) and w = scratch width (m).

The friction factor (F$_{ab}$) was determined following the procedures described by Rovani et al. [30]. The F$_{ab}$ calculation followed Equation (2) [31].

$$F_{ab} = \frac{\text{area of groove} - \text{area of pile} - \text{up}}{\text{area of groove}} \tag{2}$$

The surface topography before and after the nitriding treatments was determined using an optical interferometer (Talysurf CCI Lite, M12-3993-03, green light, Taylor Hobson, Leicester, UK). The scratch cross-sections were determined by setting a 0.8 × 0.8 mm² measurement area using 1024 × 1024-pixel resolution. Three measurements per load/sample condition combination were carried out in the scratch central region. The scratch 3D topography and the surrounding flat surface were measured. Next, the mean plane was defined based on the unscratched area only. Finally, the groove and pile-up areas were calculated using the "profile extraction" tool by TalyMap® (Taylor Hobson, Leicester, UK).

2.7. Electrochemical Techniques

For all localized corrosion tests, the samples were first mounted in epoxy resin, aiming to expose the region of interest with a 2.2 cm^2 fixed area. All potentials refer to the reference electrode: saturated calomel (SCE). A graphite bar was the counter electrode. All tests were evaluated in a NaCl 3.56 wt.% aerated solution at 20 ± 2 °C and started after the open circuit potential (OCP) stabilization for an hour. Each experiment was repeated at least three times in order to check the reproducibility, and the scatter in the data was reported.

2.7.1. Linear Polarization

Tests were performed in a glass cell filled with 150 mL solution. The potentiodynamic scanning (1 mV/s) started at -0.2 V vs. OCP and ended when reaching the transpassivation region (5000 µA/cm^2 [32]). The main corrosion parameters, such as corrosion potential (E_{corr}) and corrosion current density (i_{corr}) in the Tafel region, passivation current density (i_{pass}), and pitting potential (E_{pit}), were determined according to the literature [10,13,32,33].

2.7.2. Potentiostatic Polarization

Potentiostatic polarization tests at 1.0 V were performed to evaluate susceptibility to crevice corrosion. This potential was chosen for comparison purposes since it is at the end of the untreated IN 718 passivation region. A Teflon crevice-former presenting three 2.25×6.25 mm^2 rectangular areas each was used. The contact between the crevice-former and the mounted sample surface was reached by applying 0.5 N torque with a nylon screw. The 3D topography of the crevice corrosion marks was analyzed using 3D interferometry.

3. Results
3.1. X-ray Diffraction and Thickness Layer Analysis

The nitriding process formed a homogeneous and continuous layer (Figure 1). As reported in several studies [1,3,4,15,16,18,19], the nitriding treatments generate an outstanding superficial nitride layer (Figure 1). Growth in layer thickness with increasing temperature, as widely reported in the literature [7,14,18,19], was observed. However, the increase was insignificant in this case when considering the standard deviation in Figure 1. The high dispersion in the N400 condition is associated with the noteworthy differences in the layer thickness influenced by the crystallographic orientations of the austenitic microstructure (Figure 1a), which has also been reported in other studies [6,16]. The arrows in Figure 1 indicate the presence of intermetallic precipitates, mainly composed of Ni, Al, Ti, and Nb. Their semi-quantitative compositions were identified using SEM-EDX, for which the values (wt.%) ranged from 0.99–50.89% for Ni, 0.27–0.44% for Al, 1.02–5.65% for Ti, and 4.74–72.65% for Nb. Those intermetallic compounds, known as γ'/γ'', resulted from the IN 718 precipitation hardening treatment.

Figure 2 shows SEM cross-section micrographs on the left side, their respective EDX line scans, and the EDX maps for nitrogen on the right side. The nitrogen concentration (wt.%) as a function of depth from the surface shows a plateau trend, which correlates with the thickness of the nitride layer for both samples. The nitrogen content ranges from 2 to 10 wt.% in both nitrided layers (Figure 2). These results are consistent with the nitrogen distribution obtained through EDX mapping.

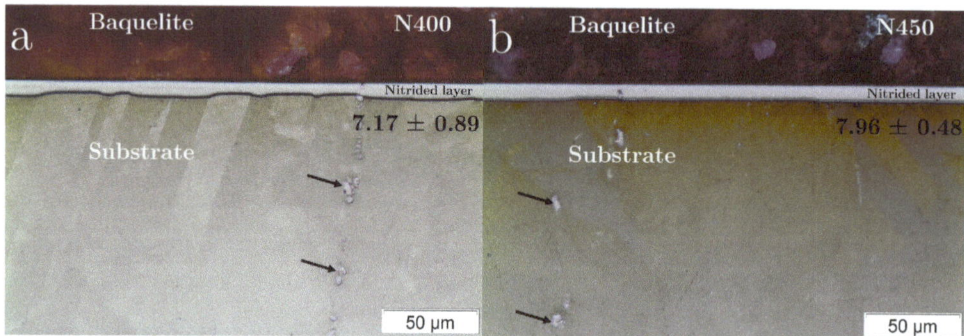

Figure 1. Cross-section OM micrographs of the nitrided layers at (**a**) 400 °C and (**b**) 450 °C. Nitrided layer thickness values correspond to the white layer that stands out from the substrate.

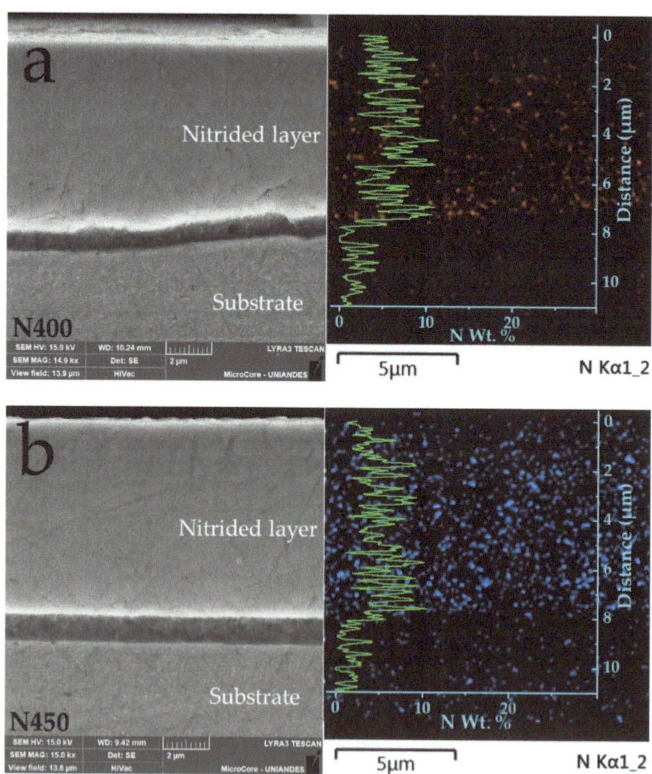

Figure 2. Cross-section SEM micrographs, their EDX line scans, and mapping for nitrogen (wt.%): plasma nitriding at (**a**) 400 °C and (**b**) 450 °C.

Figure 3 shows the X-ray diffraction spectra of the material, as received from the manufacturer, and the nitrided conditions. Nitriding at 400 °C generated a layer composed of N-expanded austenite (γ_N), chromium nitride (CrN), and austenite (γ, with increased full width at half maximum (FWHM)), which agrees with several works on IN 718 treated with LTPN [1,3,6,7,14,15]. However, nitriding at a higher temperature (450 °C) formed deleterious phases that were identified as CrN without the γ_N phase, as reported in other studies, and this increased the temperature or treatment time [4,6,7,14–16,18]. The broader, asymmetric, and lower intensity austenite peaks γ in the N400 and N450 spectra do not represent substrate influence since the XRD penetration depths are less than the nitrided layer thickness (details in Section 2.3) [3,15].

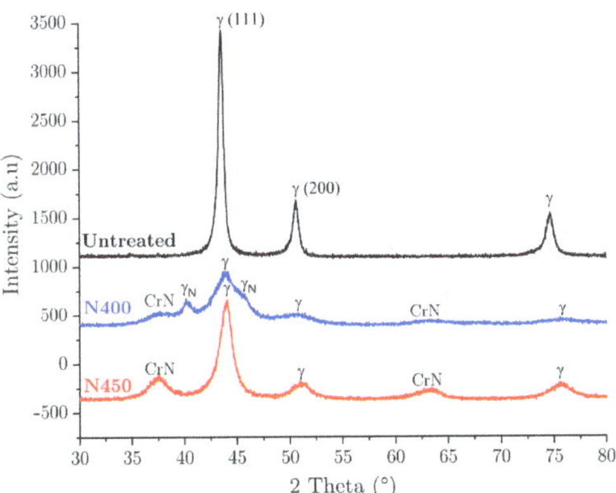

Figure 3. X-ray diffraction patterns.

The lattice parameters of intermetallic precipitates, such as $\gamma'(Ni_3(Al,Ti))$ and $\gamma''(Ni_3Nb)$, are similar to those of the γ matrix and, therefore, cannot be resolved under XRD (Untreated: Figure 3) [3,15]. Regarding this finding, Tao et al. [15] identified the formation of N-modified precipitation-strengthening phases using grazing-angle XRD and TEM analysis. The appearance of nitrided γ'/γ'' or N-modified intermetallic precipitates was not confirmed in this study.

3.2. Indentation Hardness and Scratch Resistance

Indentation hardness was measured on the top of the sample surfaces. The nitrided samples showed higher hardness profiles when compared to the as-received IN 718 (Figure 4a). The indentation size of both N400 and N450 (Figure 4c,d) was reduced when compared to the untreated samples (Figure 4b). Increased hardness on the surface nitrided at 400 °C is explained by the formation of the γ_N, for which nitrogen saturation increased crystalline lattice stress [3,4,6,7,15,16]. However, the presence of CrN precipitates in the N400 layer also contributes [3,14,15]. On the other hand, the hardness of N450 is related to hard compound-layer (CrN) formation; the distribution of this within the layer prevents free dislocation movement, which has been extensively studied [3,7,14,16,18,19]. It is relevant to highlight that the Berkovich tip penetration depth remained within the nitrided layers (Figure 4a) but beyond 10% thickness [28]. Consequently, the substrate influenced the indentation measurement in both of the nitriding surfaces.

Moreover, the micrographs in Figure 4c–d exhibited the topography of the grain boundaries on the nitrided surfaces in contrast with the untreated sanded topography (Figure 4b). Further topographic information can be found in Table 4. The parabolic profile of the indentation hardness profiles as a function of depth (Figure 4a) responded to the significant effect of surface roughness due to nitriding treatment, as reported in other studies [3,14,15,18,28]. In addition, the indentations in the rougher nitrided surfaces also showed the non-negligible dispersion of the indentation hardness depth curves.

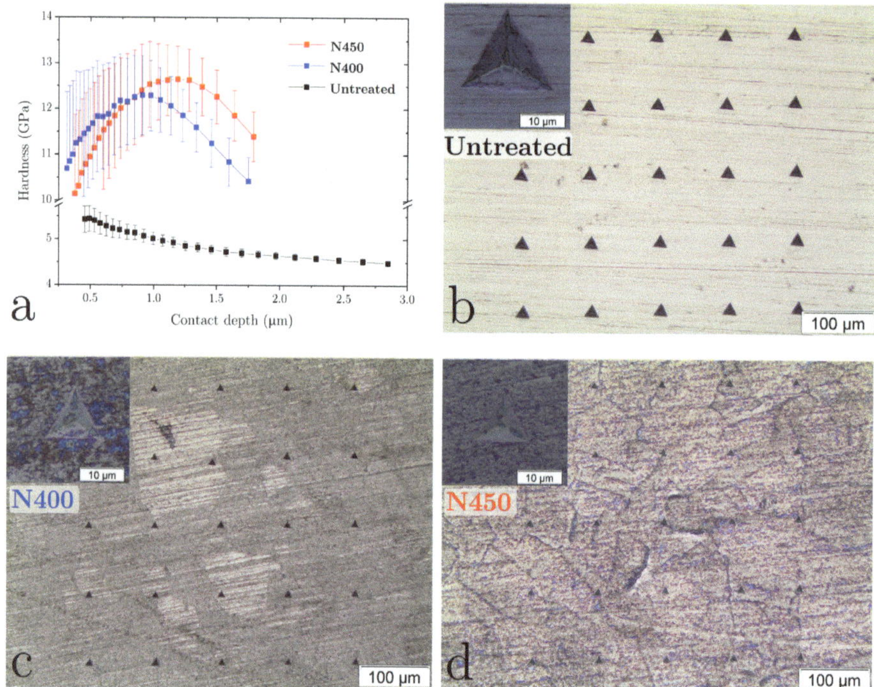

Figure 4. (**a**) Indentation hardness curves as a function of the indentation depth (QCSM method) (Error bars represent standard deviation). Instrumented nano-indentation matrix with enlarged indentation mark: (**b**) untreated; (**c**) N400; (**d**) N450.

Table 4. Roughness results (±represents standard deviation): S_q (root mean square height), S_{sk} (skewness), S_{ku} (kurtosis), S_z (maximum height), S_{dq} (root mean square surface slope), and S_{dr} (developed interfacial area ratio).

Condition	S_q (µm)	S_{sk}	S_{ku}	S_z (µm)	S_{dq}	S_{dr} (%)
Untreated	0.0127 ± 0.0009	−0.19 ± 0.03	4.14 ± 0.09	0.151 ± 0.006	0.0116 ± 0.0008	0.0067 ± 0.0009
N400	0.0148 ± 0.0002	0.11 ± 0.08	4.58 ± 0.24	0.228 ± 0.011	0.0135 ± 0.0004	0.0091 ± 0.0005
N450	0.0146 ± 0.0002	−0.02 ± 0.03	3.58 ± 0.06	0.174 ± 0.004	0.0138 ± 0.0001	0.0095 ± 0.0002

Scratch resistance was evaluated by producing constant load scratches from 2 N to 8 N on the sample surfaces. The first column in Figure 5 shows the topography results obtained using 3D interferometry before the test. The results of roughness characterization are presented below (Table 4). The second column in Figure 5 corresponds to the 3D reconstruction of a scratch section. The reconstruction of the mean cross-section profile along the scratch is shown in the third column. The results shown (Figure 5) correspond to a representative sample of each condition evaluated at 8 N. The same procedure was adopted for all loads (2–8 N), the results of which are summarized in the graphs below (Figure 6).

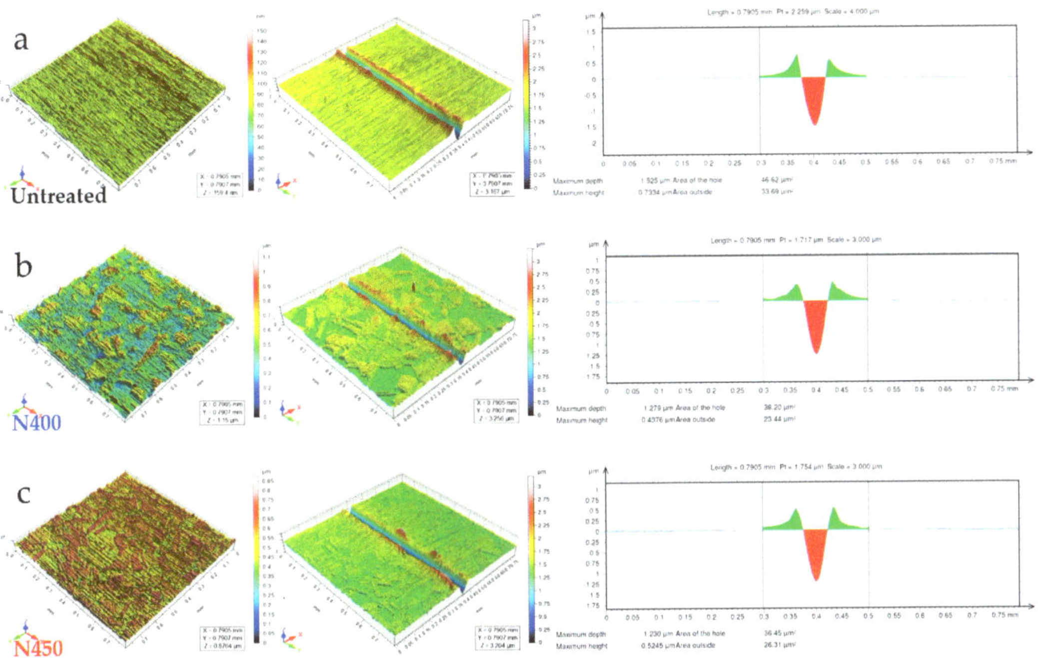

Figure 5. A sequence of surface analysis using 3D interferometry before and after the scratch tests: initial topography and the reconstruction of scratches and their cross-sections: (**a**) untreated, (**b**) N400, (**c**) N450 (the results presented in this figure correspond to the 8 N load).

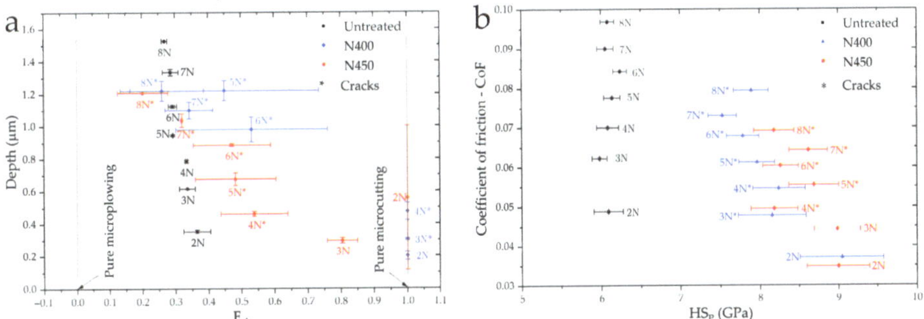

Figure 6. (**a**) Scratch depth as a function of abrasion factor (F_{ab}). (**b**) Coefficient of friction as a function of the scratch hardness (HS_P). Error bars represent confidence intervals.

Table 4 shows the most relevant roughness parameters describing the topographic changes due to the nitriding process. The S_q, S_z, S_{dq}, and S_{dr} values of N400 and N450 confirmed a general increase in roughness. S_{dq} is a general measure of the slopes that compose the surface. S_{dr} represents the percentage of the additional surface area contributed to by the texture in comparison with an ideal plane of the measurement region. Both results are coherent with the grain boundary topography observed in the reconstruction in Figure 5b,c. In particular, the surfaces nitrided at 450 °C are characterized as being normally distributed since S_{sk} and S_{ku} are close to 0.00 and 3.00, respectively. Thus, the topography developed in N400 is characterized by the predominance of peak structures ($S_{sk} > 0$), which agrees with the presence of excessively high peaks ($S_{ku} > 3.00$) and the maximum surface heights (highest S_z).

The graph below (Figure 6a) depicts the scratch depth (in the cross-section) as a function of the abrasion factor (F_{ab}) in all sample conditions and loads. The F_{ab} values indicate the predominant scratch wear mechanism. As shown in Figure 6a, a zero (0) value indicates pure microploughing wear, while a value of one (1) represents pure microcutting wear (Figure 6a). During the scratch tests, the friction coefficient (CoF) was recorded, and the results (as a function of scratch hardness (HS_P)) are represented in Figure 6b for all conditions.

All samples showed increased scratch penetration as a function of load increase (Figure 6a). Regarding IN 718 (untreated), we observed that F_{ab} hardly depends on the load, keeping values around 0.25–0.4, showing a tendency for plastic deformation mechanisms without crack formation (microploughing). The nitrided surfaces presented cracks at the sides of the scratches from the 3 N load for N400 and from 4 N in N450. At low loads, the pure microcutting mechanism operates in both nitrided conditions, within the 2–4 N range for N400 and only 2 N in N450. From the loads mentioned, a mixed mechanism predominates with a plowing tendency when the load grows. When considering the penetration depth, the substrate did not influence the tests up to 4 and 5 N in N400 and N450, respectively. This occurred because the scratch depth did not exceed 10% of the nitride layer thickness [28]. Consequently, from those loads upwards, the substrate influenced the abrasion response on the nitrided surfaces.

At similar loads, the CoF (Figure 6b) was always lower for the nitrided surfaces when compared to the untreated ones. In addition, the CoF values were lower in the N450 condition when compared to N400. The surface hardening effect (Figure 6b) as a result of nitriding was observed since the nitrided surfaces presented higher scratch hardness values (HS_p) than those observed on the untreated surface. Additionally, the N450 condition showed a tendency for higher HS_p mean values when compared to N400.

3.3. Linear Polarization

The most representative polarization curves of each condition are presented in Figure 7. The most relevant corrosion parameters are shown in Table 5. The tests revealed that the general corrosion resistance, measured based on the i_{corr}, E_{corr}, and C_o parameters, is comparable between the untreated and the N450 conditions. The N400 samples showed E_{corr} values close to that of the untreated material and higher i_{corr} and C_o values. An increase of several orders of magnitude was observed for current density from $E > 0.25$ V in the nitrided conditions (Figure 7). Regarding N450, the appearance of bubbles on the surface of the samples was observed due to water oxidation reactions and the hydrolysis of the alloy metals [10,33]. Simultaneously, on the auxiliary electrode's surface, bubbles ascribed to oxygen and water reduction reactions were also observed from $E > 0.25$ [33]. The phenomena described did not occur when the current density increased in the N400 condition ($E > 0.30$ V). Therefore, it is possible to affirm that in the 0.65–1.00 V potential range for N400, second passivation occurred—actually, a pseudo-passivation—since the current was ~100 μA/cm^2 (Table 5).

The micrographs (Figure 8) evidenced that the passivation layer failure occurred mainly due to the localized pitting corrosion in all samples. For this reason, the passive layer breakdown potential was named E_{pit} in the curves of Figure 7 and Table 5. The N450 nitriding revealed a sharp decrease in pitting resistance, observed as (i) a significantly lower E_{pit} value when compared to the untreated and N400 samples (Table 5); (ii) the formation of larger pits (Figure 8c), and (iii) greater pit nucleation density. On the other hand, the N400 samples showed slightly lower E_{pit} values than those presented by the untreated IN 718 (Table 5). On both nitrided surfaces, the pits were observed as starting mainly at the grain boundaries and were larger than those found in IN 718. However, the N400 condition showed lesser pit density when compared to the untreated and N450 conditions.

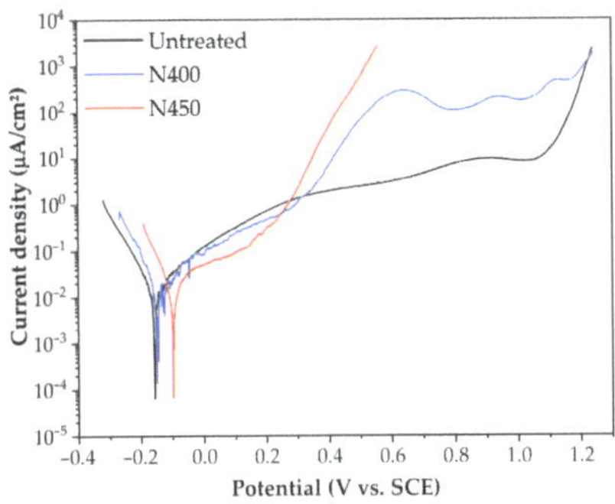

Figure 7. Polarization curves of untreated and nitrided samples (3.56 wt.% NaCl, 20 ± 2 °C, 1 mV/s).

Table 5. Summary of E_{corr}, i_{corr}, C_o, i_{pass}, and E_{pit} parameter values (± represents standard deviation).

Parameter	Untreated	N400	N450
E_{corr} (V)	-0.174 ± 0.037	-0.14 ± 0.02	-0.13 ± 0.03
i_{corr} ($\mu A/cm^2$)	0.015 ± 0.002	0.025 ± 0.001	0.015 ± 0.003
C_o (mm/year) $\times 10^{-4}$	1.57 ± 0.24	2.55 ± 0.12	1.59 ± 0.35
Passivation region (V)	0.00–1.00	0.00–0.25 * 0.65–1.00	0.00–0.25
i_{pass} ($\mu A/cm^2$)	4.827 ± 3.432	0.26 ± 0.15 * 176 ± 44	0.141 ± 0.091
E_{pit} (V)	1.131 ± 0.004	1.109 ± 0.005	0.26 ± 0.042

*—Pseudo-passivation.

Figure 8. Micrographs of surfaces after the linear polarization tests: (**a**) Untreated, (**b**) N400 and (**c**) N450.

3.4. Crevice Corrosion

The potentiostatic polarization curves with crevice formers are shown in Figure 9 (representative curve of each sample). The N400 condition showed lower current densities than the untreated material, presenting improvements in crevice-corrosion resistance. However, the N450 condition showed very high current densities, reaching 50,000 $\mu A/cm^2$ from the beginning of the polarization. Such densities indicate very high corrosion rates that go over the potentiostat tolerance limit, provoking the suspension of the measurement.

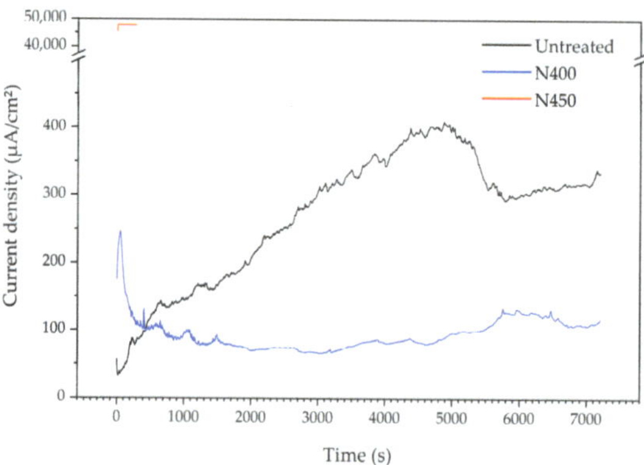

Figure 9. Potentiostatic polarization curves (1.00 V) in tests with crevice-former.

Crevice corrosion marks are seen in Figure 10. A reconstruction of the micrographs of the corrosion marks obtained using optical microscopy is shown in Figure 10 (center). On the left of Figure 10, the details of a corroded region analyzed using 3D interferometry can be observed. The micrographs of the untreated and N400 conditions revealed that the corrosion occurred preferentially on the boundaries of the crevice-former. The 3D interferometry analysis only evidenced a severe and profound attack in the untreated material (Figure 10a). The N400 samples did not present deep crevice corrosion but had extended corrosion in the surrounding areas of the crevice-former's contour (Figure 10b).

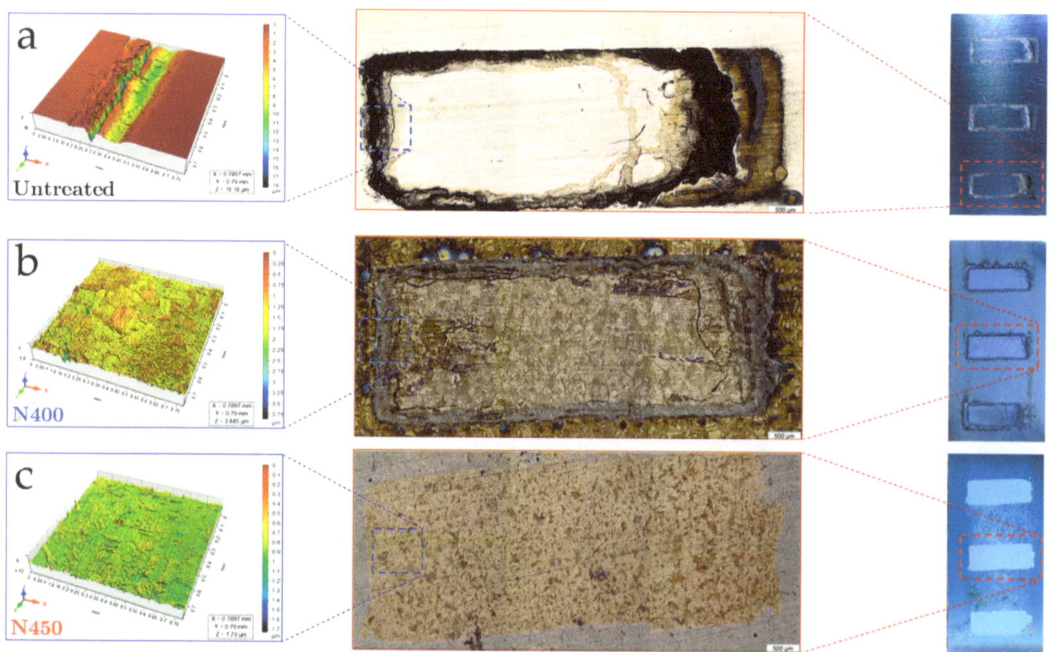

Figure 10. Potentiostatic polarization curves (1.00 V) of tests with crevice former: (**a**) untreated, (**b**) N400, and (**c**) N450.

The N450 sample tests were interrupted due to high current densities: only a few seconds were enough to generate corrosion on the whole surface, as observed in the macrography (Figure 10c) (right side). Pitting corrosion was not observed inside or outside the crevice areas in any sample. This finding confirms that the potentiostatic test at 1.00 V exclusively evaluated the behavior of the material concerning the localized crevice corrosion.

4. Discussion

4.1. X-ray Diffraction and Thickness Layer Analysis

The modification of surfaces of Inconel alloys using nitriding has been developed through different techniques (Table 1). The main common microstructural changes observed in Inconel alloy nitriding studies are (i) the dislocation, loss of symmetry, and generalized widening of the $\gamma(111)$ and $\gamma(200)$ crystallographic peaks in the XRD spectrum (Figure 3); (ii) the presence of CrN and γ_N peaks (Figure 3); (iii) the formation of a nitride layer differentiated from the substrate (Figure 1).

Regarding the first change, (i) the peak characteristics indicate an alteration in the crystalline lattice parameter due to the introduction of nitrogen via microstresses, non-homogeneous chemicals, and microstructural composition, or the residual stress related to chemical gradients and crystalline lattice anisotropic expansion [3,6–8,10,11,13,34]. The peaks referring to the γ_N phase (Figure 3) appear due to the formation of the nitrogen interstitial solid solution in the austenitic structure of Inconel alloys, as reported by [1,3,6,7,14,15]. Regarding the second change (ii), the simultaneous presence of peaks referring to the CrN phases in N400 and the disappearance of the γ_N peaks in N450 indicate γ_N phase degradation in CrN + γ. This was observed when IN 718 was nitrided at higher temperatures ($\geq$450 °C) [4,6,7,14,18,19] and also for longer periods (>8 h) [6,19]. A more detailed discussion on the formation of the γ_N phase and CrN is given elsewhere [3,6,15,34].

Finally (iii), the formation of a surface layer well differentiated from the substrate (Figures 1 and 2) was already reported in several studies [1,3,4,15,16,18,19]. According to Kovací et al. [18] and Aw, Batchelor, and Loh [19], a separate layer occurs as a consequence of the high speed of the sputtering-redeposition process and the fast reaction between N and Cr, which favor the formation of CrN instead of γ_N. The preferential electrochemical attack of the layer-substrate interface (Figures 1 and 2) may be associated with CrN nucleation, as reported in [18,19]. According to Kovací et al. [18], this does not allow the N to penetrate deeply into the surface and does not allow the diffusion layer to become too narrow. Only two nitriding studies on IN 718 have evaluated nitrogen distribution in the nitrided layer. Mondragon-Rodriguez et al. [3] observed a uniform distribution of N in the nitrided layer at 450 °C in IN 718, similar to Figure 2. Already in the work of Zhang et al. [1], the N distribution followed a linear decreasing trend from the surface to a depth of ~50% of the thickness of the nitrided layer. The remaining 50% of the layer exhibited an approximately constant N value (2–3 at.%). The discrepancies with Zhang et al. [1] may be attributed to the effect of residual stresses from the surface previously treated by UNSM from nitrogen diffusion.

On the other hand, other authors also evidenced differences in layer thickness (as observed in Figure 1a, N400), attributed to the presence of slip lines or crystallographic orientations within the layer [6,14] and in the substrate [5,16]. In both situations, these lines appear due to the difficulty of the γ-FCC structure to deform and the formation of pile-up gaps and dislocations resulting from the introduction of nitrogen [16,34]. At higher nitriding temperatures (in this case, 450 °C, Figure 1b), extensive CrN precipitation and γ_N degradation occurred, promoting lattice relaxation and achieving a uniform layer thickness in N450.

4.2. Indentation Hardness and Scratch Resistance

Upwards of a two-fold increase in hardness due to the nitriding treatment (Figure 4) is in agreement with all studies evaluating nitride Inconel alloy microhardness [3,4,6,7,14–16,19]. The severe distortion of the crystalline reticulate by N diffusion promotes solid solution strengthening. Moreover, the intrinsic higher hardness of CrN [1,4] and their distribution in the layer enables the precipitation hardening mechanism [7,14,16,18,19]. However, increased nitriding temperature did not cause significant differences in hardness despite the layer composition in the N450 condition (CrN + γ). According to Tao et al. [15], the close hardness values ($HV_{0.01}$: 1570 and 1671) in nitriding treatments at 400 °C for 20 h and 450 °C for 20 h resulted from significant substrate influence (thinner layers of 4 and 7 μm, respectively). In this study, the similar indentation hardness values at a contact depth of <1 μm are due to the effect of high roughness (Table 4) [28]. In contact, depths of >1 μm are already influenced by the substrate (penetration in excess of 10% of the thickness), and values that were slightly higher were observed in the N450 condition, similar to those obtained in [15].

Increased roughness due to nitriding treatment (Table 4) is a well-known and widely reported finding [1,3,7,11,15,17,18,34]. Regarding Inconel alloys, the studies found in the literature [1,3,7,15,18,34] have evaluated the R_a parameter. However, a broader analysis of topographic changes, such as the one in this study (Table 4), was never reported (to the author's knowledge). In addition, to increase mean roughness (S_q, S_z), the nitriding treatment increased the root mean square surface slope (S_{dq}) and the real contact area (S_{dr}). The untreated surface was characterized by the predominance of deep valley structures, which were confirmed by the $S_{sk} < 0$ and $S_{ku} > 3.00$ parameters. The higher temperature (450 °C) promoted uniformity in the distribution of the peaks and valleys (S_{sk} and S_{ku} close to 0.00 and 3.00; a normal distribution). Nitriding at 400 °C was more severe in promoting topographic changes, differentiated by the predominance of peaky structures ($S_{sk} > 0$, $S_{ku} > 3.00$, highest Sz). According to Kovací et al. [18], the increased process parameters (time and temperature) intensify ion bombarding and the fast cathodic pulverization, which cause changes to the surface texture. In addition to sputtering effects, other studies [3,15,17,34] reported that increased roughness is mainly related to nitrogen diffusion (causing microstructure expansion or swelling). These reasons justify the more significant topographic changes in N400 than in N450 due to the formation of the γ_N + CrN layer.

Increased surface roughness is usually related to a higher wear volume and friction coefficient (CoF) due to raised contact areas between the tribological pairs [1,35]. Kapoor et al. [35] established that the wear volume should increase with the roughness slope and decrease with increases in hardness. Thus, a higher CoF would be expected, and the scratch wear resistance in nitriding conditions would decrease based solely on the topographic changes already discussed (Table 4, Figure 5). However, over a two-fold hardness increase (Figure 4) on the nitrided surfaces predominated regarding tribological behavior. Zhang et al. [1] observed that the increase in hardness (1.9 times) due to surface modification (ultrasonic nanocrystal) was compensated for by an increase in roughness (Ra increase of five times), resulting in similar performances from the point of view of the wear for IN 718 (polished surface). This study obtained improved scratch resistance with nitriding treatment, evidenced by the decreased penetration depth, increased scratch hardness (HS_P), and reduced CoF (Figure 6). These results agree with the studies put forward by Kovací et al. [18], Maniee et al. [7], and Aw, Batchelor, and Loh [19], in which nitrided IN 718 presented a lower CoF and lower wear when compared to the as-received material in dry pin-on-disk wear tests (Table 1). Thus, in this study, the nitriding treatments improved the surface mechanical resistance and remarkably influenced the scratch resistance. The roughness and topographical changes after nitriding did not negatively affect tribological performance.

However, little difference was observed in the scratch depth between N400 and N450, which agrees with the similar indentation hardness and layer thickness. On the other hand, the N450 samples showed lower CoF and higher HS_P values than N400 in loads of $\geq$5 N (Figure 6b). Both results can be explained, as proposed by Aw, Batchelor, and Loh [19], since, for equivalent hardness, the negative effect of the topography of surfaces nitrided at 400 °C predominates (peak predominance). Nitriding resulted in wear micromechanism changes when compared to the untreated material (Figure 6a). The adhesive-abrasive combined mechanisms in untreated and nitrided IN 718 (using different techniques) were identified by [1,14,16]. All those authors identified lower adhesive wear contributions on nitrided surfaces, with a consequent reduction in CoF and wear rate in dry tests. Likewise, in scratch tests, mixed micromechanisms prevailed in the IN 718 and nitrided conditions (Figure 6a), in which microcutting predominated with a tendency to crack in low load conditions (2–4 N). However, despite the gain in scratch wear resistance with nitriding, microcutting + cracks might provoke higher losses due to wear with increased time/distance/severity in the tribological test, as pointed out by Zhang et al. [1].

4.3. Linear Polarization

The linear polarization curves corresponding to IN 718 before and after nitriding are presented in Figure 7. The best performance of the N450 condition, when compared to the N400 condition up to the 0.20 V potential (that is, in E_{corr}, i_{corr}, and C_0), might be related to the better stability of the oxide layer formed on the surface composed of CrN (N450), as observed by Xue et al. [4] in salt bath-nitrided IN 718. Further studies are required to confirm this hypothesis. However, significant differences were observed for the Tafel region in both nitriding conditions when compared to the untreated IN 718. E_{pit} values close to the untreated condition (Table 5 and lower pit nucleation density in N400 (Figure 8b) indicate some improvements in pitting resistance in a NaCl (3.56 wt.%) solution. This result is ascribed to the γ_N phase, which demonstrated resistance to localized corrosion in Inconel [7,14–16] and other stainless-steel alloys [8–13,20] due to nitrogen enrichment. Nitrogen plays a role in neutralizing local pit acidification in anodic areas; further explanations can be found in other sources [23–26]. The polarization results revealed that CrN precipitation at 400 °C for 4 h (XRD results, Figure 3) is still insufficient to cause substantial damage to corrosion performance. These results agree with those reported in previous studies [15,16].

An increased treatment temperature from 400 to 450 °C resulted in a passivation region and pitting corrosion resistance reduction. Chrome depletion in many matrix areas, resulting from CrN formation, might cause such detrimental corrosion effects. Similar results were previously reported by [7,15,16] when they obtained a nitride layer (without γ_N) due to an increase in nitriding temperature (Table 1). Otherwise, the preferential pit nucleation at the grain boundaries of the nitrided surfaces (Figure 8) was already reported by Xue et al. [4] for IN 718 treated by liquid nitriding. According to [4], this results from the concentration of lattice defects, such as dislocation at the grain boundaries, which reduces the corrosion potential when compared to neighboring grains, forming galvanic micropiles. This effect was also reported by Calabokis et al. [10] in nitrided UNS S32507 steel and by Berton et al. [13] in nitrided AISI 409.

4.4. Crevice Corrosion

Few studies in the literature have evaluated crevice corrosion in Ni-based alloys [21,22]. In fact, the research into Inconel alloy nitriding (Table 1) does not assess crevice corrosion. This study is the first to approach crevice corrosion in these nitrided alloys, according to the author's knowledge. Miller and Lillard [22] evaluated the crevice corrosion behavior of IN 625 in synthetic seawater. Through potentiostatic tests in a potential lower than the transpassivation potential, the authors of [22] found out that, regardless of the potential applied, a more profound attack occurs closer to the crevice opening (as shown in Figure 10a,b). According to Miller and Lillard [22], corrosion always starts inside the crevice, with a superficial attack. Then, over time, the attack moves toward the opening, where it propagates and deepens.

The results in Figures 9 and 10 showed that the formation of a layer composed of γ_N and CrN (N400) increases crevice corrosion resistance in IN 718. Nitriding at 450 °C harmed the resistance seen in the excessive current density. These results agree with the reduction in the passivation region and higher current densities in the linear polarization tests (N450: Figure 7, Table 5). Both results are due to the presence of Cr-depleted zones as a consequence of the transformation $\gamma_N \rightarrow CrN + \gamma$, already reported.

In fact, nitrided stainless-steel crevice corrosion evaluation is also relatively scarce in the literature [9,10,20]. Studies by Fossati et al. [9] and Bottoli et al. [20] reported crevice corrosion in modified AISI 316L via gaseous nitriding treatments [20] and glow-discharge nitriding [9], and Calabokis et al. [10] evaluated it in LTPN UNS S32750 duplex stainless steel. Studies [9,10,20] evidenced an improvement in crevice corrosion resistance due to the formation of γ_N. Nitrogen positively affects the passive layer resistance and its repassivation, which has been confirmed by several studies [23–26]. In agreement with this study, CrN precipitation in AISI 316L was seen to be harmful to crevice corrosion [9] due to the formation of Cr-depleted zones around the CrN precipitates.

Therefore, the results confirmed that a layer composed of CrN + γ (without γ_N) significantly reduced localized pitting and crevice corrosion resistance in IN 718. An improvement in the crevice corrosion resistance of IN 718 was obtained after nitriding at 400 °C, which was confirmed by a reduction in current density in potentiostatic tests (Figure 9) and a reduction in crevice corrosion depth (Figure 10).

5. Conclusions

The LTPN treatment produced layers with thicknesses of 7.17 ± 0.89 µm (400 °C) and 7.96 ± 0.48 µm (450 °C) composed of an N-expanded austenite phase (γ_N) and CrN at 400 °C, and CrN + γ at 450 °C.

Surface hardening due to nitriding is associated with the solid-solution strengthening and precipitation-hardening mechanisms, with a maximum indentation hardness of 12 GPa (400 °C) and 12.5 GPa (450 °C), respectively, when compared to 5 GPa for substrate hardness. Surface hardening prevails in the tribological response in scratch tests against the effect of an increase in roughness (higher S_q and S_z) and topographic changes (S_{ku}, S_{sk}, S_{dq}, and S_{dr}) resulting from nitriding.

The nitriding condition at 450 °C showed slightly higher scratch resistance and a lower CoF compared to the N400 condition, with significant damage to localized pitting and crevice corrosion resistance when compared to IN 718. Those results are a consequence of the CrN + γ hard layer, which generates chromium-depleted regions around the CrN precipitates.

Even with CrN precipitation, the γ_N formation in N400 results in better performance regarding localized crevice corrosion (corrosion wear not measurable and a lower current density) without hampering the electrochemical behavior (E_{corr}, i_{corr}, C_0, and i_{pass}) of the untreated material.

Author Contributions: Y.N.d.l.R.: Conceptualization, methodology, validation, formal analysis, investigation, writing—original draft, writing—review and editing, and project administration. O.P.C.: Conceptualization, methodology, validation, formal analysis, investigation, writing—original draft, writing—review and editing and project administration. V.B.-B.: Writing—review and editing, funding acquisition, resources, and project administration. C.L.T.: resources, project administration, and funding acquisition. P.C.B.: Validation, methodology, formal analysis, investigation, writing—original draft and writing—review and editing and funding acquisition. All authors have read and agreed to the published version of the manuscript.

Funding: The APC and the research was funded by the Fundación Universitaria Los Libertadores—Colombia (FULL) (Project N° ING-08-23). The author Y. Nuñez acknowledges the Coordenação de Aperfeiçoamento de Pessoal de Nível Superior—Brazil (CAPES) (grant number 88887.484833/2020-00). The author O. P. Calabokis acknowledges the Fundação de Apoio à Educação, Pesquisa e Desenvolvimento Científico e Tecnológico da Universidade Tecnológica Federal do Paraná—Brazil (FUNTEF-PR) for the scholarship ACT NO. 03/2020. The author P. C. Borges receives a research scholarship from Conselho Nacional de Desenvolvimento Científico e Tecnológico—Brazil (CNPq) (Process 308716/2021-3).

Data Availability Statement: The data that support this work and its results are not available to be shared because they are under confidentiality agreements. Access to the data can be requested through an official document.

Acknowledgments: Acknowledgments to the Materials Characterization Multiuser Laboratory—UTFPR (CMCM) for the SEM-EDS and X-ray diffraction services. The authors thank the company *Arotec S.A. Indústria e Comércio* for the PMI analysis. The authors thank the Laboratory of Nanomechanical Properties of Surfaces and Thin Films of the Federal University of Paraná (LABNANO-UFPR) for the nano-indentation analysis performed.

Conflicts of Interest: The authors declare no conflict of interest.

References

1. Zhang, H.; Qin, H.; Ren, Z.; Zhao, J.; Hou, X.; Doll, G.L.; Dong, Y.; Ye, C. Low-Temperature Nitriding of Nanocrystalline Inconel 718 Alloy. *Surf. Coat. Technol.* **2017**, *330*, 10–16. [CrossRef]
2. Kruk, A.; Wusatowska-Sarnek, A.M.; Ziętara, M.; Jemielniak, K.; Siemiątkowski, Z.; Czyrska-Filemonowicz, A. Characterization on White Etching Layer Formed During Ceramic Milling of Inconel 718. *Met. Mater. Int.* **2018**, *24*, 1036–1045. [CrossRef]
3. Mondragón-Rodríguez, G.C.; Torres-Padilla, N.; Camacho, N.; Espinosa-Arbeláez, D.G.; de León-Nope, G.V.; González-Carmona, J.M.; Alvarado-Orozco, J.M. Surface Modification and Tribological Behavior of Plasma Nitrided Inconel 718 Manufactured via Direct Melting Laser Sintering Method. *Surf. Coat. Technol.* **2020**, *387*, 125526. [CrossRef]
4. Xue, L.; Wang, J.; Li, L.; Chen, G.; Sun, L.; Yu, S. Enhancement of Wear and Erosion-Corrosion Resistance of Inconel 718 Alloy by Liquid Nitriding. *Mater. Res. Express* **2020**, *7*, 096510. [CrossRef]
5. Kopec, M.; Gorniewicz, D.; Kukla, D.; Barwinska, I.; Jóźwiak, S.; Sitek, R.; Kowalewski, Z.L. Effect of Plasma Nitriding Process on the Fatigue and High Temperature Corrosion Resistance of Inconel 740H Nickel Alloy. *Arch. Civ. Mech. Eng.* **2022**, *22*, 57. [CrossRef]
6. Jing, Y.; Jun, W.; Tan, G.; Ji, X.; Fan, H. Phase Transformations during Low Temperature Nitrided Inconel 718 Superalloy. *ISIJ Int.* **2016**, *56*, 1076–1082. [CrossRef]
7. Maniee, A.; Mahboubi, F.; Soleimani, R. Improved Hardness, Wear and Corrosion Resistance of Inconel 718 Treated by Hot Wall Plasma Nitriding. *Met. Mater. Int.* **2020**, *26*, 1664–1670. [CrossRef]
8. Borgioli, F. The "Expanded" Phases in the Low-Temperature Treated Stainless Steels: A Review. *Metals* **2022**, *12*, 331. [CrossRef]
9. Fossati, A.; Borgioli, F.; Galvanetto, E.; Bacci, T. Corrosion Resistance Properties of Glow-Discharge Nitrided AISI 316L Austenitic Stainless Steel in NaCl Solutions. *Corros. Sci.* **2006**, *48*, 1513–1527. [CrossRef]
10. Palma Calabokis, O.; Núñez de la Rosa, Y.; Lepienski, C.M.; Perito Cardoso, R.; Borges, P.C. Crevice and Pitting Corrosion of Low Temperature Plasma Nitrided UNS S32750 Super Duplex Stainless Steel. *Surf. Coat. Technol.* **2021**, *413*, 127095. [CrossRef]
11. Núñez de la Rosa, Y.E.; Palma Calabokis, O.; Borges, P.C.; Ballesteros Ballesteros, V. Effect of Low-Temperature Plasma Nitriding on Corrosion and Surface Properties of Duplex Stainless Steel UNS S32205. *J. Mater. Eng. Perform.* **2020**, *29*, 2612–2622. [CrossRef]
12. Núñez, Y.; Mafra, M.; Morales, R.E.; Borges, P.C.; Pintaude, G. The Effect of Plasma Nitriding on the Synergism between Wear and Corrosion of SAF 2205 Duplex Stainless Steel. *Ind. Lubr. Tribol.* **2020**, *72*, 1117–1122. [CrossRef]
13. Berton, E.M.; das Neves, J.C.K.; Mafra, M.; Borges, P.C. Quenching and Tempering Effect on the Corrosion Resistance of Nitrogen Martensitic Layer Produced by SHTPN on AISI 409 Steel. *Surf. Coat. Technol.* **2020**, *395*, 125921. [CrossRef]
14. Singh, V.; Meletis, E.I. Synthesis, Characterization and Properties of Intensified Plasma-Assisted Nitrided Superalloy Inconel 718. *Surf. Coat. Technol.* **2006**, *201*, 1093–1101. [CrossRef]
15. Tao, X.; Kavanagh, J.; Li, X.; Dong, H.; Matthews, A.; Leyland, A. An Investigation of Precipitation Strengthened Inconel 718 Superalloy after Triode Plasma Nitriding. *Surf. Coat. Technol.* **2022**, *442*, 128401. [CrossRef]
16. Skobir Balantič, D.A.; Donik, Č.; Podgornik, B.; Kocijan, A.; Godec, M. Improving the Surface Properties of Additive-Manufactured Inconel 625 by Plasma Nitriding. *Surf. Coat. Technol.* **2023**, *452*, 129130. [CrossRef]
17. Chollet, S.; Pichon, L.; Cormier, J.; Dubois, J.B.; Villechaise, P.; Drouet, M.; Declemy, A.; Templier, C. Plasma Assisted Nitriding of Ni-Based Superalloys with Various Microstructures. *Surf. Coat. Technol.* **2013**, *235*, 318–325. [CrossRef]
18. Kovací, H.; Ghahramanzadeh, H.; Albayrak, Ç.; Alsaran, A.; Çelik, A. Effect of Plasma Nitriding Parameters on the Wear Resistance of Alloy Inconel 718. *Metal Sci. Heat Treat.* **2016**, *58*, 470–474. [CrossRef]
19. Aw, P.K.; Batchelor, A.W.; Loh, N.L. Structure and Tribological Properties of Plasma Nitrided Surface Films on Inconel 718. *Surf. Coat. Technol.* **1997**, *89*, 70–76. [CrossRef]
20. Bottoli, F.; Jellesen, M.S.; Christiansen, T.L.; Winther, G.; Somers, M.A.J. High Temperature Solution-Nitriding and Low-Temperature Nitriding of AISI 316: Effect on Pitting Potential and Crevice Corrosion Performance. *Appl. Surf. Sci.* **2018**, *431*, 24–31. [CrossRef]
21. Mulford, S.J.; Tromans, D. Crevice Corrosion of Nickel-Based Alloys in Neutral Chloride and Thiosulfate Solutions. *Corrosion* **1988**, *44*, 891–900. [CrossRef]
22. Miller, D.M.; Lillard, R.S. An Investigation into the Stages of Alloy 625 Crevice Corrosion in an Ocean Water Environment: Initiation, Propagation and Repassivation in a Remote Crevice Assembly. *J. Electrochem. Soc.* **2019**, *166*, C3431–C3442. [CrossRef]
23. Baba, H.; Katada, Y. Effect of Nitrogen on Crevice Corrosion in Austenitic Stainless Steel. *Corros. Sci.* **2006**, *48*, 2510–2524. [CrossRef]
24. Mudali, U.K.; Dayal, R.K. Influence of Nitrogen Addition on the Crevice Corrosion Resistance of Nitrogen-Bearing Austenitic Stainless Steels. *J. Mater. Sci.* **2000**, *35*, 1799–1803. [CrossRef]

25. Levey, P.R.; Van Bennekom, A. A Mechanistic Study of the Effects of Nitrogen on the Corrosion Properties of Stainless Steels. *Corros. Sci.* **1995**, *51*, 911–921. [CrossRef]
26. Li, H.B.; Jiang, Z.H.; Yang, Y.; Cao, Y.; Zhang, Z.R. Pitting Corrosion and Crevice Corrosion Behaviors of High Nitrogen Austenitic Stainless Steels. *Int. J. Miner. Metall. Mater.* **2009**, *16*, 517–524. [CrossRef]
27. Henke, B.; Gullikson, E.; Davis, J. The Center for X-ray Optics (CXRO) X-ray Attenuation Length Calculator. Available online: https://henke.lbl.gov/optical_constants/atten2.html (accessed on 22 April 2023).
28. Li, X.; Bhushan, B. A Review of Nanoindentation Continuous Stiffness Measurement Technique and Its Applications. *Mater. Charact.* **2002**, *48*, 11–36. [CrossRef]
29. *ASTM G171-03*; Standard Test Method for Scratch Hardness of Materials Using a Diamond Stylus. ASTM International: West Conshohocken, PA, USA, 2017.
30. Rovani, A.C.; Breganon, R.; de Souza, G.S.; Brunatto, S.F.; Pintaúde, G. Scratch Resistance of Low-Temperature Plasma Nitrided and Carburized Martensitic Stainless Steel. *Wear* **2017**, *376–377*, 70–76. [CrossRef]
31. Franco, L.A.; Sinatora, A. Material Removal Factor (Fab): A Critical Assessment of Its Role in Theoretical and Practical Approaches to Abrasive Wear of Ductile Materials. *Wear* **2017**, *382–383*, 51–61. [CrossRef]
32. *ASTM G61-86*; Standard Test Method for Conducting Cyclic Potentiodynamic Polarization Measurements for Localized Corrosion Susceptibility of Iron-, Nickel-, or Cobalt-Based Alloys. ASTM International: West Conshohocken, PA, USA, 2018. [CrossRef]
33. Esmailzadeh, S.; Aliofkhazraei, M.; Sarlak, H. Interpretation of Cyclic Potentiodynamic Polarization Test Results for Study of Corrosion Behavior of Metals: A Review. *Prot. Met. Phys. Chem. Surf.* **2018**, *54*, 976–989. [CrossRef]
34. Sudha, C.; Anand, R.; Thomas Paul, V.; Saroja, S.; Vijayalakshmi, M. Nitriding Kinetics of Inconel 600. *Surf. Coat. Technol.* **2013**, *226*, 92–99. [CrossRef]
35. Kapoor, A.; Johnson, K.L.; Williams, J.A. A Model for the Mild Ratchetting Wear of Metals. *Wear* **1996**, *200*, 38–44. [CrossRef]

metals

Article

The Effect of Cathodic Cage Plasma TiN Deposition on Surface Properties of Conventional Plasma Nitrided AISI-M2 Steel

Luiz Henrique Portela de Abreu [1], Muhammad Naeem [2,*], Renan Matos Monção [1], Thercio H. C. Costa [3], Juan C. Díaz-Guillén [4], Javed Iqbal [5,6] and Rômulo Ribeiro Magalhães Sousa [1,7]

1 Advanced Materials Interdisciplinary Laboratory (LIMAV), Postgraduate Program in Materials Science and Engineering, Federal University of Piaui (UFPI), Teresina 64000-000, PI, Brazil; luizhabreu92@gmail.com (L.H.P.d.A.); renanmatos2010@hotmail.com (R.M.M.); romulorms@gmail.com (R.R.M.S.)
2 Department of Physics, Women University of Azad Jammu and Kashmir, Bagh 12500, Pakistan
3 Mechanical Department, Federal University of Rio Grande do Norte (UFRN), Natal 59000-000, RN, Brazil; thercioc@gmail.com
4 CONACYT-Corporación Mexicana de Investigación en Materiales, Saltillo 25290, Mexico; jcarlos@comimsa.com
5 Department of Physics, University of Azad Jammu and Kashmir, Muzaffarabad 13100, Pakistan; javedkiqbal@gmail.com
6 Department of Physics and Astronomy, University of Latvia, Raina bulvaris 19, LV-1586 Riga, Latvia
7 Mechanical Departament, Federal University of Piaui (UFPI), Teresina 64000-000, PI, Brazil
* Correspondence: mnaeem@wuajk.edu.pk

Citation: de Abreu, L.H.P.; Naeem, M.; Monção, R.M.; Costa, T.H.C.; Díaz-Guillén, J.C.; Iqbal, J.; Sousa, R.R.M. The Effect of Cathodic Cage Plasma TiN Deposition on Surface Properties of Conventional Plasma Nitrided AISI-M2 Steel. *Metals* **2022**, *12*, 961. https://doi.org/10.3390/met12060961

Academic Editors: Shinichiro Adachi, Francesca Borgioli, Thomas Lindner and Marcello Cabibbo

Received: 9 May 2022
Accepted: 31 May 2022
Published: 2 June 2022

Publisher's Note: MDPI stays neutral with regard to jurisdictional claims in published maps and institutional affiliations.

Abstract: In this study, a combination of conventional plasma nitriding and cathodic cage plasma deposition (CCPD) at different temperatures (400 and 450 °C) is implemented to enhance the surface properties of AISI-M2 steel. This combination effectively improves the surface hardness and the formation of a favorable hardness gradient toward the core, which would benefit the load-bearing capacity of substrate. The duplex-treated samples exhibit iron nitrides (Fe_4N, $Fe_{2-3}N$) and titanium nitride (TiN) phases. The thickness of the hard-TiN layer is 1.35 and 2.37 μm, whereas the combined thickness of the hard film and diffusion layer is 87 and 124 μm, for treatment at 400 and 450 °C, respectively. The wear rate and friction coefficient are dramatically reduced by duplex treatment. The oxidative wear mechanism and adhesive wear mechanism are dominant for duplex-treated samples. This study suggests that the cathodic cage plasma deposition technique can attain a combination of hard film and diffusion layer. The plasma nitriding before CCPD is beneficial for attaining an adequate nitrogen diffusion layer thickness. The drawbacks of conventional TiN film deposition, such as "egg-shell" problems, can be removed.

Keywords: cathodic cage plasma deposition; hardness; wear resistance; titanium nitride

1. Introduction

AISI-M2 is molybdenum-containing high-speed steel widely used in several cold work applications such as punching, pressing, and forming tools. The performance of high-speed steel tools can be upgraded by surface modification either by thermochemical diffusion techniques or the deposition of hard coatings. As a result, the machining performance of tools can be increased, reducing personnel cost and maintenance expenditure [1].

Among thermochemical diffusion techniques, plasma nitriding is widely applied to enhance the tribo-mechanical properties of tool steels by altering their microstructure [2]. In this process, nitrogen atoms are diffused in the surface and near-surface zone, and thus, surface mechanical and tribological properties are changed. Commonly, the nitrided steel contains a compound zone (white layer) and diffusion zone [3,4]. The white layer contains iron nitrides formed on the top surface of the nitrided sample, while the diffusion zone is due to interstitially dissolved nitrogen atoms or precipitated iron nitrides [3]. The presence of a white layer can improve the surface hardness of tool steel, but it is not beneficial for tribological properties. The white layer is brittle, and thus it is cracked down during sliding contact, generating hard abrasive particles on the sliding path, and thus it can damage the wear resistance [5,6]. White layer formation, which is unsuitable for industrial application, can be avoided by fine-tuning control parameters such as temperature, gas composition, and treatment time [7]. On the other hand, hard film deposition such as titanium nitride by physical vapor deposition (PVD) or chemical vapor deposition (CVD) is used to improve the surface properties of steel. TiN coating exhibits high hardness, low friction coefficient, and superior wear resistance [8–10]. Unfortunately, such hard coatings have poor adhesion with the substrate due to a significant difference in hardness between coating and substrate [11]. Thus, the combination of the thermochemical diffusion technique and hard film deposition is beneficial.

In the conventional plasma-nitriding (CPN) technique, the substrate is kept at the cathodic potential, and the reactor body acts as an anode. Its working is based on the sputtering re-deposition model [12]: the bombardment of incident ions on cathodic samples, removal of sample material, and reaction with nitrogen gas in plasma to form iron nitrides. These nitrides are re-condensed on the sample's surface, and nitrogen atoms subsequently diffuse in the sample. Here, sample material is sputtered and re-deposited, and thus, only nitrides of sample materials can be formed. Approximately two decades ago [13,14], the cathodic cage plasma-nitriding technique (CCPN) was introduced. In CCPN, samples are kept at floating potential and covered with a cathodic metallic screen (active screen or cathodic cage) [13]. Ions directly bombard the cathodic cage, and the material of the cathodic cage is sputtered and re-deposited on the surface of the sample. Thus, nitrides of any material can also be deposited on samples, and due to this fact, it is referred to as cathodic cage plasma deposition (CCPD). This method is widely used to deposit numerous materials such as iron, chromium, aluminum, nickel, niobium, copper, silver, and titanium [15–19]. As in cathodic cage plasma deposition, simultaneous diffusion of nitrogen atoms and deposition of cage material occurs, and thus, it is predicted to be more valuable to deposit titanium nitride hard film.

Several authors reported the duplex treatment (plasma nitriding and TiN film deposition) on steel samples [20–22]. They found that duplex treatment can create a hardness gradient among TiN film and sample, and thus adhesion of hard film can be enhanced. However, they used PVD, CVD techniques for TiN film deposition, and a separate system is required for nitriding and TiN film deposition. Thus, this combination is relatively expensive. Besides this, the TiN film has low adhesion with substrate and shows an "egg-shell" problem [23].

In this study, samples are nitrided by the CPN technique, and subsequently, TiN is deposited by the CCPD technique. First, the samples are initially nitrided by CPN to create an effective diffusion zone [8] and a hardness gradient before deposition of TiN. Then, the TiN is deposited on pre-nitrided samples at different temperatures, 400 °C and 450 °C, by the CCPD technique.

2. Experimental Details

AISI-M2 high-speed steel sample (nominal composition provided in Table 1 with a cylindrical geometry (diameter 2 cm and thickness 1 cm) were considered. Before plasma treatment, samples were polished using silicon carbide papers of multiple sizes (300–2500). Then, they were mirror polished with alumina powder, cleaned in an ultrasonic bath, and rinsed in flowing water.

Table 1. Chemical composition (wt.%) of AISI-M2 high-speed steel samples.

Si	Mn	Co	W	V	Mo	Cr	C	Fe
0.45	0.3	1	6.1	1.9	5	4.1	0.8	Balance

The CPN treatment was carried out in a laboratory-scale plasma-nitriding reactor made of austenitic stainless steel (304) with a cylindrical shape having 30 cm diameter and 30 cm height. The reactor's body was grounded, which works as an anode, whereas the sample holder works as a counter electrode (cathode). The power of 400 W is kept constant during processing. The details of a CPN reactor are described elsewhere [24]. The titanium nitride is deposited by a CCPD system equipped with a titanium cathodic cage. This system is developed by covering the cathodic base plate of CPN with a titanium cathodic cage and keeping samples on the ceramic insulator plate [24]. The cathodic cage was biased with a high-voltage source (2 A and 1200 V). The cathodic cage comprises two concentric cylinders. The inner cylinder is 35 mm in height and 45 mm in diameter, and the outer cylinder is 45 mm in height and 65 mm in diameter, as described earlier [25]. The samples are first treated by CPN to create an effective diffusion layer and subsequently treated by CCPD to deposit a hard titanium nitride layer. Thus, this combination of treatments is named duplex plasma treatment. The labeling of samples and detailed processing conditions are given in Table 2.

Table 2. Labelling of samples and corresponding treatment conditions.

Conventional Plasma Nitriding				Cathodic Cage Plasma Deposition				Sample Labeling
Gases (sccm)	Pressure (Pa)	Time (h)	Temperature (°C)	Gases (sccm)	Pressure (Pa)	Time (h)	Temperature (°C)	
Quenching and tempering								Base
30 H$_2$_10N$_2$	350	4	400					CPN
30 H$_2$_10N$_2$	350	4	400	30 H$_2$ _ 10N$_2$	350	4	400	Duplex 400
30 H$_2$_10N$_2$	350	4	400	30 H$_2$ _ 10N$_2$	350	4	450	Duplex 450

The X-ray diffraction (XRD) data were obtained by an X-ray diffractometer (Malvern Panalytical LTD, Malvern , UK)Empyrean, radiation Cu-Kα (λ = 1.54 Å)) operated with a copper target tube at a voltage of 45 kV, current of 40 mA, scanning angle of (2θ) from 30° to 90°, and a stepping angle of 0.013°. It was equipped with Soller slits, which consist of large numbers of parallel plates in the plane of diffraction to limit the spread of incident and diffracted X-ray beams. Additionally, it contained graded multilayer Goebel mirrors. The phases formed on the near-surface zone and deep inside the sample were assessed by grazing incidence (0.3, 1, 3, and 5°) and conventional X-ray diffraction, respectively. The XPert High Score Plus software was used for the phase's identification. The microhardness of samples was evaluated by a Vickers microhardness tester. The equipment used to perform the microhardness testing was a Shimadzu microhardness tester (model HMV 2000). The load used in each analysis was 50 gf. For the superficial microhardness analysis, 10 indentations were made along the sample surface, while for the preparation of the microhardness profile along with the nitrided layer, the samples were cut, sanded with silicon carbide papers, embedded in resin, and indentations were made along its cross-section in a total of 8 indentations. The surface and cross-sectional scanning electron microscope (SEM) analyses were carried out by Tescan Mira 3, operated with 12 kV voltage. The titanium composition in films was assessed by energy dispersive spectroscopy (EDS) (Bruker, model X flash, Billerica, MA, USA). The EDS analysis was performed by 15 kV accelerating voltage over 100 s. The presence of nitrogen in deposited films was examined by wavelength-dispersive spectroscopy (WDS) [26]. The WDS conditions were 20 kV beam voltage, 14.5 μm work distance, dwell time of 2.5 s, and beam intensity close to 20 nA. For nitrogen analysis, a Bruker detector with Crystal BRML80-Ka1 signal was used. Optical microscopy was also used in the cross-sectional observation of deposited layers. The wear resistance was assessed by a ball-on-disc wear tester containing a SiC ball of 6 mm diameter and a normal load of 1 N. A sliding speed of 5 cm/s and wear track radius of 7 mm were used.

3. Results and Discussion

The hardness profiles of base material (quenched and tempered as described elsewhere [27]) and conventional plasma-nitrided (CPN) and duplex-treated samples (base sample is first treated in CPN, and then TiN is deposited by CCPD at 400 °C and 450 °C) are shown in Figure 1. The first point in the hardness profile corresponds to the hardness on the sample's surface. It shows that the hardness of the base sample (~590 $HV_{0.05}$) increases up to 1000, 1467, and 1716 $HV_{0.05}$ for the CPN sample and duplex-treated samples at 400 °C and 450 °C, respectively. Thus, the hardness is improved by duplex treatment and is significantly higher than only nitrided M2 steel samples. In addition, the hardness profile shows an increase up to a higher depth. Our results show that a favorable hardness gradient is formed along with the sample's depth, which can be ascribed to the diffusion of nitrogen atoms up to higher depth during CPN and subsequent diffusion during CCPD. This can be further confirmed by the cross-sectional SEM analysis presented later.

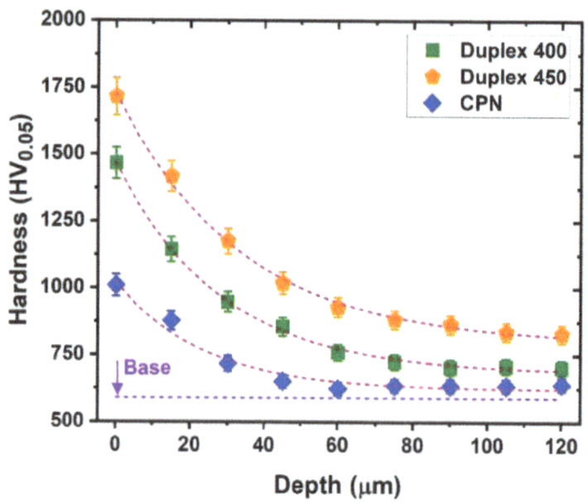

Figure 1. Hardness profile of conventional plasma-nitrided (CPN) and duplex-treated samples with substrate temperature for TiN deposition at 400 °C and 450 °C.

The conventional X-ray diffraction pattern of a base sample and CPN and duplex-treated samples at 400 °C and 450 °C is plotted in Figure 2. The base material (Figure 2a) comprises two phases: martensite, and carbide. The α-Fe martensite phase appears from the steel matrix, whereas the carbide phase (M6C) comes from the combination of alloying elements (where M = Mo, W, Cr, V) with carbon [28]. The M6C carbide phase has a face-centered cubic (FCC) structure [29]. The nitrided sample (Figure 2b) shows the presence of iron nitrides Fe_4N (COD: 96-900-4226) and $Fe_{2-3}N$ (96-152-5732) phases. The samples nitrided in CPN and TiN-deposited by CCPD show (Figure 2c,d) the presence of iron nitrides Fe_4N (COD: 96-900-4226) and $Fe_{2-3}N$ (96-152-5732) and less intense peaks of titanium nitride TiN (96-101-1101). However, it is quite challenging to identify the phases of nitrided M2 steel due to the formation of nitrides of substrate alloying elements and conversion of carbides (primary and secondary) into carbo-nitrides—or only nitrides—as reported in the literature [28]. The phases of many nitrides appear close to the carbides, and thus, diffraction peaks of nitrides are not distinguishable and overlap. Thus, the intensity of background peaks is enhanced [28]. This change in intensity and broadness of background peaks reveals the formation of iron nitrides in addition to background peaks. As nitrided samples are post-treated by Ti cathodic cage, the peaks corresponding to TiN are less intense. This shows that a thin TiN layer is formed. Thus, the film is characterized by grazing incidence XRD using different incident angles 0.3–5°, as presented in Figure 3. The top layer is mainly composed of TiN, and no intense background peaks corresponding to iron or alloying element nitrides appear. However, when shifting to the core of the sample (higher incidence angles), alloying element carbides and iron nitrides appear again. While comparing the intensity of background peaks at the same incidence angle, at 400 °C, peaks are more intense than at 450 °C. This reveals that a thicker TiN layer is formed in this condition, confirmed by cross-sectional SEM analysis. No intense peaks corresponding to Ti or Ti_2N are observed; thus, the deposited films are stoichiometric. The mechanism behind the formation of TiN by CCPD can be ascribed to the well-known mechanisms of CCPN. The working principles of both CCPN and CCPD are similar, and models proposed for CCPN can be applied to CCPD. The only difference in these systems is that under a certain processing pressure range, the additional hollow cathode effect arises through the holes of the cathodic cage, which can increase the intensity of discharge [30]. As proposed in several models, such as Saeed et al. [31], Hubbard et al. [32], Gallo et al. [33], and Fraczek et al. [34], primarily energetic ions sputter titanium atoms from the cathodic cage, subsequently react with active nitrogen atoms, and form nitrides of metal atoms, such as TiN in this case. These nitrides are re-deposited on the substrate, and afterward, nitrogen atoms are diffused in the substrate.

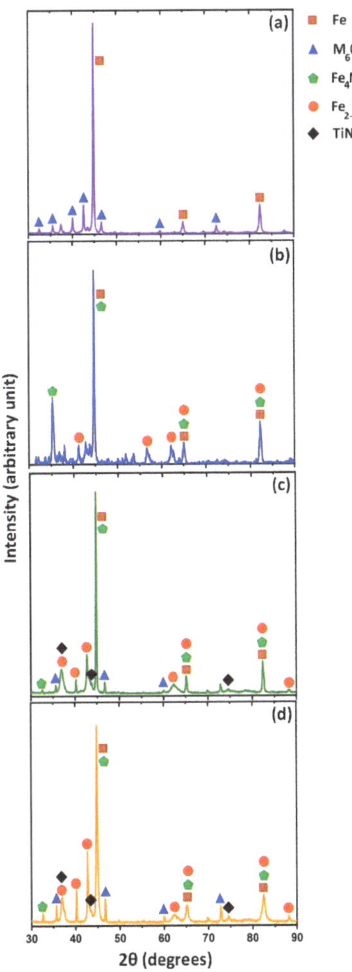

Figure 2. Conventional X-ray diffraction of (**a**) base sample, (**b**) CPN, and duplex-treated samples with substrate temperature for TiN deposition at (**c**) 400 °C and (**d**) 450 °C.

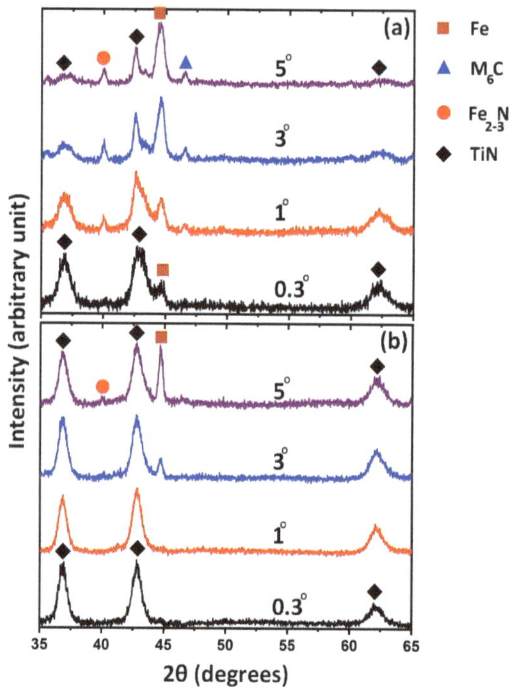

Figure 3. Grazing incidence X-ray diffraction of duplex-treated samples with substrate temperature for TiN deposition at (**a**) 400 °C and (**b**) 450 °C.

The surface SEM images of base material and nitrided and duplex-treated samples are shown in Figure 4. The base surface (Figure 4a) contains polishing scratches induced by the mechanical polishing of samples before plasma processing. The surface of the nitrided sample (Figure 4b) contains uniformly distributed polygonal iron nitride particles, following the literature [7]. The surface of the duplex-treated samples (Figure 4c,d) contains uniformly dispersed spherical nanoparticles agglomerated to coralloid granular nitrides, as shown in high-resolution images. These particles are mainly composed of Ti and N elements observed by EDS analysis (spectra are not presented here). The particles are more agglomerated for the sample treated at higher temperatures [35]. The deposition of Ti can be ascribed to the cathodic cage sputtering and its re-deposition on the substrate.

The thickness of the deposited layer is assessed by cross-sectional optical microscopy, as presented in Figure 5a,b,d. It reveals thicknesses of 71 μm, 98 μm, and 124 μm for nitrided and duplex-treated samples at 400 °C and 450 °C. The thickness increases with processing temperature due to better diffusion of nitrogen atoms in the steel matrix. This layer thickness is the combined thickness of the TiN and diffusion layer, and layers are not distinguishable due to resolution limitations. Therefore, the samples were further etched to observe the TiN layer and observed by high-resolution cross-sectional SEM analysis, as presented in Figure 5c,e. The titanium nitride layer in the top zone is confirmed by titanium elemental mapping, as presented in Figure 5b. This shows that the thickness of the TiN layer is 1.35 μm and 2.37 μm for treatment at 400 °C and 450 °C, respectively. The increase in thickness of TiN layer with temperature might be due to the higher sputtering from the cathodic cage [36]. The typical EDS spectrum of duplex-treated samples with TiN deposition at 400 °C is shown in Figure 6, which shows the presence of Ti and N on the surface. This is further clarified by cross-sectional elemental line scan analysis and WDS, as presented in Figure 7. This shows that titanium is mainly present in the top region of treated samples and is quite comparable to the thickness of the TiN layer measured by cross-sectional SEM images. However, nitrogen is presented up to a higher depth due to the diffusion during the pre-nitriding process and CCPD. This diffusion of nitrogen atoms to a higher depth is the cause of the hardness gradient between a sample and its TiN layer, and thus, this film is predicted to be valuable for wear resistance. Usually, in conventional techniques, the titanium nitride layer is not so adhesive with the substrate, and it can be avoided by CCPD. Additionally, the pre-nitriding process is responsible for effective diffusion zone production, and thus, a better hardness gradient with the substrate can be produced, as observed in Figure 1.

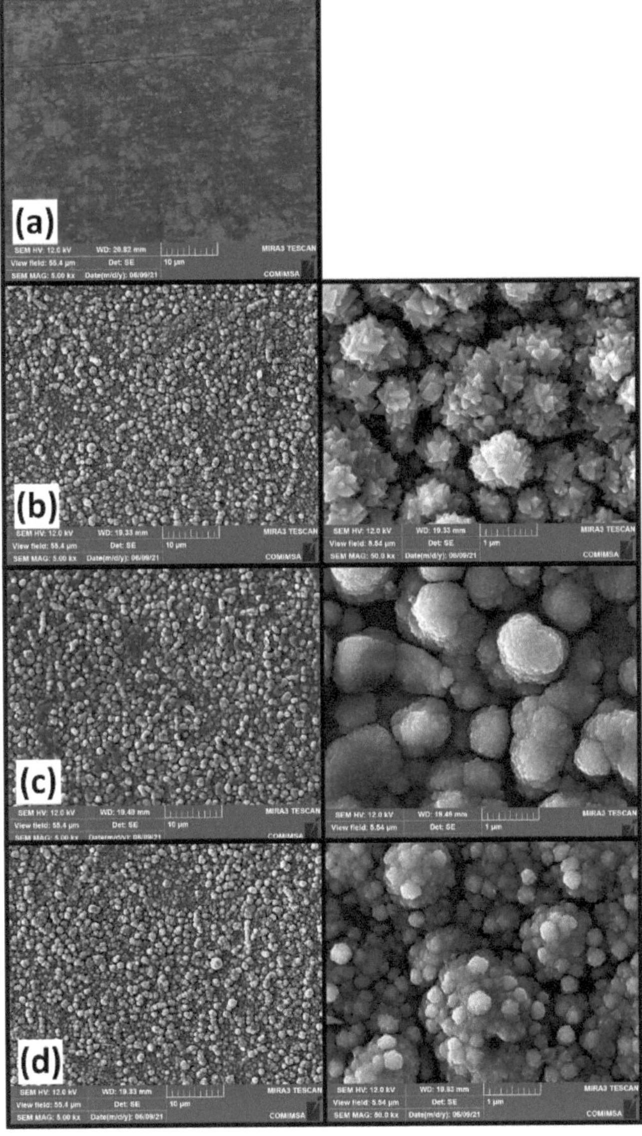

Figure 4. SEM micrographs of (**a**) base sample and (**b**) CPN and duplex-treated samples with substrate temperature for TiN deposition at (**c**) 400 °C and (**d**) 450 °C.

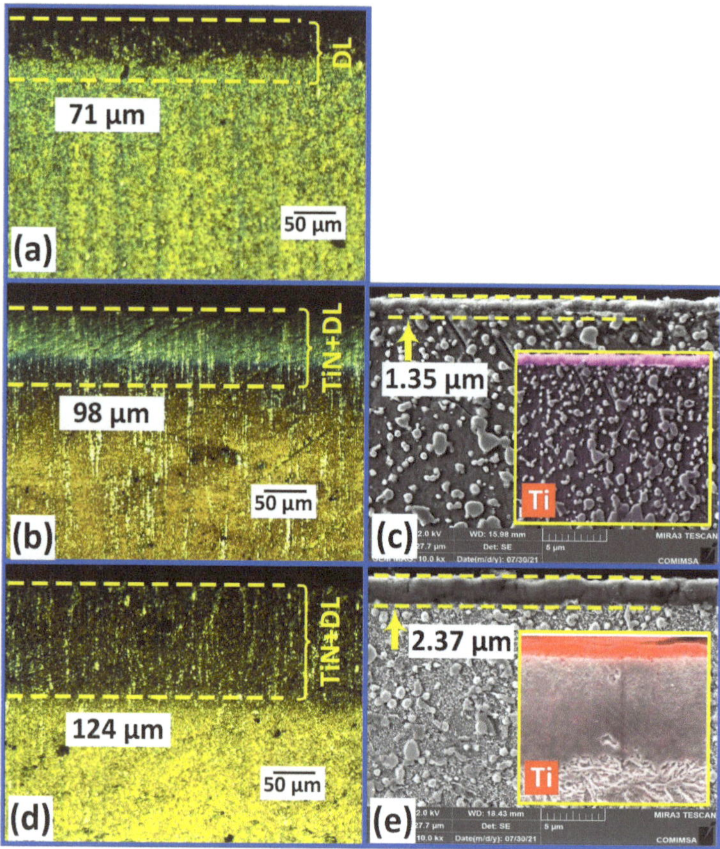

Figure 5. Cross-sectional optical micrograph of (**a**) nitrided sample, optical micrographs with SEM images of duplex-treated samples with substrate temperature for TiN deposition at (**b**,**c**) 400 °C and (**d**,**e**) 450 °C.

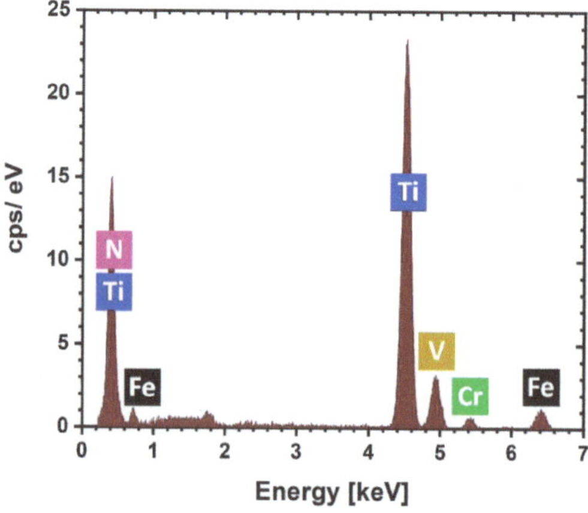

Figure 6. Typical EDS spectrum of duplex-treated samples with TiN deposition temperature of 400 °C.

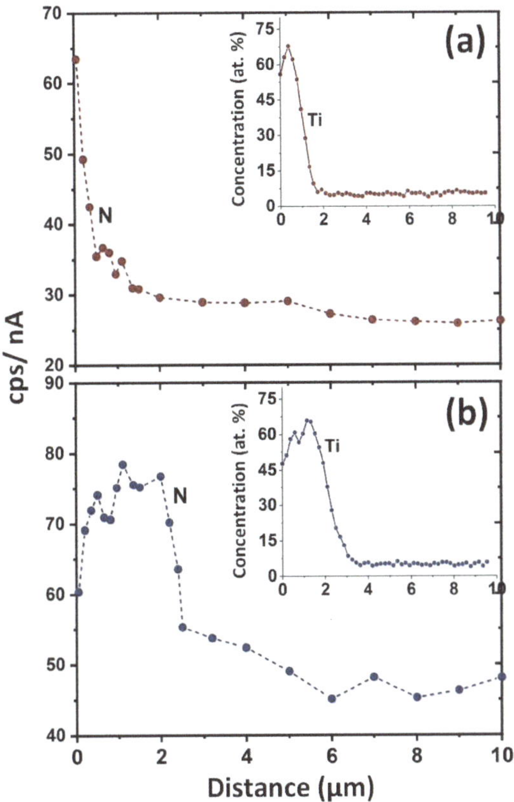

Figure 7. WDS profile (for N detection) and EDS line scan (for Ti detection) of duplex-treated samples with substrate temperature for TiN deposition at (**a**) 400 °C and (**b**) 450 °C.

The performance of deposited films in sliding contact parts applications is assessed by a ball-on-disc wear tester and worn tracks SEM images are presented in Figure 8. It shows that base sample (Figure 8a) is severely worn, and tracks are wider than treated samples. The abrasion grooves appear, and the surface contains metallic debris and microcracks. These grooves and microcracks appear due to the low hardness of the base material, and thus, micro-ploughing and plastic deformation is induced during sliding contact of the ball. The tracks show that the abrasive wear mechanism contributes to the base sample. The worn track of the nitrided sample (Figure 8b) is relatively narrow and smoother than the base material due to increased hardness and resistance to plastic deformation. The duplex-treated sample's surface in Figure 8c,d shows that the surface is smooth—free of cracks and metallic debris. This shows the existence of an adhesive wear mechanism. The surface is hard enough for treated samples to prevent plastic deformation during sliding contact against the ball. The oxygen elemental mapping reveals that the track is oxidized, and an oxidative wear mechanism is also present. The titanium elemental mapping shows that the entire surface of the sample, including the worn surface, contains homogeneous titanium distribution. The titanium nitride layer is not removed during sliding contact against the ball. This is primarily due to the hybrid mechanism of CCPD, which involves the deposition of titanium, and the thermochemical diffusion process. This hybrid mechanism produces TiN film with good adherence to the substrate. Additionally, the plasma pre-nitriding process also produces an effective diffusion layer, increasing adhesion with the substrate. Thus, the wear resistance is increased by duplex plasma treatment.

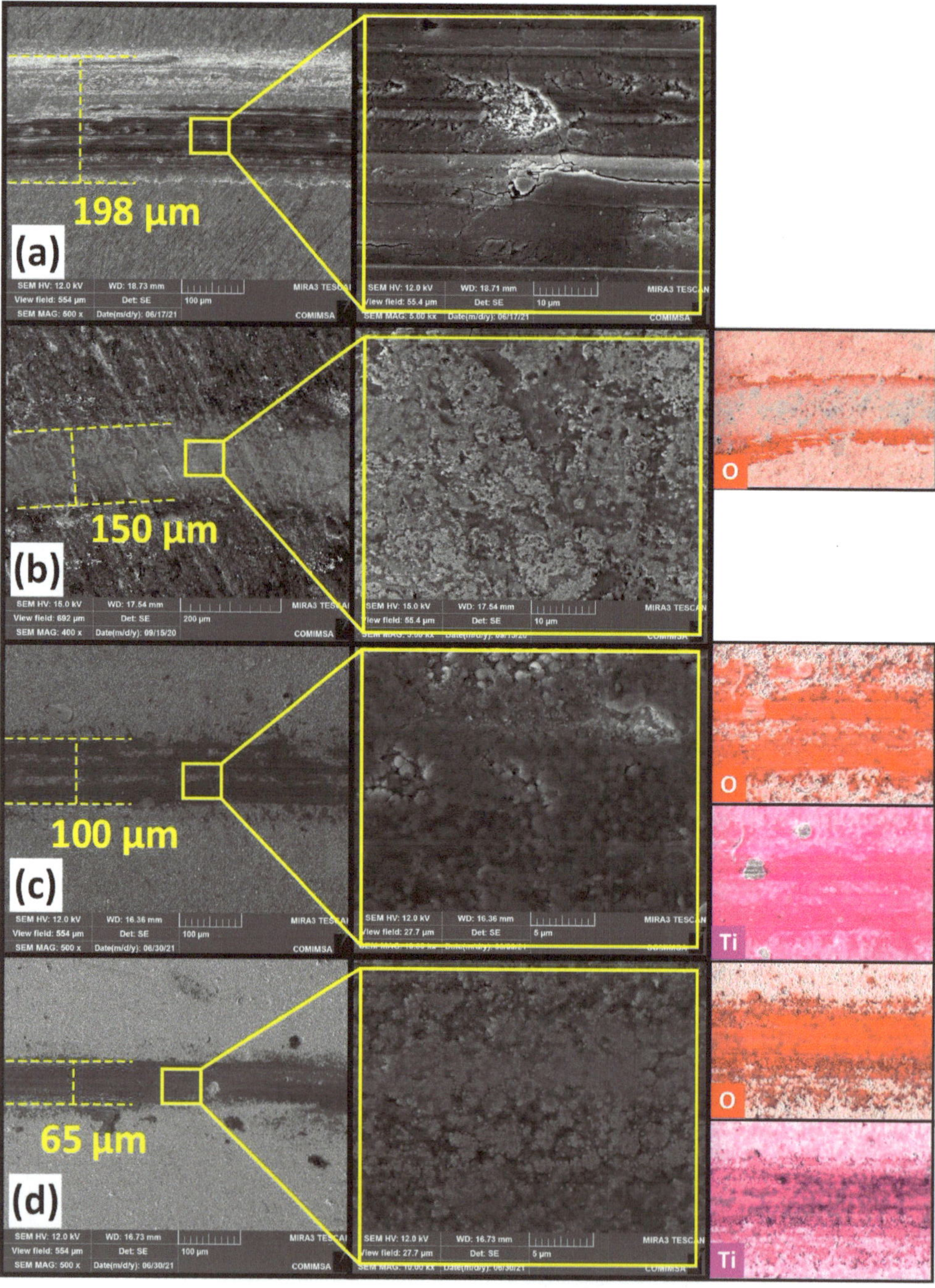

Figure 8. SEM images along with corresponding elemental mapping of (**a**) base sample and (**b**) CPN and duplex-treated samples with substrate temperature for TiN deposition at (**c**) 400 °C and (**d**) 450 °C.

The friction coefficients of the base material and nitrided and duplex-plasma-treated samples with sliding distance are compared in Figure 9, and corresponding averages are plotted in Figure 10. The friction coefficient of the base material fluctuates with the sliding distance, and average values are higher than the treated samples. This can be ascribed to the low hardness of the base sample, and thus, severe wear as described earlier [37]. The variation of friction coefficient with change in sliding distance is probably caused by the formation and subsequent removal of the oxide layer on further sliding [37]. This is supported by the surface SEM image of the base sample, in which bright and dark regions appear due to oxidation of the surface and later spalling. The friction coefficient of treated samples shows relatively smooth and low values due to sufficient hardness and low wear rate of treated samples. This is also supported by surface SEM images, in which the surface of treated samples is smooth, clean, and free of imperfection. The wear rates of the base sample and nitrided and duplex-treated samples are compared in Figure 10. This shows a significant decrease in wear rate by duplex plasma treatment due to sufficient hardness of the sample and sufficient adhesion with the substrate, due to which hard TiN film is not detached from the substrate during sliding contact.

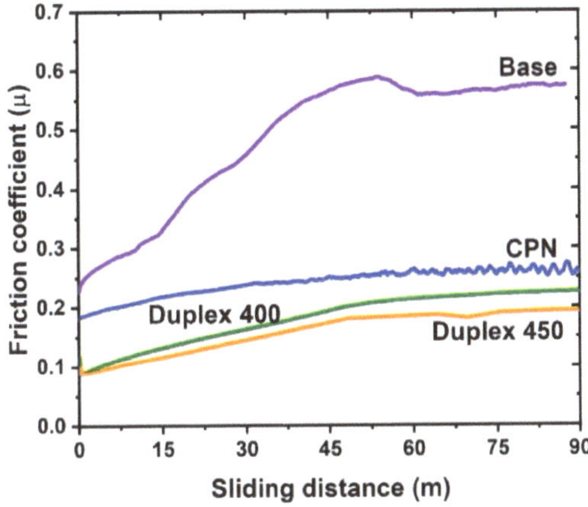

Figure 9. Variation of friction coefficient with sliding distance of base sample, CPN, and duplex-treated samples with substrate temperature for TiN deposition at 400 °C and 450 °C.

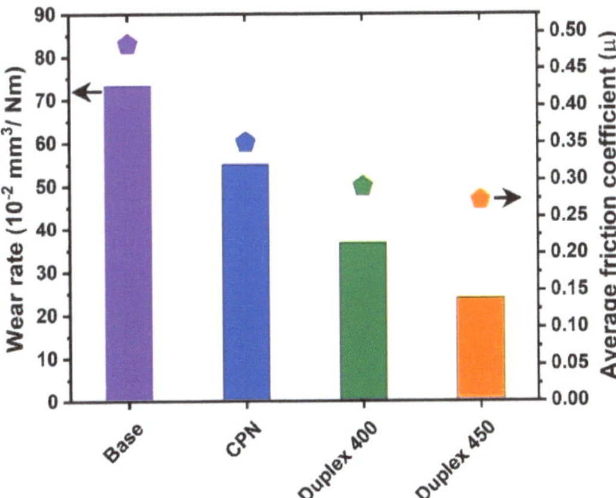

Figure 10. Wear rate and average friction coefficient of base sample and CPN and duplex-treated samples with substrate temperature for TiN deposition at 400 °C and 450 °C.

4. Conclusions

Nitriding of AISI-M2 steel is carried out in a conventional plasma nitriding system and subsequently, cathodic cage plasma deposition with a titanium cage is used to improve surface properties. The hardness of the base sample (590 $HV_{0.05}$) is increased by duplex treatment, and the maximum hardness of ~1716 $HV_{0.05}$ is attained by cathodic cage plasma deposition at 450 °C. Besides the increase in surface hardness, a favorable hardness gradient with depth is obtained by duplex treatment due to the formation of the diffusion layer during plasma nitriding as well as cathodic cage plasma deposition. The grazing-incidence-angle X-ray diffraction depicts titanium nitride formed on the nitrided sample during CCPD. The combined thicknesses of titanium nitride and diffusion layer are 87 and 124 μm for samples treated at 400 °C and 450 °C, respectively. The individual thickness of the titanium nitride layer is 1.53 μm and 2.37 μm. Furthermore, duplex treatment reduces the wear rate and friction coefficient significantly due to the formation of the hard titanium nitride layer in the near-surface zone and favorable hardness gradient, which increases hard film adhesion with the substrate. As a result, the wear tracks are relatively narrower and shallower for treated samples and adhesive, and the oxidative wear mechanism is dominant for treated samples in contrast with abrasive and oxidative wear in the base sample.

Author Contributions: Conceptualization, M.N., L.H.P.d.A., R.R.M.S., Methodology, R.M.M., T.H.C.C., J.C.D.-G., writing orignal draft, M.N., Writing review and editing, J.C.D.-G., R.M.M. and J.I. Supervision, R.R.M.S., T.H.C.C., Formal analysis, J.I., R.M.M., J.C.D.-G. All authors have read and agreed to the published version of the manuscript.

Funding: This study was financed in part by the "Coordenação de Aperfeiçoamento de Pessoal de Nível Superior (CAPES), Brazil" Finance Code 001. L. Henrique is thankful to "Universidade Federal do Piaui"and "Instituto Federal do Amazonas" for partial financial support in this research work. Javed Iqbal is thankful to European Regional Development Fund (Project No. 1.1.1.5/19/A/003) for the development of quantum optics and photonics lab at the university of Latvia.

Institutional Review Board Statement: Not applicable.

Informed Consent Statement: Not applicable.

Data Availability Statement: All data generated or analyzed during this study are included in this published article.

Conflicts of Interest: The authors declare no conflict of interest.

References

1. Libório, M.S.; Praxedes, G.B.; Lima, L.L.F.; Nascimento, I.G.; Sousa, R.R.M.; Naeem, M.; Costa, T.H.; Alves, S.M.; Iqbal, J. Surface modification of M2 steel by combination of cathodic cage plasma deposition and magnetron sputtered MoS_2-TiN multilayer coatings. *Surf. Coat. Technol.* **2020**, *384*, 125327. [CrossRef]
2. Nascimento, I.O.; Naeem, M.; Freitas, R.S.; Nascimento, R.M.; Viana, B.C.; Sousa, R.R.M.; Feitor, M.C.; Iqbal, J.; Costa, T.H.C. Comparative study of structural and stoichiometric properties of titanium nitride films deposited by cathodic cage plasma deposition and magnetron sputtering. *Eur. Phys. J. Plus* **2022**, *137*, 319. [CrossRef]
3. daS Rocha, A.; Strohaecker, T.; Hirsch, T. Effect of different surface states before plasma nitriding on properties and machining behavior of M2 high-speed steel. *Surf. Coat. Technol.* **2003**, *165*, 176–185. [CrossRef]
4. Bonu, V.; Srinivas, G.; Kumar, V.P.; Joseph, A.; Narayana, C.; Barshilia, H.C. Temperature dependent erosion and Raman analyses of arc-deposited H free thick DLC coating on Cr/CrN coated plasma nitrided steel. *Surf. Coat. Technol.* **2022**, *436*, 128308. [CrossRef]
5. Das, K.; Alphonsa, J.; Ghosh, M.; Ghanshyam, J.; Rane, R.; Mukherjee, S. Influence of pretreatment on surface behavior of duplex plasma treated AISI H13 tool steel. *Surf. Interfaces* **2017**, *8*, 206–213. [CrossRef]
6. Alsaran, A.; Altun, H.; Karakan, M.; Celik, A. Effect of post-oxidizing on tribological and corrosion behaviour of plasma nitrided AISI 5140 steel. *Surf. Coat. Technol.* **2004**, *176*, 344–348. [CrossRef]
7. Díaz-Guillén, J.C.; Naeem, M.; Hdz-García, H.M.; Acevedo-Davila, J.L.; Díaz-Guillén, M.R.; Khan, M.A.; Iqbal, J.; Mtz-Enriquez, A.I. Duplex plasma treatment of AISI D2 tool steel by combining plasma nitriding (with and without white layer) and post-oxidation. *Surf. Coat. Technol.* **2020**, *385*, 125420. [CrossRef]
8. Balashabadi, P.; Larijani, M.M.; Shokri, A.A.; Jafari-Khamse, E.; Seyedi, H.; Eshghi, S. The effect of bias voltage on microstructure and hardness of TiN films grown by ion coating deposition. *Eur. Phys. J. Plus* **2015**, *130*, 1–10. [CrossRef]

9. Guha, S.; Das, S. Investigation over effect of different carbon content on various properties of titanium carbon nitride (TiCN) coating grown on Si (100) substrate by chemical vapor deposition (CVD) process. *Eur. Phys. J. Plus* **2022**, *137*, 363. [CrossRef]

10. Smagoń, K.; Stach, S.; Ţălu, Ş.; Arman, A.; Achour, A.; Luna, C.; Ghobadi, N.; Mardani, M.; Hafezi, F.; Ahmadpourian, A.; et al. Studies of the micromorphology of sputtered TiN thin films by autocorrelation techniques. *Eur. Phys. J. Plus* **2017**, *132*, 1–15. [CrossRef]

11. Libório, M.S.; Almeida, E.O.; Alves, S.M.; Costa, T.H.C.; Feitor, M.C.; Nascimento, R.M.; Sousa, R.R.M.; Naeem, M.; Jelani, M. Enhanced surface properties of M2 steel by plasma nitriding pre-treatment and magnetron sputtered TiN coating. *Int. J. Surf. Sci. Eng.* **2020**, *14*, 288–306. [CrossRef]

12. Edenhofer, B. Physical and Metallurgical Aspects of Ionitriding. Pt. 1 and Pt. 2. In *Heat Treatment Metals*; Wolfson Heat Treatment Centre: Birmingham, UK, 1974.

13. Li, C.; Georges, J.; Li, X. Active screen plasma nitriding of austenitic stainless steel. *Surf. Eng.* **2002**, *18*, 453–457. [CrossRef]

14. Li, C.; Bell, T. Sliding wear properties of active screen plasma nitrided 316 austenitic stainless steel. *Wear* **2004**, *256*, 1144–1152. [CrossRef]

15. Lin, K.; Li, X.; Tian, L.; Dong, H. Active screen plasma surface co-alloying of 316 austenitic stainless steel with both nitrogen and niobium for the application of bipolar plates in proton exchange membrane fuel cells. *Int. J. Hydrogen Energy* **2015**, *40*, 10281–10292. [CrossRef]

16. Naeem, M.; Shafiq, M.; Zaka-ul-Islam, M.; Bashir, M.I.; Díaz-Guillén, J.C.; Lopez-Badillo, C.M.; Zakaullah, M. Novel duplex cathodic cage plasma nitriding of non-alloyed steel using aluminum and austenite steel cathodic cages. *J. Alloys Compd.* **2017**, *721*, 307–311. [CrossRef]

17. Fernades, F.; Filho, E.R.; Souza, I.; Nascimento, I.; Sousa, R.; Almeida, E.; Feitor, M.; Costa, T.; Naeem, M.; Iqbal, J. Novel synthesis of copper oxide on fabric samples by cathodic cage plasma deposition. *Polym. Adv. Technol.* **2020**, *31*, 520–526. [CrossRef]

18. Morell-Pacheco, A.; Kim, H.; Wang, T.; Shiau, C.H.; Balerio, R.; Gabriel, A.; Shao, L. Ni coating on 316L stainless steel using cage plasma treatment: Feasibility and swelling studies. *J. Nucl. Mater.* **2020**, *540*, 152385. [CrossRef]

19. Da Silva, S.S.; Bottoni, R.; Gontijo, L.C.; Ferreira, S.O. Plasma deposition of titanium nitride thin films under the effect of hollow cathode length in cathodic cage. *Matéria* **2017**, *22*, 1–12.

20. Podgornik, B.; Vižintin, J.; Wänstrand, O.; Larsson, M.; Hogmark, S.; Ronkainen, H.; Holmberg, K. Tribological properties of plasma nitrided and hard coated AISI 4140 steel. *Wear* **2001**, *249*, 254–259. [CrossRef]

21. Sprute, T.; Tillmann, W.; Grisales, D.; Selvadurai, U.; Fischer, G. Influence of substrate pre-treatments on residual stresses and tribo-mechanical properties of TiAlN-based PVD coatings. *Surf. Coat. Technol.* **2014**, *260*, 369–379. [CrossRef]

22. De Las Heras, E.; Egidi, D.A.; Corengia, P.; González-Santamaría, D.; García-Luis, A.; Brizuela, M.; López, G.A.; Martinez, M.F. Duplex surface treatment of an AISI 316L stainless steel; microstructure and tribological behaviour. *Surf. Coat. Technol.* **2008**, *202*, 2945–2954. [CrossRef]

23. Bashir, M.I.; Shafiq, M.; Naeem, M.; Zaka-ul-Islam, M.; Díaz-Guillén, J.C.; Lopez-Badillo, C.M.; Zakaullah, M. Enhanced surface properties of aluminum by PVD-TiN coating combined with cathodic cage plasma nitriding. *Surf. Coat. Technol.* **2017**, *327*, 59–65. [CrossRef]

24. Sousa, R.R.M.D.; Moura, Y.J.L.; Sousa, P.A.O.D.; Medeiros Neto, J.Q.; Costa, T.H.D.C.; Alves Junior, C. Nitriding of AISI 1020 steel: Comparison between conventional nitriding and nitriding with cathodic cage. *Mater. Res.* **2014**, *17*, 708–713. [CrossRef]

25. Barbosa, M.G.C.; Viana, B.C.; Santos, F.E.P.; Fernandes, F.; Feitor, M.C.; Costa, T.H.C.; Naeem, M.; Sousa, R.R.M. Surface modification of tool steel by cathodic cage TiN deposition. *Surf. Eng.* **2021**, *37*, 334–342. [CrossRef]

26. Diaz-Guillen, J.C.; Naeem, M.; Acevedo-Davila, J.L.; Hdz-Garcia, H.M.; Iqbal, J.; Khan, M.A.; Mayen, J. Improved Mechanical Properties, Wear and Corrosion Resistance of 316L Steel by Homogeneous Chromium Nitride Layer Synthesis Using Plasma Nitriding. *J. Mater. Eng. Perform.* **2020**, *29*, 877–889. [CrossRef]

27. Araujo, A.G.F.; Naeem, M.; Araujo, L.N.M.; Costa, T.H.C.; Khan, K.H.; Díaz-Guillén, J.C.; Iqbal, J.; Liborio, M.S.; Sousa, R.R.M. Design, manufacturing and plasma nitriding of AISI-M2 steel forming tool and its performance analysis. *J. Mater. Res. Technol.* **2020**, *9*, 14517–14527. [CrossRef]

28. Moreno-Bárcenas, A.; Alvarado-Orozco, J.M.; Carmona, J.G.; Mondragón-Rodríguez, G.C.; González-Hernández, J.; García-García, A. Synergistic effect of plasma nitriding and bias voltage on the adhesion of diamond-like carbon coatings on M2 steel by PECVD. *Surf. Coat. Technol.* **2019**, *374*, 327–337. [CrossRef]

29. Serna, M.M.; Rossi, J.L. MC complex carbide in AISI M2 high-speed steel. *Mater. Lett.* **2009**, *63*, 691–693. [CrossRef]

30. Alves, C., Jr.; De Araújo, F.O.; Ribeiro, K.J.B.; Da Costa, J.A.P.; Sousa, R.D.; De Sousa, R.S. Use of cathodic cage in plasma nitriding. *Surf. Coat. Technol.* **2006**, *201*, 2450–2454. [CrossRef]

31. Saeed, A.; Khan, A.W.; Jan, F.; Abrar, M.; Khalid, M.; Zakaullah, M. Validity of "sputtering and re-condensation" model in active screen cage plasma nitriding process. *Appl. Surf. Sci.* **2013**, *273*, 173–178. [CrossRef]

32. Hubbard, P.; Partridge, J.G.; Doyle, E.D.; McCulloch, D.G.; Taylor, M.B.; Dowey, S.J. Investigation of nitrogen mass transfer within an industrial plasma nitriding system I: The role of surface deposits. *Surf. Coat. Technol.* **2010**, *204*, 1145–1150. [CrossRef]

33. Gallo, S.C.; Dong, H. On the fundamental mechanisms of active screen plasma nitriding. *Vacuum* **2009**, *84*, 321–325. [CrossRef]

34. Fraczek, T.; Ogorek, M.; Skuza, Z.; Prusak, R. Mechanism of ion nitriding of 316L austenitic steel by active screen method in a hydrogen-nitrogen atmosphere. *Int. J. Adv. Manuf. Technol.* **2020**, *109*, 1357–1368. [CrossRef]

35. Rad, H.F.; Amadeh, A.; Moradi, H. Wear assessment of plasma nitrided AISI H11 steel. *Mater. Des.* **2011**, *32*, 2635–2643.

36. Hoshiyama, Y.; Chiba, K.; Maruoka, T. Effect of Active Screen Plasma Nitriding on Mechanical Properties of Spheroidal Graphite Cast Iron. *Metals* **2021**, *11*, 412. [CrossRef]
37. Zhang, F.; Yan, M. Microstructure and wear resistance of in situ formed duplex coating fabricated by plasma nitriding Ti coated 2024 Al alloy. *J. Mater. Sci. Technol.* **2014**, *30*, 1278–1283. [CrossRef]

Article

Electrochemical Corrosion Behavior of Ti-N-O Modified Layer on the TC4 Titanium Alloy Prepared by Hollow Cathodic Plasma Source Oxynitriding

Jiwen Yan [1], Minghao Shao [2], Zelong Zhou [1], Zhehao Zhang [3], Xuening Yi [1], Mingjia Wang [1], Chengxu Wang [1], Dazhen Fang [3], Mufan Wang [1], Bing Xie [2], Yongyong He [3,*] and Yang Li [1,*]

[1] School of Nuclear Equipment and Nuclear Engineering, Yantai University, Yantai 264005, China
[2] School of Electromechanical Automobile Engineering, Yantai University, Yantai 264005, China
[3] State Key Laboratory of Tribology, Tsinghua University, Beijing 100084, China
* Correspondence: heyy@mail.tsinghua.edu.cn (Y.H.); liyang@ytu.edu.cn or metalytu@163.com (Y.L.)

Abstract: TC4 alloy is widely used in dental implantation due to its excellent biocompatibility and low density. However, it is necessary to further improve the corrosion resistance and surface hardness of the titanium alloy to prevent surface damage that could result in the release of metal ions into the oral cavity, potentially affecting oral health. In this study, Ti-N-O layers were fabricated on the surface of TC4 alloy using a two-step hollow cathode plasma source oxynitriding technique. This resulted in the formation of TiN, Ti_2N, TiO_2, and nitrogen-stabilized $\alpha(N)$-Ti phases on the TC4 alloy, forming a Ti-N-O modified layer. The microhardness of the samples treated with plasma oxynitriding (PNO) was found to be 300–400% higher than that of untreated (UN) samples. The experimental conditions were set at 520 °C, and the corrosion current density of the PNO sample was measured to be 7.65×10^{-8} A/cm^2, which is two orders of magnitude lower than that of the UN sample. This indicates that the PNO-treated TC4 alloy exhibited significantly improved corrosion resistance in the artificial saliva solutions.

Keywords: TC4 alloy; oxynitriding; electrochemical testing; artificial saliva; EIS; XPS

check for updates

Citation: Yan, J.; Shao, M.; Zhou, Z.; Zhang, Z.; Yi, X.; Wang, M.; Wang, C.; Fang, D.; Wang, M.; Xie, B.; et al. Electrochemical Corrosion Behavior of Ti-N-O Modified Layer on the TC4 Titanium Alloy Prepared by Hollow Cathodic Plasma Source Oxynitriding. *Metals* **2023**, *13*, 1083. https://doi.org/10.3390/met13061083

Academic Editors: Sundeep Mukherjee and Elisabetta Petrucci

Received: 1 April 2023
Revised: 28 May 2023
Accepted: 2 June 2023
Published: 7 June 2023

1. Introduction

Titanium alloys are widely used for manufacturing medical implants in humans due to their exceptional properties, such as high specific strength, fatigue strength, and biocompatibility [1,2]. These properties make them an ideal choice for implants that require durability and the ability to withstand the corrosive environment of the human body. When exposed to air at ambient temperatures, titanium and its alloys form a surface oxide layer. However, this naturally occurring oxide layer can degrade in certain corrosive environments, leading to its dissolution [3]. The natural oxide film is susceptible to dissolution in reducing or complex media and can rapidly decompose in acidic or fluoride solutions [4,5]. Increased acidity of the medium or the presence of ions (F^-, Cl^-, SO_4^{2-}) can increase the corrosiveness of the medium, reduce the protective properties of the oxide passivation layer, or accelerate its degradation rate [6,7]. While titanium alloys are vulnerable to fluoride ions, other active anions, such as Cl^-, can also harm the oxide layer of titanium alloys. For instance, Cl^- present in artificial saliva can be adsorbed on the oxide film, leading to its decomposition [8]. In successful cases involving surgical and medical instruments, various surface engineering technologies, such as physical vapor deposition (PVD), chemical vapor deposition (CVD), and plasma nitriding (PN), have been employed to enhance biocompatibility, antibacterial effects, and high corrosion resistance in the human body [9,10]. Among these techniques, PN is extensively used as a surface modification method capable of forming relatively thick, complex, adherent films with corrosion resistance [11–13]. TiN coating, in particular, exhibits good biocompatibility and holds promise for various applications in human implants [14]. However, in the harsh oral corrosion environment, a TiN protective layer alone cannot effectively prevent oral bacterial infection, consequently impacting its antibacterial activity. Wang et al. demonstrated that TiN-coated samples on pure titanium experienced erosion and damage when exposed to fluoride-containing artificial saliva [15]. Therefore, further improvements in the performance of the TiN-modified layer are necessary.

Wierzchoń et al. [16] conducted a study in which they utilized a plasma-assisted oxidation-nitridation process to prepare a TiO_2 + Ti_2N + Ti(N) diffusion layer, resulting in improved corrosion and wear resistance of the substrate material. Furthermore, the presence of nano-crystalline titanium oxide (rutile) demonstrated enhanced biological performance when titanium and its alloys came into contact with blood. The formation of TiNxOy metal compounds involves the replacement of nitrogen atoms with oxygen atoms in the face-centered cubic (fcc) lattice of TiN, causing lattice distortion and the introduction of defects, ultimately improving the properties of the material [17]. The incorporation of oxygen into the cationic fcc sublattice plays a key role in this transformation. Banakh et al. [18] found that titanium nitride coatings were deposited on commercially pure titanium and observed significant bioactivity. In another study, Albayrak et al. [19] investigated the nitriding and oxynitride behavior of CP-Ti samples through the anodization of the TiO_2 layer after the plasma nitriding process. The porous oxynitride samples showed better corrosion resistance.

The hollow cathode discharge effect plays a crucial role in enhancing the efficiency of ionization and creating a denser plasma through multiple electron–gas collisions [20–22]. This effect is particularly advantageous for the nitriding of titanium alloys, which require high diffusion temperatures. The use of high-density plasma allows higher and more controlled processing temperatures to be achieved in a shorter period [23]. In the nitriding furnace, a hollow cathode device is employed to generate a hollow cathode discharge, serving as a powerful working medium and an effective heat source for heating the workpiece [24]. Plasma nitriding technology offers numerous benefits, including rapid nitriding speed, high efficiency, energy saving, and the production of high-quality, corrosion-resistant nitride layers at low temperatures [25,26]. In this study, a two-step hollow cathodic plasma source oxynitride was employed to prepare a Ti-N-O modified layer on a TC4 alloy substrate. The surface treatment involved plasma nitriding followed by plasma oxynitride, resulting in the formation of a nitride layer that subsequently transformed into a Ti-N-O modified layer. By utilizing this duplex technology, it is anticipated that high-quality composite modified layers can be achieved, leading to improved corrosion resistance of titanium alloys in artificial saliva solutions.

2. Materials and Methods

2.1. Preparation of Ti-N-O Compound Layer

The material used for the experiments was TC4 alloy bolts, size M8 × 13 mm. The samples were acid washed with HF-HNO$_3$ (45 mL HF (49%) + 205 mL HNO$_3$ (65%) + 750 mL deionized water) solution to wash off the oil stains and oxide on the surface. The duration of the acid wash was ten minutes. Finally, the samples were cleaned separately using an ultrasonic treatment for 10 min with acetone and anhydrous ethanol. The samples were dried and loaded into the nitriding equipment.

The self-developed experimental hollow cathode device is shown in Figure 1. This hollow cathode device intensified the collision between the plasma and the TA2 screen, thereby increasing the plasma concentration. This enabled the sample to achieve plasma nitriding rapidly. The parameters for the surface treatment of samples are shown in Table 1.

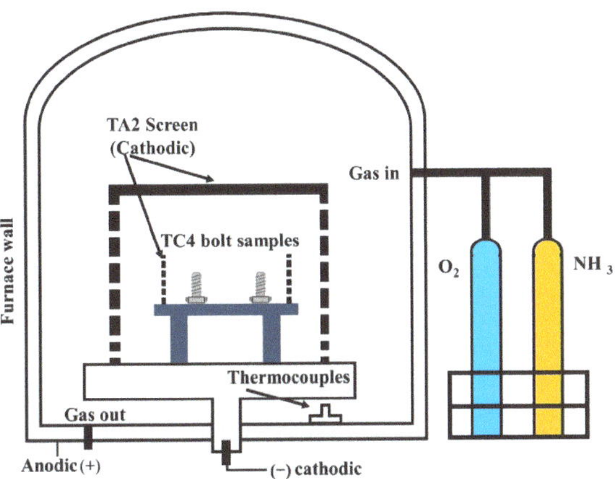

Figure 1. Structure diagram of hollow cathode device.

Table 1. Parameters for surface treatment of samples.

Sample	Temperature	Heating Time (min)	Step 1 Nitriding (min) NH$_3$	Step 2 Oxynitriding (min) NH$_3$:O$_2$ (9:1)
UN	-	-	-	-
PNO-500	500 °C	120	60	10
PNO-520	520 °C	120	60	10
PNO-540	540 °C	120	60	10

2.2. Performance Test and Organization Observation

X-ray diffraction (XRD) analysis was performed using a D8ADVANCE X-ray diffractometer system made by BRUKER with Cu-Kα radiation. The analysis involved scanning in the 2θ range of 20°–90° to evaluate the crystal structure of UN and PNO samples. The microstructure of the sample section was observed using a Zeiss Axio Observer 3M metallographic microscope, and the surface hardness was measured with an HXD-1000TM/LCD microhardness tester. X-ray photoelectron spectroscopy (XPS) analysis of the surface chemical composition was conducted using a PHI Quantra II system from JAPAN, and data were processed using CasaXPS software. The analysis used peak calibration with energy C1s (284.8 eV). For electrochemical testing, a CS310 electrochemical workstation was used with a three-electrode system. An artificial saliva composition was used (see Table 2) and maintained at 37 °C and pH 6.65 ± 0.01. The titanium bolt was connected to a silver wire and immersed in the artificial saliva as a working electrode. The contact area between the silver wire and the bolt was excluded from the analysis (illustrated in Figure 1), and the surface area of the electrode was approximately 215 mm^2. A reference electrode of Ag/AgCl was used, and a platinum sheet was the auxiliary electrode. An open circuit potential test was performed on the sample for 10 min. Electrochemical impedance spectroscopy (EIS) measurement was conducted using a sinusoidal potential perturbation of 10 mV in the frequency range of 10^{-2}–10^5 Hz. On the basis of the EIS results, Nyquist and Bode plots of the sample were obtained, and a dynamic potential polarization test was performed in the range of 0.6–1.5 V (relative to the reference electrode) at a scanning rate of 0.5 mV/s. Three replicate experiments were carried out, and the experimental data were fitted using Zview-2 and Origin 2018 software.

Table 2. Chemical compositions of artificial saliva.

Na$_2$S	Mg$_2$P$_2$O$_7$	CaCL$_2$	KCL	NaCL	Mucin	CO(NH$_2$)$_2$	Na$_2$HPO$_4$	Distilled Water
0.0008 g	0.0008 g	0.3000 g	0.2 g	0.2 g	2.0 g	0.5 g	0.3 g	0.5 L

3. Results and Analysis

3.1. Analysis of Organization and Physical Structure

3.1.1. The Thickness of The Composite Layer of The PNO Samples

The cross-sectional morphology of the PNO samples after corrosion with the Kroll solution is shown in Figure 2. The PNO samples had a prominent bright white layer on the surface, which was not affected by the corrosive solution. The Ti-N-O composite layer of titanium was superior to the substrate in terms of corrosion resistance. Observations revealed a light-yellow layer above the white and bright layer on the PNO sample. The reason behind this phenomenon is the formation of TiN, which presents a characteristic golden-yellow hue resulting from the diffusion of nitrogen into the TC4 alloy [27]. For the PNO samples, the thickness of the composite layer was approximately 5.5 µm, 5.7 µm, and 10.0 µm, respectively.

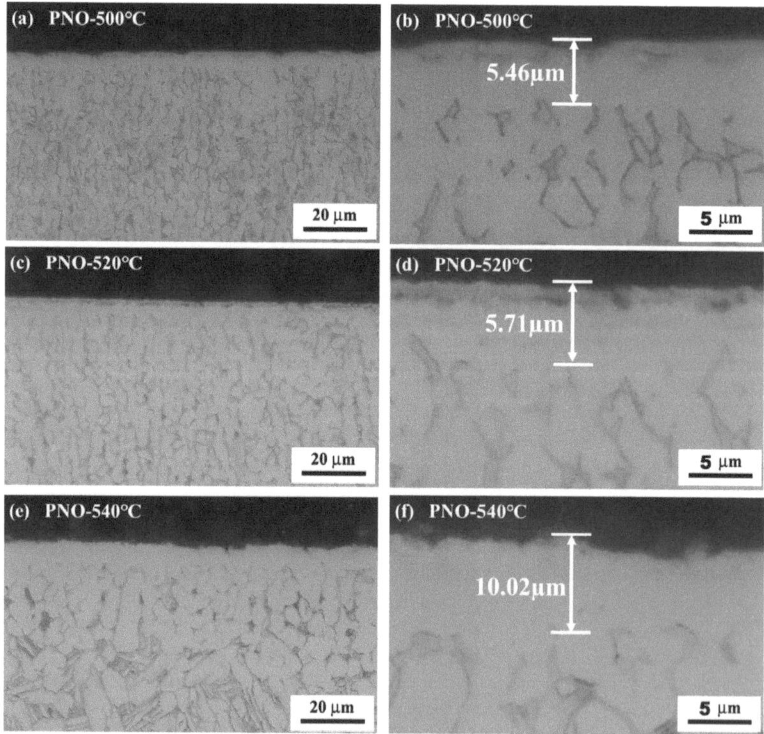

Figure 2. Cross-sectional micrographs of UN and PNO samples.

3.1.2. Phase Determination

Figure 3 shows the XRD patterns of the UN and the PNO samples. The UN sample consisted of α-Ti with a close-packed hexagonal lattice (hcp) (100) orientation and β-Ti with a body-centered cubic structure (bcc) (101) orientation. For the PNO samples compared with the original sample, the α-Ti peak was weakened. The β-Ti was unchanged, while the diffraction peaks of TiN (PDF#38-1420) and Ti$_2$N (PDF#76-0198) and some TiNx peaks appeared. In addition, both the α-Ti (100) and (101) diffraction peaks (PDF # 89-3725) shifted to a low angle, which indicates that the crystal lattice was distorted during the process of oxynitride [28,29]. The Ti_2N phase was generated at a lower temperature of 500 °C, while the intensity of the TiN peak increased at higher temperatures of 520 and 540 °C. The main phases of the PNO samples were TiN and Ti_2N.

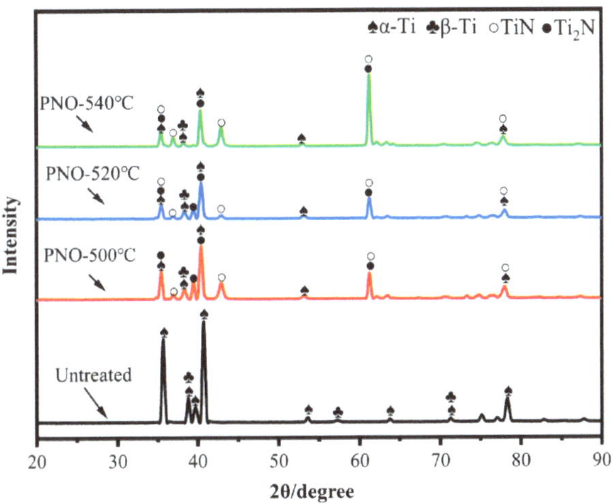

Figure 3. X-ray diffraction patterns of UN and PNO samples.

3.1.3. XPS Analysis of PNO-520 Samples before Artificial Saliva Corrosion

As shown in Figure 4, the Ti2p peak spectrum suggested that several components were present, including TiO_2, TiN, TiON, and Ti_2O_3. By analyzing the binding energy, we can distinguish between the different components. Specifically, the presence of TiN can be identified by the binding energies of 454.98 and 460.61 eV, while TiON in TiN_xO_y can be represented by the binding energies of 455.99 and 461.60 eV [30,31]. The binding energy at 458.28 (standard value 458.3) and 464.16 eV (average value 464.19 eV) correspond to TiO_2. This could be attributed to the distortion of the lattice structure in TiO_2 caused by N-doping during plasma source oxynitride, which shifts the energy levels of the orbitals [32]. The binding energies of Ti_2O_3 were 456.96 and 462.87 eV [33]. The binding energy points of 395.8 and 396.98 eV in N1s corresponded to TiON and TiN, respectively. Three peaks were found in O1s, namely TiO_2, Ti_2O_3, and TiON, with binding energies of 529.76, 529.8, and 531.34 eV, respectively [34].

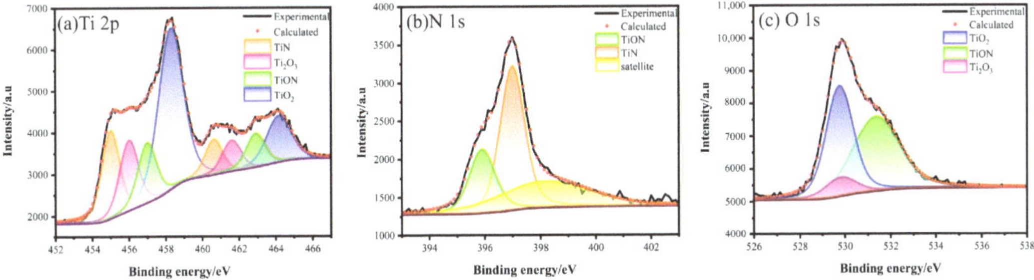

Figure 4. XPS spectra of Ti2p, N1s, and O1s of PNO−520 sample.

3.2. Analysis of Mechanical Properties

As shown in Figure 5, the surface hardness of the PNO samples was significantly improved compared with the matrix hardness of TC4 alloy (300.0 $HV_{0.1}$). The surface hardness values of the PNO samples were 798.3 $HV_{0.1}$, 982.1 $HV_{0.1}$, and 1142.7 $HV_{0.1}$, respectively. The results obtained from XRD indicated that TiN and Ti_2N phases were on the surface of the PNO samples. The primary factor behind the increased surface hardness was the presence of the TiN and Ti_2N phases [35,36]. As their content increases, so does the hardness of the surface. The microhardness change is associated with the treatment temperature. With the rise in the oxynitride temperature, the depth of the diffusion layer increased gradually. The diffusion layers of the PNO-500, PNO-520, and PNO-540 samples were 45 µm, 48 µm, and 75 µm, respectively. Nitrogen and titanium form an interstitial solid solution in the diffusion layer, improving the substrate's surface hardness.

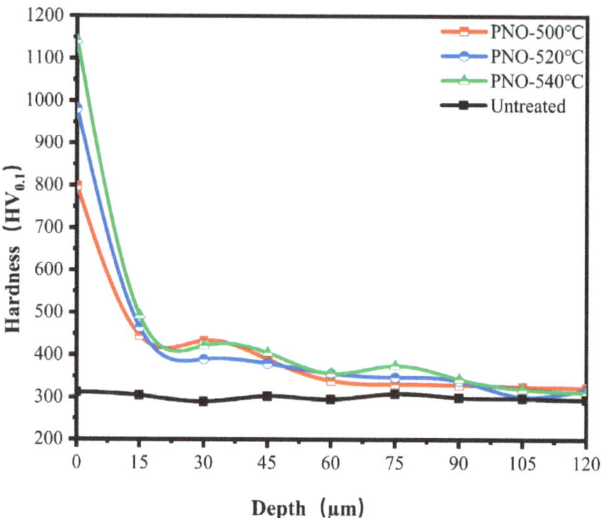

Figure 5. Microhardness depth distribution of UN and PNO samples.

3.3. Analysis of Corrosion Resistance Properties

3.3.1. Analysis of Polarization Curves

When titanium alloys are implanted into the human body as dental implants, the saliva in the human mouth can corrode and degrade implanted titanium alloy teeth. The corrosion of this process is simulated through an electrochemical corrosion test. In the electrochemical corrosion experiment of artificial saliva, bubbles can be observed on the sample's surface. This phenomenon indicates that a hydrogen evolution reaction occurs during the corrosion process (1) [37,38].

$$2Ti + 6H^+ \rightarrow 2Ti^{3+} + 3H_2 \uparrow \tag{1}$$

As shown in Figure 6, in the AB anodic region, the titanium alloy dental implant sample oxide began to dissolve, and the activation polarization affected the dissolution rate. The current density of the PNO sample changed very little at a potential of -0.2–0.1 V and hardly changed with the potential. In addition, the BC region was the active–passive transition region, where the formation of TiO_2 passivation film could prevent the further dissolution of the base material [39]. The PNO samples had a common feature: the passivation film was generated at almost the same current density. By contrast, the UN sample generated passivation film at a much higher current density [40]. At 0.1–0.2 V, the passivation film dissolved, at which time the current density and potential increased.

At point D, there was a sharp decrease in current density and a corresponding increase in potential, leading to the regeneration of the passivation film [41]. The potential for further growth in EF regional voltage is still significant, while the current remains relatively unchanged. The surface of the sample generated a dense passivation film, indicating that it was in a stable passivation stage. The anodic polarization suggested that the passivation film on the PNO samples underwent a repassivation process involving forming, dissolving, and regenerating the indicated morphology of the anode region, suggesting the presence of a protective layer. By contrast, the cathode region produced predominantly hydrogen. Electrochemical polarization characteristic tests showed that the corrosion resistance of the PNO samples was better than that of the UN sample in the human oxidation potential (-58–212 mV) range.

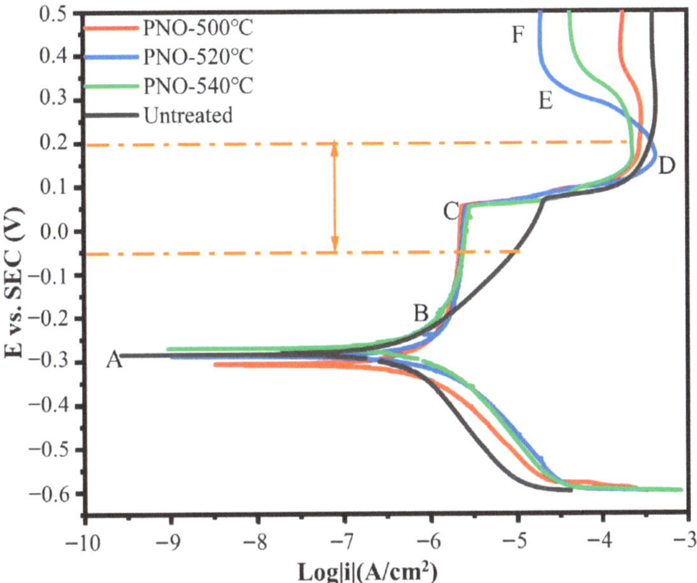

Figure 6. Potentiodynamic polarization curves of UN and PNO samples in artificial saliva.

The polarization voltage, corrosion current density, and polarization resistance were obtained through fitting using Cview-2 software. As shown in Table 3, the corrosion current density calculated by the software was based on the Stern–Geary Equation (2) [42,43]:

$$I_{corr} = \frac{\beta_a \times \beta_c}{(\beta_a + \beta_c)R_p} = \frac{b_a \times b_C}{2303 \times (b_a + b_C)R_p} \tag{2}$$

Table 3. The I_{corr}, E_{corr}, and R_p of different samples.

Sample	E_{corr} (V)	I_{corr} (A/cm²)	R_p ($\Omega \cdot cm^2$)
UN	-0.330	2.78×10^{-6}	9.35×10^3
PNO-500	-0.294	1.55×10^{-6}	1.70×10^4
PNO-520	-0.289	7.65×10^{-8}	3.40×10^5
PNO-540	-0.265	9.35×10^{-7}	2.78×10^4

Table 3 shows the data polarization potential (E_{corr}), corrosion current density (I_{corr}), and polarization resistance (R_p) corresponding to the polarization curves. E_{corr} reflects the thermodynamic tendency for corrosion to occur [44]. The higher the value of E_{corr}, the lower the corrosion tendency of the sample [45]. According to the E_{corr}, the UN sample was

-0.289 V. As the temperature increased, the E_{corr} of PNO-540 was -0.265 V. It can be seen that temperature had a particular influence on the E_{corr} of the TC4 alloy dental implant bolts. The corrosion resistance of a material can be judged by its I_{corr}. The lower the I_{corr} value, the better the corrosion resistance of the material [46]. After the treatment, the I_{corr} and R_p of the PNO-520 sample were 7.65×10^{-8} A/cm^2 and 3.40×10^5 $\Omega \cdot$cm^2, respectively. The I_{corr} and R_p of the PNO-540 sample were 9.35×10^{-7} A/cm^2 and 2.78×10^4 $\Omega \cdot$cm^2, respectively. The I_{corr} value of the PNO-520 sample was lower than the I_{corr} values of the PNO-500 and PNO-540 samples. Similarly, the R_p value of the PNO-520 sample was one order of magnitude higher than those of the PNO-500 and PNO-540 samples. A sufficiently thick Ti-N-O layer was found to reduce the I_{corr} values of the substrate. However, if the nitride layer is too thin, it reduces the resistance to Cl$^-$ ions in the solution, thereby reducing the I_{corr} values [47]. Moreover, when the thickness of Ti-N-O was high enough to prevent the adsorption of Cl$^-$ ions in artificial saliva, the surface roughness became the main factor that affected the corrosion resistance of the sample. During the process of hollow cathodic plasma source oxynitride, the sample was bombarded by the plasma. The higher the temperature, the more active the plasma, and the rougher the surface morphology of the sample [48]. Thus, it was observed that the corrosion resistance of the PNO-540 sample deteriorated as the surface roughness increased. Therefore, the PNO-520 sample exhibited the best corrosion resistance. Bao et al. [49] prepared TiNxOy coating on the TC4 surface through physical vapor deposition. The I_{corr} of the coating sample in an acidic solution (pH = 5.2) was only 6.0×10^{-8} A/cm^2. Du et al. [50] produced various Ti-N-O coatings with an I_{corr} range of 100–500 nA/cm^2 and an R_p range of 70–300 k$\Omega \cdot$cm^2. Meanwhile, Subramanian et al. [51] manufactured titanium oxynitride coatings with an I_{corr} of 0.67×10^{-7} A/cm^2 and a V_{corr} of 3.1×10^{-3} mm/a. The I_{corr} of the PNO-520 sample prepared by our process was 7.65×10^{-8} A/cm^2, and the R_p was 3.40×10^5 $\Omega \cdot$cm^2. The above results are similar, indicating that the Ti-N-O modified layer prepared by our process improved the corrosion resistance of the sample.

3.3.2. XPS Analysis of PNO-520 Samples after Artificial Saliva Corrosion

To elucidate the factors affecting the corrosion resistance of PNO samples in artificial saliva and their passivation film repassivation mechanism, XPS analysis was conducted on PNO-520 corrosion specimens (Figure 7). A Ti_2O_3 peak observed at the 456.2 and 461.81 eV binding energies in the Ti2p spectrum suggested that oxidation reactions occurred on the surface of the titanium alloy. The fitted peaks of Ti2p were all paired. The observed binding energies of $Ti2p^{3/2}$ = 455.12, 457.2, and 458.46 eV, and $Ti2p^{1/2}$ = 460.76, 463.31, and 464.48 eV indicated the presence of TiN, TiON, and TiO_2. N1s correspond to TiON and TiN with binding energies of 395.99 and 397.06 eV (standard binding energy: 397.00 eV) with an overall decrease in the area of the peaks. The O1s orbital exhibited three closely fitted peaks. The binding energy of 529.92 eV is a prominent characteristic of TiO_2. The binding energy of 531.09 eV meant that the hydration reaction of TiO_2 in artificial saliva generated OH$^-$. The binding energy at 531.96 eV represented H_2O as the bound water absorbed by the sample oxide film [52]. The O1s peak showed that the passivation film mainly comprised TiO_2, a small amount of Ti(OH)x, and an oxide film combined with water. The presence of these oxides improves the microstructure's uniformity and improves the sample's corrosion resistance [53].

3.3.3. Electrochemical Impedance Spectroscopy Analysis of Different Samples

Figure 8 shows the Nyquist and Bode plots of TC4 alloy in human artificial saliva. The Nyquist curve represents the variation of impedance among different titanium alloy samples. The capacitance vs. frequency curve exhibited a flat, incomplete resistance arc across the entire frequency range. Moreover, the shape of the arc suggested the involvement of electrochemical charge transfer [49]. With an increase in temperature, the transfer of charges at the surface of the electrode exhibits greater intricacy, causing an increase in resistance and an enhancement in corrosion resistance [11,54]. As a consequence, a greater scope of the capacitive arch arises. This tendency holds ultimately. The convex peak on the Bode phase angle diagram and the linear change in the Bode impedance diagram in the low-frequency region range (10^{-2}–10^1 Hz) with a slope close to -1 reflect the capacitive behavior at the interface between the passivated film and the artificial saliva with the typical characteristics of an ideally polarized electrode [49,55]. In comparison with the PNO samples, the UN sample showed considerably lower values for the capacitive arc radius, as well as for the Bode phase angle and impedance at low and medium frequencies. This suggests that the UN sample exhibits more pronounced ion transport and electrode reaction processes [6,56]. The correlation between measured data and simulated data is shown in Figure 8a. The measured data matched well with the simulated data. The corrosion resistance of the UN sample was worse than that of the PNO samples. This showed that the corrosion resistance of the specimens was in the order of PNO-520 > PNO-500 > PNO-540 > UN.

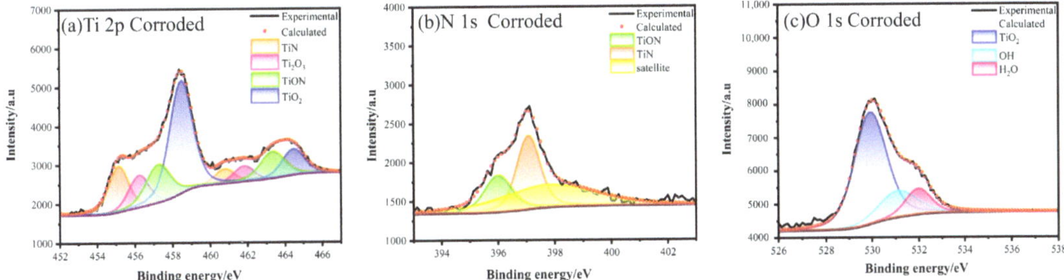

Figure 7. XPS Spectra of Ti2p, N1s, and O1s after corrosion of PNO-520 sample in artificial saliva.

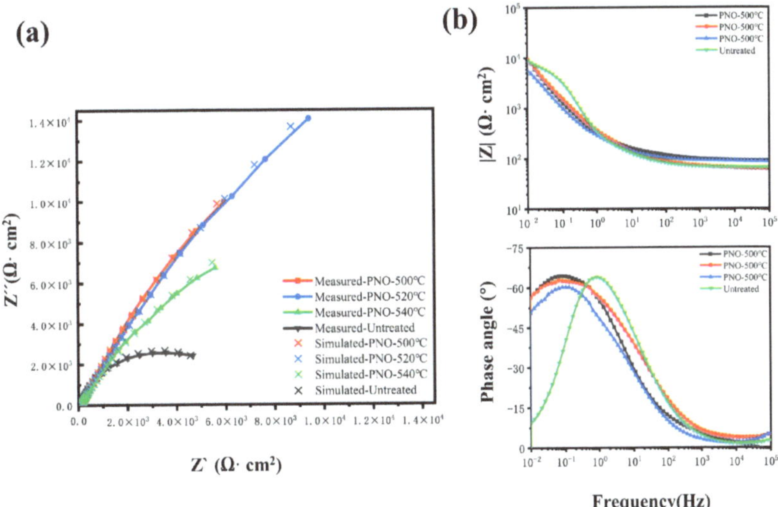

Figure 8. Electrochemical impedance spectroscopy analysis of different samples in artificial saliva: (**a**) Nyquist diagram, (**b**) Bode diagram.

3.3.4. The Mechanism of The Unique Repassivation Process Phenomenon in PNO Samples

According to the EIS curve data, the equivalent circuit was fitted to the sample with Zview-2 software. The corrosion state of the model in the artificial saliva was explained with a three-dimensional stereogram, as shown in Figure 9, in which the UN and PNO samples corresponded to the $R_s((C_fR_f)(CPE_{dl}R_{ct}))$, and $R_s(CPE_{dl}R_{ct})$ models, respectively.

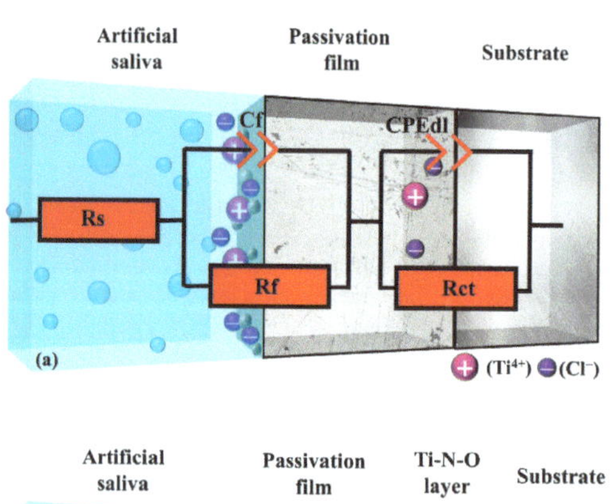

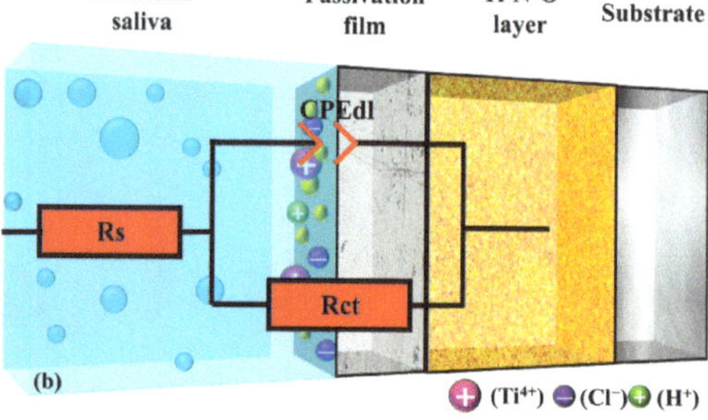

Figure 9. The equivalent circuit diagram in artificial saliva: (**a**) UN sample, (**b**) PNO sample.

The CPE impedance is defined as $Z_C = 1/[C(j\omega)^n]$, where n represents the deviation from the actual capacitive behavior. It is the CPE exponent with values between 0 and 1. When n = 1, the normal phase angle element is an ideal capacitor. On the other hand, when n = 0, the normal-phase angle element is pure resistance.

The size of the circuit parameters was simulated using Zview-2 software. The correlation between the measured data and the calculated data is shown in Figure 8a. The measured data matched well with the simulated data. The UN sample was a non-dense porous electrode circuit with a double-time constant that denoted the solution resistance. R_f represents the charge transfer reaction resistance. C_f denotes the bilayer capacitance. R_{ct} and CPE_{dl} represent the resistance and capacitance of the passivated film [57]. As can be seen in Table 4, the passivation film R_{ct} (6.50×10^3 $\Omega\cdot cm^2$) of the UN sample was more significant than the charge transfer resistance R_f (86.98 $\Omega\cdot cm^2$). This indicates that the passivation film's generation hindered the charge transfer from the artificial saliva to the titanium alloy substrate. At this time, corrosion mainly occurred in the interface between the artificial saliva and the multilayer pores of the passivation film (Figure 9a). The above behavior (Figure 9a) is called crevice corrosion, in which the artificial saliva can penetrate the TC4 alloy substrate through defects, crevices, and pores [58].

Table 4. Electrochemical EIS circuit fitting parameters for UN and PNO samples in artificial saliva.

	R_s ($\Omega\cdot cm^2$)	C_f ($F\cdot cm^{-2}$)	R_f ($\Omega\cdot cm^2$)	CPE_{dl} ($F\cdot cm^{-2}$)	R_{ct} ($\Omega\cdot cm^2$)
UN	65.35	1.21×10^{-3}	86.98	2.7×10^{-4}	6.50×10^3
PNO-500	126.5			4.71×10^{-4}	8.47×10^4
PNO-520	90.33			4.27×10^{-4}	1.36×10^5
PNO-540	92.93			7.09×10^{-4}	6.18×10^4

Studies have shown that the soluble chlorine compounds formed in various types of artificial body fluids due to the presence of Cl^- reduce the stability of the passivation film on the surface of a TC4 alloy, accelerate the dissolution of the anode, and reduce the corrosion resistance [59,60]. When titanium alloy is implanted as a dental implant bolt in the human mouth, the presence of Cl^- in the saliva accelerates the reaction under the corrosion of oral saliva. The artificial saliva chosen for this experiment was a weak acidic solution with a pH of 6.65 ± 0.01. The following crevice corrosion reactions (3) and (4) [60] occurred:

$$Ti + 4Cl^- \rightarrow TiCl_4 + 4e^- \tag{3}$$

$$TiCl_4 + 2H_2O \rightarrow TiO_2 + 4Cl^- + 4H^+ \tag{4}$$

The PNO samples were single-time constant dense multiphase structured electrode circuits. R_s represented the solution resistance, and R_{ct} and CPE_{dl} represented the nitrogen oxide resistance and the double-layer capacitance, respectively. The fitted parameters are shown in Table 4, which indicates that the PNO-520 sample resistance R_{ct} (1.36×10^5 $\Omega \cdot cm^2$) was two orders of magnitude larger than the initial sample resistance R_{ct} (6.43×10^3 $\Omega \cdot cm^2$). The PNO-500 (8.47×10^4 $\Omega \cdot cm^2$) and PNO-540 (6.18×10^4 $\Omega \cdot cm^2$) sample resistance values were one order of magnitude higher than the initial resistance value. Figure 7c depicts the XPS detection analysis results, which show that upon exposure to an artificial environment, the material's surface undergoes corrosion. This process leads to the formation of bound water and hydroxyl groups on its surface. The growth of TiO_2 was also observed to occur concomitantly with hydration. Ti-OH groups were formed on the oxidized surface OH^- groups, which could attract cations (Mg^{2+}, Ca^{2+}, K^{1+}, Na^+), and the following reaction occurred (5) [59,61]:

$$TiO_2 + H_2O + 4H^+ + 3e^- \rightarrow Ti(OH)_3^+ \tag{5}$$

The experiments demonstrated that within a given solution, the titanium alloy's corrosion byproduct Ti^{3+} can be converted to Ti^{4+}, which serves as a substance that decelerates corrosion, thereby prompting the resumption of passivation [40].

The repassivation process of the passive film is shown in Figure 10 and can be divided into three steps: (6), (7), and (8) [8,62]:

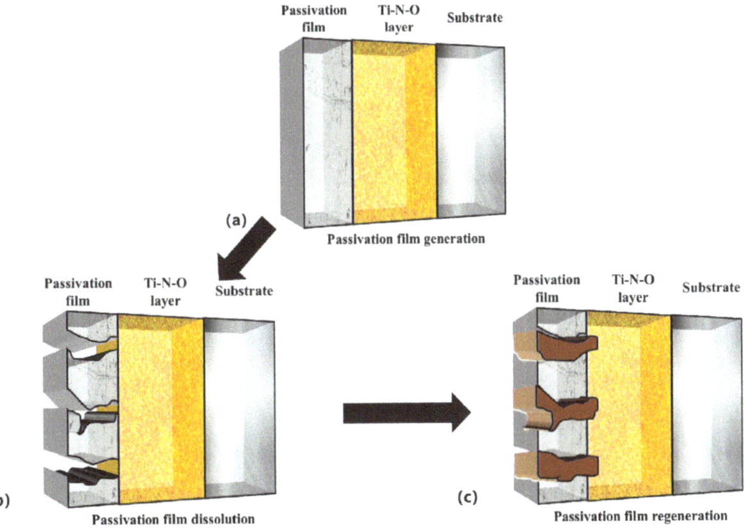

Figure 10. Three-dimensional schematic diagram of dissolution and regeneration of Ti-N-O modified layer passivation film in artificial saliva.

Passivation was observed in the PNO samples exposed to a corrosive artificial saliva environment (Figure 6, −0.3 V–0.05 V). A passivation film was formed on their surfaces (Figure 10a). The analysis (Figure 7) indicated that the film was composed of TiO_2 and Ti_2O_3.

Secondly, the chemical stability of Ti-N-O, the intermediate product of the titanium-nitrogen compound and titanium-oxygen compound, is not as good as titanium oxide. Since the Ti-N bond dissociation energies are lower than those of the Ti-O bond, the intermediate product TiNxOy partially dissolves. Additionally, according to reaction (6), the passivation film undergoes a hydration reaction, leading to the further dissolution of the film (see Figure 10b).

Finally, the PNO samples showed secondary passivation behavior (0.2–0.5 V). The repassivation mechanism of the passivation film of the Ti-N-O modified layer gave the sample durable and adequate corrosion resistance under the artificial saliva. The PNO samples formed a layer of oxide and $TiN + Ti_2N + \alpha$-Ti(N) nitrides. The nitrides and oxides became a corrosion-resistant barrier layer, effectively protecting the substrate [63–66].

$$Ti^{3+} + xH_2O \rightarrow TiO_x + 2xH^+ + (2x-3)e^- \tag{6}$$

$$2TiN_xO_y + 2H_2O \rightarrow 2Ti^{3+} + (1+y)O_2 + xN_2 + 4H^+ + 10e^- \tag{7}$$

$$2TiN + 2yH_2O \rightarrow 2TiN_xO_y + (1-x)N_2 + 4yH^+ + 4ye^- \tag{8}$$

The above analysis shows that the PNO samples have a more vital ability to inhibit the penetration of corrosion ions than the UN sample. The repassivation process of the PNO sample's passive film improves its corrosion resistance in artificial saliva. This is attributed to the dense structures of the nitrides Ti_2N and TiN, which prevent the invasion of saliva, and the superior corrosion resistance of the oxide TiO_2, which hampers active ion diffusion. As a result, the electrochemical corrosion effect was reduced, and contact between the substrate and artificial saliva was effectively blocked, reducing charge transfer [67]. The Ti-N-O modified layer with a dense structure, fast passivation ability, and high chemical stability was responsible for the improved corrosion resistance of the samples [68].

4. Conclusions

The corrosion resistance of TC4 titanium alloy in artificial saliva was evaluated by comparing UN and PNO samples in terms of the phase type, surface hardness, depth of diffusion layer, and EIS testing. The key findings of this study are as follows:

(1) By employing a two-step hollow cathodic plasma source oxynitride technique at temperatures of 500, 520, and 540 °C, a modified Ti-N-O layer was successfully formed on the surface of the TC4 alloy, enhancing its properties. The PNO sample consisted of TiO_2, TiN, and Ti_2N layers and a-Ti(N) phase, forming the Ti-N-O modified layer.

(2) The composite layer thickness of the PNO samples ranged from 5.5 to 10.0 μm, with corresponding hardness values of 798.3 to 1142.7 $HV_{0.1}$. Specifically, for the PNO-520 sample, the composite layer had a thickness of approximately 6.0 μm, and the diffusion layer extended beyond 48 μm. The surface hardness of this sample was measured at 982.1 $HV_{0.1}$, which was approximately three times higher than the hardness of the substrate.

(3) The PNO sample exhibited a significantly stronger ability to inhibit the penetration of corrosion ions in the artificial saliva solutions compared with the UN sample. The impedance measurement of the PNO-520 sample yielded a value of 1.36×10^5 $\Omega \cdot cm^2$, which was two orders of magnitude higher than that of the UN sample (6.50×10^3 $\Omega \cdot cm^2$). Furthermore, the corrosion current density of the PNO-520 sample was 7.65×10^{-8} A/cm^2, lower than that of the UN sample (2.78×10^{-6} A/cm^2). The unique repassivation process of the passivating film in the PNO sample, involving generation, dissolution, and regeneration, effectively enhanced its corrosion resistance in harsh environments.

Author Contributions: Conceptualization, Y.H. and Y.L.; methodology, J.Y.; software, M.S. and M.W. (Wang Mufan); validation, J.Y., Z.Z. (Zelong Zhou) and Z.Z. (Zhehao Zhang); formal analysis, X.Y. and B.X.; investigation, J.Y., D.F. and Z.Z. (Zelong Zhou); resources, Z.Z. (Zhehao Zhang); data curation, M.W. (Mingjia Wang) and C.W.; writing—original draft preparation, J.Y.; writing—review and editing, Y.L.; visualization, J.Y. and X.Y.; supervision, Y.L.; project administration, M.W. (Mingjia Wang); funding acquisition, Y.L. and M.W. (Mingjia Wang). All authors have read and agreed to the published version of the manuscript.

Funding: This research was funded by the National Natural Science Foundation of China, grant number 52175192, and the Shandong Natural Science Foundation, grant number ZR2022ME211.

Institutional Review Board Statement: Not applicable.

Informed Consent Statement: Not applicable.

Data Availability Statement: Not applicable.

Conflicts of Interest: The authors declare no conflict of interest.

References

1. Siony, N.; Vuong, L.; Lundaajamts, O.; Kadkhodaei, S. Computational design of corrosion-resistant and wear-resistant titanium alloys for orthopedic implants. *Mater. Today Commun.* **2022**, *33*, 104465. [CrossRef]
2. Gurel, S.; Yagci, M.B.; Bal, B.; Canadinc, D. Corrosion behavior of novel Titanium-based high entropy alloys designed for medical implants. *Mater. Chem. Phys.* **2020**, *254*, 123377. [CrossRef]
3. Omarov, S.; Nauryz, N.; Talamona, D.; Perveen, A. Surface Modification Techniques for Metallic Biomedical Alloys: A Concise Review. *Metals* **2022**, *13*, 82. [CrossRef]
4. Krishnan, V.; Krishnan, A.; Remya, R.; Ravikumar, K.K.; Nair, S.A.; Shibli, S.M.; Varma, H.K.; Sukumaran, K.; Kumar, K.J. Development and evaluation of two PVD-coated beta-titanium orthodontic archwires for fluoride-induced corrosion protection. *Acta Biomater* **2011**, *7*, 1913–1927. [CrossRef]
5. Atapour, M.; Pilchak, A.L.; Shamanian, M.; Fathi, M.H. Corrosion behavior of Ti–8Al–1Mo–1V alloy compared to Ti–6Al–4V. *Mater. Des.* **2011**, *32*, 1692–1696. [CrossRef]
6. Lario, J.; Vicente Escuder, Á.; Segovia, F.; Amigó, V. Electrochemical corrosion behavior of Ti–35Nb–7Zr–5Ta powder metallurgic alloys after Hot Isostatic Process in fluorinated artificial saliva. *J. Mater. Res. Technol.* **2022**, *16*, 1435–1444. [CrossRef]
7. Yu, F.; Addison, O.; Davenport, A. Temperature-Dependence Corrosion Behavior of Ti6Al4V in the Presence of HCl. *Front. Mater.* **2022**, *9*, 880702. [CrossRef]
8. Pohrelyuk, I.M.; Fedirko, V.M.; Tkachuk, O.V.; Proskurnyak, R.V. Corrosion resistance of Ti–6Al–4V alloy with nitride coatings in Ringer's solution. *Corros. Sci.* **2013**, *66*, 392–398. [CrossRef]
9. Grigoriev, S.; Sotova, C.; Vereschaka, A.; Uglov, V.; Cherenda, N. Modifying Coatings for Medical Implants Made of Titanium Alloys. *Metals* **2023**, *13*, 718. [CrossRef]
10. Naeem, M.; Awan, S.; Shafiq, M.; Raza, H.A.; Iqbal, J.; Díaz-Guillén, J.C.; Sousa, R.R.M.; Jelani, M.; Abrar, M. Wear and corrosion studies of duplex surface-treated AISI-304 steel by a combination of cathodic cage plasma nitriding and PVD-TiN coating. *Ceram. Int.* **2022**, *48*, 21473–21482. [CrossRef]
11. Adachi, S.; Ueda, N. Wear and Corrosion Properties of Cold-Sprayed AISI 316L Coatings Treated by Combined Plasma Carburizing and Nitriding at Low Temperature. *Coatings* **2018**, *8*, 456. [CrossRef]
12. Hoja, S.; Steinbacher, M.; Zoch, H.W. Compound layer design for deep nitrided gearings. *Metals* **2020**, *10*, 455. [CrossRef]
13. Sharifahmadian, O.; Zhai, C.; Hung, J.; Shineh, G.; Stewart, C.A.C.; Fadzil, A.A.; Ionescu, M.; Gan, Y.; Wise, S.G.; Akhavan, B. Mechanically robust nitrogen-rich plasma polymers: Biofunctional interfaces for surface engineering of biomedical implants. *Mater. Today Adv.* **2021**, *12*, 100188. [CrossRef]
14. Liu, J.; Liu, D.; Li, S.; Deng, Z.; Pan, Z.; Li, C.; Chen, T. The effects of graphene oxide doping on the friction and wear properties of TiN bioinert ceramic coatings prepared using wide-band laser cladding. *Surf. Coat. Technol.* **2023**, *458*, 129354. [CrossRef]
15. Wang, X.; Bai, S.; Li, F.; Li, D.; Zhang, J.; Tian, M.; Zhang, Q.; Tong, Y.; Zhang, Z.; Wang, G.; et al. Effect of plasma nitriding and titanium nitride coating on the corrosion resistance of titanium. *J. Prosthet. Dent.* **2016**, *116*, 450–456. [CrossRef] [PubMed]
16. Wierzchoń, T.; Czarnowska, E.; Grzonka, J.; Sowińska, A.; Tarnowski, M.; Kamiński, J.; Kulikowski, K.; Borowski, T.; Kurzydłowski, K.J. Glow discharge assisted oxynitriding process of titanium for medical application. *Appl. Surf. Sci.* **2015**, *334*, 74–79. [CrossRef]
17. Behzadi, P.; Badr, M.; Zakeri, A. Duplex surface modification of pure Ti via thermal oxidation and gas nitriding: Preparation and electrochemical studies. *Ceram. Int.* **2022**, *48*, 34374–34381. [CrossRef]
18. Banakh, O.; Moussa, M.; Matthey, J.; Pontearso, A.; Cattani-Lorente, M.; Sanjines, R.; Fontana, P.; Wiskott, A.; Durual, S. Sputtered titanium oxynitride coatings for endosseous applications: Physical and chemical evaluation and first bioactivity assays. *Appl. Surf. Sci.* **2014**, *317*, 986–993. [CrossRef]

19. Albayrak, C.; Hacisalihoglu, I.; Vangolu, S.Y.; Alsaran, A. Tribocorrosion behavior of duplex treated pure titanium in Simulated Body Fluid. *Wear* **2013**, *302*, 1642–1648. [CrossRef]
20. Zhang, C.W.; Wen, K.; Gao, Y. Columnar and nanocrystalline combined microstructure of the nitrided layer by active screen plasma nitriding on surface-nanocrystalline titanium alloy. *Appl. Surf. Sci.* **2023**, *617*, 156614. [CrossRef]
21. Zhang, L.; Shao, M.H.; Wang, Z.W.; Zhang, Z.H.; He, Y.Y.; Yan, J.W.; Lu, J.P.; Qiu, J.X.; Li, Y. Comparison of tribological properties of nitrided Ti-N modified layer and deposited TiN coatings on TA2 pure titanium. *Tribol. Int.* **2022**, *174*, 107712. [CrossRef]
22. de Abreu, L.H.; Naeem, M.; Monção, R.M.; Costa, T.H.C.; Díaz-Guillén, J.C.; Iqbal, J.; Sousa, R.R. The Effect of Cathodic Cage Plasma TiN Deposition on Surface Properties of Conventional Plasma Nitrided AISI-M2 Steel. *Metals* **2022**, *12*, 961. [CrossRef]
23. Domínguez-Meister, S.; Ibáñez, I.; Dianova, A.; Brizuela, M.; Braceras, I. Nitriding of titanium by hollow cathode assisted active screen plasma and its electro-tribological properties. *Surf. Coat. Technol.* **2021**, *411*, 126998. [CrossRef]
24. Borgioli, F.; Galvanetto, E.; Bacci, T. Surface Modification of Austenitic Stainless Steel by Means of Low Pressure Glow-Discharge Treatments with Nitrogen. *Coatings* **2019**, *9*, 604. [CrossRef]
25. Adachi, S.; Yamaguchi, T.; Ueda, N. Formation and Properties of Nitrocarburizing S-Phase on AISI 316L Stainless Steel-Based WC Composite Layers by Low-Temperature Plasma Nitriding. *Metals* **2021**, *11*, 1538. [CrossRef]
26. Wang, Z.W.; Li, Y.; Zhang, Z.H.; Zhang, S.Z.; Ren, P.; Qiu, J.X.; Wang, W.W.; Bi, Y.J.; He, Y.Y. Friction and wear behavior of duplex-treated AISI 316L steels by rapid plasma nitriding and (CrWAlTiSi)N ceramic coating. *Results Phys.* **2021**, *24*, 104132. [CrossRef]
27. Sun, F.; Liu, X.-L.; Luo, S.-Q.; Xiang, D.-D.; Ba, D.-C.; Lin, Z.; Song, G.-Q. Duplex treatment of arc plasma nitriding and PVD TiN coating applied to dental implant screws. *Surf. Coat. Technol.* **2022**, *439*, 128449. [CrossRef]
28. Li, Y.; Wang, Z.; Wang, L. Surface properties of nitrided layer on AISI 316L austenitic stainless steel produced by high temperature plasma nitriding in short time. *Appl. Surf. Sci.* **2014**, *298*, 243–250. [CrossRef]
29. Cherenda, N.N.; Basalai, A.V.; Shymanski, V.I.; Uglov, V.V.; Astashynski, V.M.; Kuzmitski, A.M.; Laskovnev, A.P.; Remnev, G.E. Modification of Ti-6Al-4V alloy element and phase composition by compression plasma flows impact. *Surf. Coat. Technol.* **2018**, *355*, 148–154. [CrossRef]
30. Zhao, Y.T.; Lu, M.Y.; Fan, Z.Q.; Huang, S.Q.; Huang, H. Laser deposition of wear-resistant titanium oxynitride/titanium composite coatings on Ti-6Al-4V alloy. *Appl. Surf. Sci.* **2020**, *531*, 147212. [CrossRef]
31. Ye, Q.W.; Li, Y.; Zhang, M.Y.; Zhang, S.Z.; Bi, Y.J.; Gao, X.P.; He, Y.Y. Electrochemical behavior of (Cr, W, Al, Ti, Si)N multilayer coating on nitrided AISI 316L steel in natural seawater. *Ceram. Int.* **2020**, *46*, 22404–22418. [CrossRef]
32. Jaeger, D.; Patscheider, J. A complete and self-consistent evaluation of XPS Spectra of TiN. *J. Electron Spectrosc. Relat. Phenom.* **2012**, *185*, 523–534. [CrossRef]
33. Wang, Z.G.; Zu, X.T.; Xiang, X.; Zhu, S.; Wang, L.M. Surface modification of Ti-4Al-2V alloy by nitrogen implantation. *J. Mater. Sci.* **2006**, *41*, 3363–3367. [CrossRef]
34. Li, Y.; Wang, L. Study of oxidized layer formed on aluminium alloy by plasma oxidation. *Thin Solid Film.* **2009**, *517*, 3208–3210. [CrossRef]
35. Hong, X.; Feng, K.; Tan, Y.F.; Wang, X.L.; Tan, H. Effects of process parameters on microstructure and wear resistance of TiN coatings deposited on TC11 titanium alloy by electrospark deposition. *Trans. Nonferrous Met. Soc. China* **2017**, *27*, 1767–1776. [CrossRef]
36. Mohammadi, M.; Akbari, A.; Warchomicka, F.; Pichon, L. Depth profiling characterization of the nitride layers on gas nitrided commercially pure titanium. *Mater. Charact.* **2021**, *181*, 111453. [CrossRef]
37. Guo, Y.; Fang, Y.; Dai, G.; Sun, Z.; Wang, Y.; Yuan, Q. The effect of hydrogen treatment on microstructures evolution and mechanical properties of titanium alloy fabricated by selective laser melting. *J. Alloy. Compd.* **2022**, *890*, 161642. [CrossRef]
38. Adam, D.B.; Tsai, M.-C.; Awoke, Y.A.; Huang, W.-H.; Lin, C.-H.; Alamirew, T.; Ayele, A.A.; Yang, Y.-W.; Pao, C.-W.; Su, W.-N.; et al. Engineering self-supported ruthenium-titanium alloy oxide on 3D web-like titania as iodide oxidation reaction electrocatalyst to boost hydrogen production. *Appl. Catal. B Environ.* **2022**, *316*, 121608. [CrossRef]
39. Nguyen, T.L.; Tseng, C.C.; Cheng, T.C.; Nguyen, V.; Chang, Y.H. Formation and characterization of calcium phosphate ceramic coatings on Ti-6Al-4V alloy. *Mater. Today Commun.* **2022**, *31*, 103686. [CrossRef]
40. Wei, Y.; Pan, Z.M.; Fu, Y.; Yu, W.; He, S.L.; Yuan, Q.Y.; Luo, H.; Li, X.G. Effect of annealing temperatures on microstructural evolution and corrosion behavior of Ti-Mo titanium alloy in hydrochloric acid. *Corros. Sci.* **2022**, *197*, 110079. [CrossRef]
41. Borgioli, F. From Austenitic Stainless Steel to Expanded Austenite-S Phase: Formation, Characteristics and Properties of an Elusive Metastable Phase. *Metals* **2020**, *10*, 187. [CrossRef]
42. Karimi, S.; Nickchi, T.; Alfantazi, A. Effects of bovine serum albumin on the corrosion behaviour of AISI 316L, Co–28Cr–6Mo, and Ti–6Al–4V alloys in phosphate buffered saline solutions. *Corros. Sci.* **2011**, *53*, 3262–3272. [CrossRef]
43. Li, Q.; Wei, M.; Yang, J.; Zhao, Z.; Ma, J.; Liu, D.; Lan, Y. Effect of Ca addition on the microstructure, mechanical properties and corrosion rate of degradable Zn-1Mg alloys. *J. Alloy. Compd.* **2021**, *887*, 161255. [CrossRef]
44. Luo, Y.; Wang, M.; Zhu, J.; Tu, J.; Jiao, S. Microstructure and Corrosion Resistance of Ti6Al4V Manufactured by Laser Powder Bed Fusion. *Metals* **2023**, *13*, 496. [CrossRef]
45. Shahmohammadi, M.; Sun, Y.N.; Yuan, J.C.C.; Mathew, M.T.; Sukotjo, C.; Takoudis, C.G. In vitro corrosion behavior of coated Ti6Al4V with TiO2, ZrO2, and TiO2/ZrO2 mixed nanofilms using atomic layer deposition for dental implants. *Surf. Coat. Technol.* **2022**, *444*, 128686. [CrossRef]

46. Li, Y.; Wang, Z.; Shao, M.; Zhang, Z.; Wang, C.; Yan, J.; Lu, J.; Zhang, L.; Xie, B.; He, Y.; et al. Characterization and electrochemical behavior of a multilayer-structured Ti–N layer produced by plasma nitriding of electron beam melting TC4 alloy in Hank's solution. *Vacuum* **2023**, *208*, 111737. [CrossRef]

47. Zhang, L.; Shao, M.; Zhang, Z.; Yi, X.; Yan, J.; Zhou, Z.; Fang, D.; He, Y.; Li, Y. Corrosion Behavior of Nitrided Layer of Ti6Al4V Titanium Alloy by Hollow Cathodic Plasma Source Nitriding. *Materials* **2023**, *16*, 2961. [CrossRef] [PubMed]

48. She, D.S.; Yue, W.; Fu, Z.Q.; Wang, C.B.; Yang, X.K.; Liu, J.J. Effects of nitriding temperature on microstructures and vacuum tribological properties of plasma-nitrided titanium. *Surf. Coat. Technol.* **2015**, *264*, 32–40. [CrossRef]

49. Bao, Y.C.; Wang, W.L.; Cui, W.F.; Qin, G.W. Corrosion resistance and antibacterial activity of Ti-N-O coatings deposited on dental titanium alloy. *Surf. Coat. Technol.* **2021**, *419*, 127296. [CrossRef]

50. Du, J.W.; Chen, L.; Chen, J.; Yue, J.L. Effects of additional oxygen on the structural, mechanical, thermal, and corrosive properties of TiN coatings. *Ceram. Int.* **2022**, *48*, 14432–14441. [CrossRef]

51. Subramanian, B.; Muraleedharan, C.V.; Ananthakumar, R.; Jayachandran, M. A comparative study of titanium nitride (TiN), titanium oxy nitride (TiON) and titanium aluminum nitride (TiAlN), as surface coatings for bio implants. *Surf. Coat. Technol.* **2011**, *205*, 5014–5020. [CrossRef]

52. Wang, Q.; Huang, F.; Cui, Y.-T.; Yoshida, H.; Wen, L.; Jin, Y. Influences of formation potential on oxide film of TC4 in 0.5 M sulfuric acid. *Appl. Surf. Sci.* **2021**, *544*, 148888. [CrossRef]

53. Rao, X.; Du, L.; Zhao, J.J.; Tan, X.D.; Fang, Y.X.; Xu, L.Q.; Zhang, Y.P. Hybrid TiO_2/AgNPs/g-C_3N_4 nanocomposite coatings on TC4 titanium alloy for enhanced synergistic antibacterial effect under full spectrum light. *J. Mater. Sci. Technol.* **2022**, *118*, 35–43. [CrossRef]

54. Hussein, M.A.; Yilbas, B.; Kumar, A.M.; Drew, R.; Al-Aqeeli, N. Influence of Laser Nitriding on the Surface and Corrosion Properties of Ti-20Nb-13Zr Alloy in Artificial Saliva for Dental Applications. *J. Mater. Eng. Perform.* **2018**, *27*, 4655–4664. [CrossRef]

55. Yang, X.J.; Du, C.W.; Wan, H.X.; Liu, Z.Y.; Li, X.G. Influence of sulfides on the passivation behavior of titanium alloy TA2 in simulated seawater environments. *Appl. Surf. Sci.* **2018**, *458*, 198–209. [CrossRef]

56. Borgioli, F.; Galvanetto, E.; Bacci, T. Surface Modification of a Nickel-Free Austenitic Stainless Steel by Low-Temperature Nitriding. *Metals* **2021**, *11*, 1845. [CrossRef]

57. Liu, C.L.; Wang, Y.J.; Wang, M.; Huang, W.J.; Chu, P.K. Electrochemical stability of TiO2 nanotubes with different diameters in artificial saliva. *Surf. Coat. Technol.* **2011**, *206*, 63–67. [CrossRef]

58. Wu, J.; Li, M.; Lin, C.C.; Gao, P.F.; Zhang, R.; Li, X.; Zhang, J.X.; Cai, K.Y. Moderated crevice corrosion susceptibility of Ti_6Al_4V implant material due to albumin-corrosion interaction. *J. Mater. Sci. Technol.* **2022**, *109*, 209–220. [CrossRef]

59. Li, J.; Bai, Y.; Fan, Z.D.; Li, S.J.; Hao, Y.L.; Yang, R.; Gao, Y.B. Effect of fluoride on the corrosion behavior of nanostructured Ti-24Nb-4Zr-8Sn alloy in acidulated artificial saliva. *J. Mater. Sci. Technol.* **2018**, *34*, 1660–1670. [CrossRef]

60. Chenghao, L.; Li'nan, J.; Chuanjun, Y.; Naibao, H. Crevice Corrosion Behavior of CP Ti, Ti-6Al-4V Alloy and Ti-Ni Shape Memory Alloy in Artificial Body Fluids. *Rare Met. Mater. Eng.* **2015**, *44*, 781–785. [CrossRef]

61. Mareci, D.; Chelariu, R.; Gordin, D.M.; Ungureanu, G.; Gloriant, T. Comparative corrosion study of Ti-Ta alloys for dental applications. *Acta Biomater* **2009**, *5*, 3625–3639. [CrossRef]

62. Oliveira, V.M.C.A.; Aguiar, C.; Vazquez, A.M.; Robin, A.; Barboza, M.J.R. Improving corrosion resistance of Ti–6Al–4V alloy through plasma-assisted PVD deposited nitride coatings. *Corros. Sci.* **2014**, *88*, 317–327. [CrossRef]

63. Azumi, K.; Seo, M. Changes in electrochemical properties of the anodic oxide film formed on titanium during potential sweep. *Corros. Sci.* **2001**, *43*, 533–546. [CrossRef]

64. Shen, H.Y.; Wang, L. Corrosion resistance and electrical conductivity of plasma nitrided titanium. *Int. J. Hydrog. Energy* **2021**, *46*, 11084–11091. [CrossRef]

65. Fossati, A.; Borgioli, F.; Galvanetto, E.; Bacci, T. Corrosion resistance properties of plasma nitrided Ti–6Al–4V alloy in nitric acid solutions. *Corros. Sci.* **2004**, *46*, 917–927. [CrossRef]

66. Tarnowski, M.; Borowski, T.; Skrzypek, S.; Kulikowski, K.; Wierzchoń, T. Shaping the structure and properties of titanium and Ti6Al7Nb titanium alloy in low-temperature plasma nitriding processes. *J. Alloy. Compd.* **2021**, *864*, 158896. [CrossRef]

67. Liu, J.M.; Lou, Y.X.; Zhang, C.; Yin, S.; Li, H.M.; Sun, D.Q.; Sun, X.H. Improved corrosion resistance and antibacterial properties of composite arch-wires by N-doped TiO_2 coating. *Rsc Adv.* **2017**, *7*, 43938–43949. [CrossRef]

68. Cui, W.F.; Niu, F.J.; Tan, Y.L.; Qin, G.W. Microstructure and tribocorrosion performance of nanocrystalline TiN graded coating on biomedical titanium alloy. *Trans. Nonferrous Met. Soc. China* **2019**, *29*, 1026–1035. [CrossRef]

Article

Mechanical and Magnetic Investigations of Balls Made of AISI 1010 and AISI 1085 Steels after Nitriding and Annealing

Sławomir Maksymilian Kaczmarek [1,*], Jerzy Michalski [2,*], Tadeusz Frączek [2], Agata Dudek [2], Hubert Fuks [1] and Grzegorz Leniec [1]

1 Department of Technical Physics, Faculty of Mechanical Engineering and Mechatronics, West Pomeranian University of Technology in Szczecin, al. Piastów 48, 70-311 Szczecin, Poland; hubert.fuks@zut.edu.pl (H.F.); grzegorz.leniec@zut.edu.pl (G.L.)
2 Faculty of Production Engineering and Materials Technology, Czestochowa University of Technology, Al. Armii Krajowej 19, 42-201 Czestochowa, Poland; fraczek.tadeusz@wip.pcz.pl (T.F.); dudek.agata@wip.pcz.pl (A.D.)
* Correspondence: skaczmarek@zut.edu.pl (S.M.K.); jerzymichalski987@gmail.com (J.M.)

Abstract: This paper discusses the changes in the phase composition and magnetic properties of the AISI 1010 and AISI 1085 steels that were nitrided at 570 °C in an ammonia atmosphere for 5 h and that were then annealed at 520 °C in a N_2/Ar atmosphere for 4 h. The test samples were made in the form of balls with diameters of less than 5 mm. The thickness of the obtained iron nitride layers was assessed through metallographic tests, while the phase composition was verified through X-ray tests. The magnetic properties were determined using ferromagnetic resonance (FMR) and superconducting quantum interference device (SQUID) techniques. Our research shows that, during the annealing of iron nitrides with a structure of $\varepsilon + \gamma'$, the ε phase decomposes first. As a result of this process, an increase in the content of the γ' phase of the iron nitride is observed. When the ε phase is completely decomposed, the γ' phase begins to decompose. The observed FMR signals did not come from isolated ions but from more magnetically complex systems, e.g., Fe–Fe pairs or iron clusters. Studies have shown that nitriding and annealing can be used to modify the magnetic properties of the tested steels.

Keywords: nitriding of AISI 1010 and 1085 steel balls; annealing in an inert atmosphere; phase composition; white layer; porous zone; magnetic properties; FMR spectra; FMR and SQUID susceptibility

Citation: Kaczmarek, S.M.; Michalski, J.; Frączek, T.; Dudek, A.; Fuks, H.; Leniec, G. Mechanical and Magnetic Investigations of Balls Made of AISI 1010 and AISI 1085 Steels after Nitriding and Annealing. *Metals* **2023**, *13*, 1060. https://doi.org/10.3390/met13061060

Academic Editor: Catalin Constantinescu

Received: 19 April 2023
Revised: 24 May 2023
Accepted: 29 May 2023
Published: 1 June 2023

1. Introduction

Nitriding has been used in the engineering industry for many years to improve the wear resistance of machine and tool parts [1,2]. Iron nitrides, in addition to high hardness and corrosion resistance, are characterized by very good magnetic properties. Magnetic materials are used in generators and electric motors as well as electronic and electromechanical devices; they are also used in data carriers. Alloys containing rare earth elements have excellent magnetic properties, so they are most often used in practice [3,4]. Due to limited rare earth metal resources and the intense increase in the demand for magnetic materials, research on magnetic materials without rare earth materials is fully justified. Research on the magnetic properties of iron nitrides meets these needs [5,6].

Iron nitrides belong to an important group of magnetic materials. The iron nitride γ'-Fe_4N meets the requirements for soft magnetic materials very well and, at the same time, is resistant to corrosion. The magnetization saturation of γ'-Fe_4N is slightly lower (by about 4%) than that of α-Fe, and its coercivity is negligibly small [7,8]. On the other hand, the iron nitride α''-$Fe_{16}N_2$ exhibits 30% higher magnetization saturation than α-Fe [9]. Therefore, it is a very good material for high-density magnetic storage media [10]. However, the α''-$Fe_{16}N_2$ nitride decomposes to α-Fe and γ'-Fe_4N at 200 °C and then completely converts

---

to α-Fe at 300 °C [11,12]. Its low thermal resistance reduces its attractiveness. On the other hand, the iron nitride γ'-Fe$_4$N, when heated in an inert atmosphere, is stable up to 650 °C [13]. Frączek et al. [14] showed that during the annealing of nitrided layers with the structure $\varepsilon + \gamma'$, which is present in AISI 1085 steel, ε phase decomposition ($\varepsilon \rightarrow \gamma' + \uparrow$) is observed. This takes place after annealing at 520 °C for 5 h in an inert H$_2$/N$_2$ atmosphere at a ratio of 3:1 and a pressure of 200 Pa. After another 5 h, the γ' phase also decomposes ($\gamma' \rightarrow \alpha$-Fe(N) + N$\uparrow$).

Among iron nitrides, the largest range of homogeneity in the Fe-N system is shown by the ε phase with a variable ratio of Fe:N Fe$_{2-3}$N [15]. Depending on the nitrogen concentration, the nitride ε may have ferromagnetic (ε-Fe$_x$N ($2 < x < 3$)) or paramagnetic (Fe$_2$N) properties [16]. Compared to the α''-Fe$_{16}$N$_2$ nitride, the ε-Fe$_3$N nitride has a much better thermal stability and slightly worse magnetic properties [17].

Since magnetic properties, such as coercivity, remanence, magnetic permeability, the Curie temperature, and saturation magnetization, depend on the nitrogen concentration in the Fe-N phases, the large homogeneity range of the ε phase allows its magnetic properties to be modified over a wide range [18].

The most effective method of producing iron nitrides with an assumed phase composition is gas nitriding. The composition of the obtained iron nitrides depends on the nitriding temperature and the nitrogen potential [19,20]. Arabczyk et al. [21] annealed a nanocrystalline iron catalyst in an NH$_3$/H$_2$ atmosphere at 350 °C with a varying nitrogen potential value during both nitriding and iron nitride reduction. During the process, the authors recorded a change in the magnetic permeability of the created nitride phases. Tests showed that the magnetic permeability of γ'-Fe$_4$N was 1.280 times higher than that of an iron catalyst, while that of ε-Fe$_x$N was more than 3 times lower.

The thermodynamics and kinetics of the nitriding process are exhaustively described in the literature [22,23], which makes controlling the chemical composition of iron nitrides relatively simple.

The aim of our research was to check whether annealing at 520 °C in an inert atmosphere would cause phase transformations of iron nitrides and whether, as a result of the transformations, a loss of mass in the annealed steels would be observed. We also tried to find some changes in the magnetic properties of the studied steel balls which could be attributed to the formation of new phases in the nitride layer region. As the subject of our research, we chose the AISI 1085 and AISI 1010 steels in the form of spherical balls. The balls were first nitrided, and then some of them were annealed. The balls that were thus altered were subjected to mechanical [14] and magnetic [24–26] tests.

2. Materials and Methods

2.1. Materials and Parameters of Nitriding and Annealing

AISI 1010 and AISI 1085 nonalloy steels were used in the tests. The chemical composition of these steels and the dimensions of the samples (SS—initial samples, "non modified") are listed in Table 1.

Table 1. Characteristics of the steels used. D_k—ball diameter.

Grade Steel	Sample No.	D_K (mm)	Element Contents in wt.%				
			C	Mn	Si	P	S
AISI 1010	1A; 11	3.97	0.10	0.5	0.1	0.04	0.05
AISI 1085	5A; 15	5.0	0.85	0.9	0.3	0.04	0.05

The steels were subjected to gas nitriding at 570 °C for 5 h; then, half of the samples were annealed at 520 °C for 4 h in N$_2$/Ar atmosphere at 200 Pa. The samples were weighed before and after the annealing process. The accuracy of weight measurement was as much as 10^{-5} g. The parameters of the nitriding and annealing processes are shown in Table 2.

Table 2. Parameters of nitriding and annealing processes.

Sample No.	Parameters of Nitriding				Parameters of Annealing		
	T (°C)	t (h)	Np (atm^{-05})	Inlet Atmosphere	T (°C)	t (h)	Inlet Atmosphere
1A; 5A	570	5	2.5	NH_3	-	-	-
11; 15	570	5	2.5	-	520	4	$N_2/Ar/P = 150$ Pa

2.2. Metallographic Research

Metallographic tests were carried out on the above-mentioned balls. Grinding the balls to their diameter, a layer of nitrides of actual thickness was observed and measured on metallographic microsection. In addition to the iron nitride layer, called the white layer (WL; its thickness is labeled g_{mp}; see Figure 1a), we also observed and measured the thickness of the porous zone (g_{por}; see Table 3 and comments in 3.1). Figure 1 shows a method for measuring the thickness of the iron nitride layer [14].

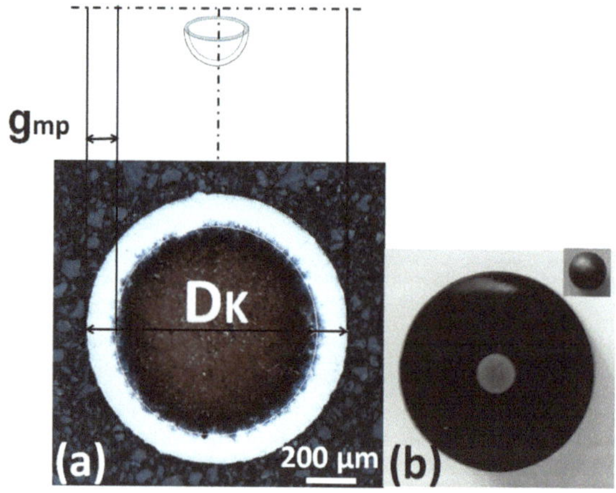

Figure 1. Diagram of measurement of the thickness of the iron nitride layer (WL) of the ball. (**a**) Gray (above) and white (below) surfaces represent the iron nitride layer, g_{mp} represents the actual thickness of the iron nitride layer (WL), and D_K represents ball diameter. (**b**) Metallographic section and appearance of the ball.

The ball diameters and thicknesses of the iron nitride layer and porous zone, observed after nitriding and annealing, are shown in Table 3.

Table 3. Sample geometry after nitriding and annealing.

Grade Steel	Sample No.	D_K (mm)	g_{mp} (μm)	g_{por} (μm)
AISI 1010	1A	3969	29 ± 1	10 ± 1
AISI 1010	11	3969	30 ± 1	12 ± 1
AISI 1085	5A	5.0	25 ± 1	10 ± 1
AISI 1085	15	5.0	25 ± 1	10 ± 1

g_{mp}—nitride layer thickness, g_{por}—thickness of the porous zone, and D_K—ball diameter.

The results of the structural X-ray examination, including the phase composition of the WL as well as the network parameters of the identified iron nitrides and their percentages in WL after nitriding and annealing, are presented in Table 4.

Table 4. The phase composition of the WL and network parameters of iron nitrides.

Grade Steel	Sample No.	PC WL	LP Fe$_4$N (Å) $a = b = c$	LP Fe$_{2-3}$N (Å) $a = b$	c	PC WL (%) Fe$_4$N	Fe$_{2-3}$N
AISI 1010	1A	Fe$_4$N-γ'; Fe$_{2-3}$N-ε	3.7986	4.6932	4.3783	4 ± 2	95 ± 2
AISI 1010	11	Fe$_4$N-γ'	3.8007	-	-	97 ± 2	-
AISI 1085	5A	Fe$_4$N-γ'; Fe$_{2-3}$N-ε	3.7978	4.6842	4.3750	13 ± 2	87 ± 2
AISI 1085	15	Fe$_4$N-γ'; Fe$_{2-3}$N-ε	3.7963	4.6998	4.3817	39 ± 2	61 ± 2

g_{mp}—WL thickness, g_{por}—thickness of the porous zone, LP—lattice parameter, and PC WL—white-layer phase composition.

2.3. Magnetic Resonance Experiment

FMR absorption spectra were recorded in the temperature range of 82–300 K using the conventional X-band spectrometer ELEXSYS E500 (BRUKER, Billerica, MA, USA) operating at f = 9.46 GHz and p = 0.62 mW of microwave power. The first derivative of the absorption spectrum was recorded as a function of the applied magnetic induction, B, which ranged from 0 to 1.4 T. The FMR susceptibility, χ_{FMR}, was calculated as a double integral of the FMR absorption spectrum depending on the temperature. The square of magnetic moment, proportional to the $\chi_{FMR} \times$ T product, and the effective position of the FMR resonant line, g_{eff}, were also analyzed. g_{eff} values were calculated from the B_{res} resonant field and the f_{res} resonant frequency using the resonance condition (g_{eff} = 71.44773 f_{res}(GHz)/B_{res}(mT)).

Static magnetic susceptibility measurements were made with the MPMS-XL7 SQUID (superconducting quantum interference device) magnetometer. Measurements were recorded in the temperature range of ~50 K to room temperature at a magnetic field of 100 Oe. Hysteresis loops were also measured to find other magnetic properties, such as H_c, coercive field, and B_r, remnant parameters.

3. Results

3.1. Phase Transformations of Iron Nitrides

On the AISI 1010 steel (sample 1A), a layer of iron nitrides with a thickness of g_{mp} = 29 $\pm$ 1 µm was formed in the nitriding process (see Table 3). In this layer, one can distinguish a lighter zone with poor etching on the substrate and a porous etching zone on the surface with a thickness of g_{por} = 10 $\pm$ 1 µm (Figure 2a).

On the diffractogram of the surface of the nitrided sample (1A), the dominant phase in the layer was the ε phase; all the characteristic lines from the ε phase were recorded in the range of the 2Θ angles studied (Figure 2b). The low intensity and the absence of the characteristic lines of the γ' {200} and γ' {111} planes indicate that the amount of the γ' phase in the layer did not exceed 5% of the mass (see Table 4).

After the annealing process, the steel structure (sample 11) did not change significantly. In the microstructure, a layer of iron nitrides with a thickness of g_{mp} = 30 $\pm$ 1 µm was visible. The thickness of the light, poorly digestible zone decreased slightly; the thickness of the porous zone (g_{por}) increased to 10 $\pm$ 1 µm, and its porosity also increased (Figure 2c). The process significantly changed the phase composition of the iron nitride layer. After the nitriding process, the iron nitride layer was a mixture of ε and γ' phases; after the annealing process, as a result of denitrification, the iron nitride layer was transformed into a single-phase γ' layer. On the diffractogram of the surface of sample 11, only two characteristic lines of γ' {111} and γ' {200} and characteristic lines of the substrate, α-Fe {110}, were identified (Figure 2d). As a result of the transformation, the lattice parameters of the formed γ'-Fe$_4$N nitride also increased (Table 4). The appearance of a characteristic line of the α-Fe substrate may be associated with the disappearance of part of the nonporous zone of the iron nitrides.

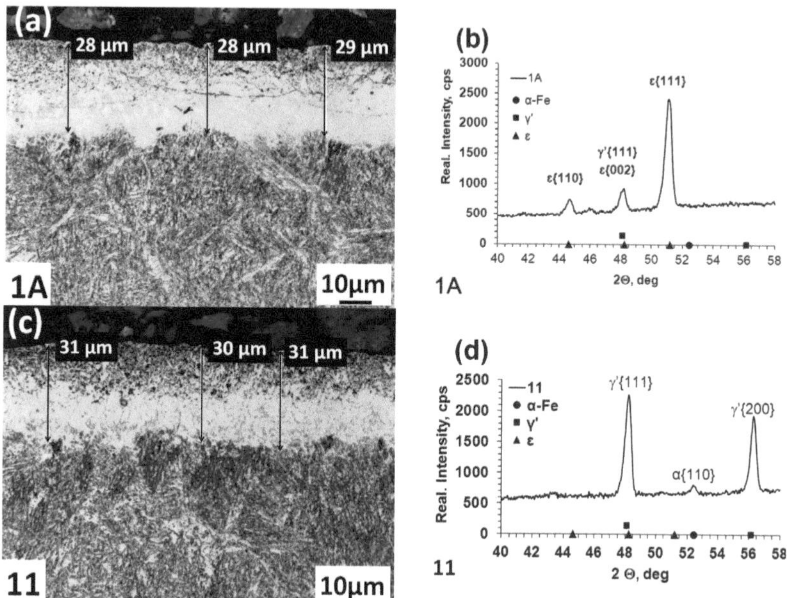

Figure 2. Microstructure of the nitrided layer on AISI 1010 steel (sample 1A) (**a**) and nitrided and annealed steel (sample 11) (**c**). Diffractograms of the surface of nitrided sample (sample 1A) (**b**) and nitrided and annealed sample (sample 11) (**d**).

On the AISI 1085 steel (sample 5A), a layer of iron nitrides with a thickness of $g_{mp} = 25 \pm 1$ μm was formed in the nitriding process (Table 3); in this layer, a lighter zone with poor etching on the substrate and a more etched porous zone with a thickness of $g_{mp} = 10 \pm 1$ μm on the surface can be distinguished (Figure 3a). On the diffractogram of the surface of the nitride sample (5A), the predominant phase in the nitride layer was the ε phase (Figure 3b); the content of the ε phase in the iron nitride layer was 87% (Table 4). All the characteristic lines of the ε phase in the range of the 2Θ angles were tested. The low intensity of the characteristic line of the γ' {111} plane and the low intensity of the characteristic line of the γ' {200} plane indicate that the amount of the γ' phase in the layer did not exceed 15% by weight.

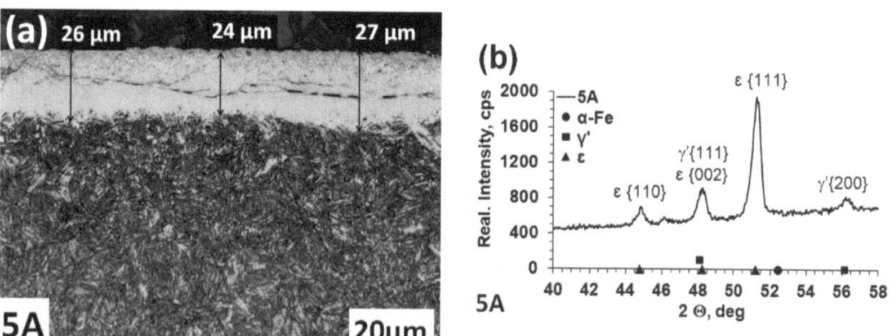

Figure 3. *Cont.*

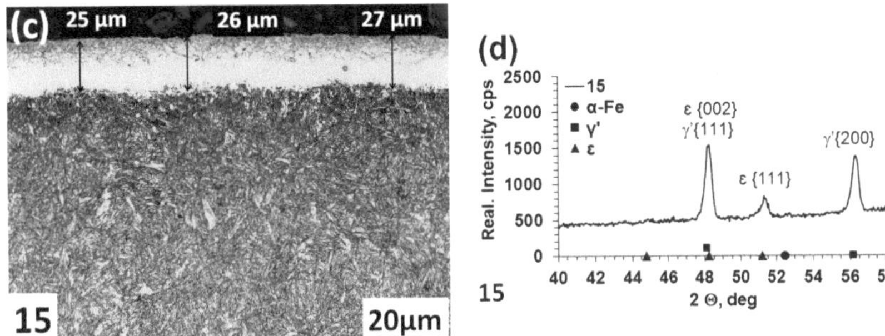

Figure 3. Microstructure of the nitrided layer on AISI 1085 steel (sample 5A) (**a**) and nitrided and annealed steel (sample 15) (**c**). Diffractograms of the surface of nitrided sample (sample 5A) (**b**) and nitrided and annealed sample (sample 15) (**d**).

After the annealing process, the steel structure (sample 15) did not change significantly. In the microstructure, there was a layer of iron nitrides with a thickness of $g_{mp} = 25 \pm 1$ μm (Table 3). The thickness of the light, poorly etched zone decreased slightly, while the thickness of the porous zone (g_{por}) increased to 12 ± 1 μm (Figure 3c). The degree of porosity of the zone increased. The process significantly changed the phase composition of the iron nitride layer. After the nitriding process, the iron nitride layer was a mixture of ε and γ' phases. During annealing, only part of the ε phase evolved into the γ' phase according to the $\varepsilon \rightarrow \gamma'\varepsilon + N\uparrow$ reaction, so the iron nitride layer was a mixture of $\varepsilon + \gamma' + \gamma'\varepsilon$ phases. In the iron nitride layer, next to the γ' phase formed by the nitriding process, there was also a $\gamma'\varepsilon$ phase formed as a result of the ε phase transformation (Figure 3d). X-ray studies of sample 15 found that the characteristic line of the ε {110} plane disappeared. The intensity of the ε {111} line decreased significantly; the intensity of the γ' {200} line and the intensity of the lines of the γ' {111} and ε {002} planes increased. In the doublet of these phases, the number of γ' phases increased. As a result of the ε-Fe$_{2-3}$N phase transformation, the content of the γ'-Fe$_4$N phase in the white layer increased. The lattice parameters of γ' phase increased slightly; the lattice parameters of the ε phase decreased (Table 4).

Figure 4 shows the change in the mass of the annealed samples after the nitriding process. The disappearance of the ε phase and the partial dissociation of the γ' phase ($\gamma' \rightarrow \alpha$-Fe(N)) explain the large weight loss in sample 11 (Figure 2d). The incomplete phase transformation of ε to γ' explains, in turn, the small loss of mass in sample 15 (Figure 3).

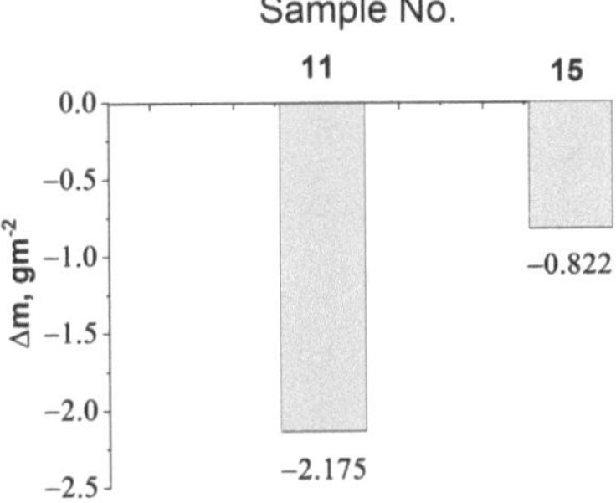

Figure 4. Change in sample weight, $\Delta m/m^2$ (samples 11 and 15), after annealing.

3.2. FMR Spectra

Investigations of the magnetic properties of AISI 1010 steel balls were carried out earlier in this paper [27]. However, these balls were subjected to nitriding processes under different thermodynamic conditions and ball diameters. The results of these and current studies confirm the independence of the conclusions about the magnetic properties of the balls with respect to their materials and geometric sizes. The magnetic properties of 1085 steel balls were analyzed by us for the first time. By analyzing the general shape of the resonant lines from Figures 5 and 6, it can be seen that "unmodified" (SS) ball materials usually have a very complex, wide, and intense FMR signal observed over the entire available temperature range of 50–300 K, which is far from the standard Lorentzian function (see Figures 5a–c and 6a,b). The AISI 1085 ball samples (SS—initial sample, 5A—nitrided sample, and 15—nitrided and annealed sample) had a diameter of 5 mm and were not suitable for measuring the temperature dependencies of the FMR spectra (diameter was too large in relation to the measuring chamber). Therefore, Figure 6a shows the FMR spectra of a ball cut from the same material but with a different diameter: AISI 1085_SS_2.5 mm. Figure 6b shows only the FMR spectra of the AISI_1085_5 mm samples measured at room temperature. The general characteristics of the spectra shown in Figure 6a correspond to those shown in Figure 6b.

Let us assume for our samples the correctness of the model of isolated Fe ions. The asymmetry of the FMR resonant signal can be explained by the very high conductivity of steel, leading to the formation of skin currents in the FMR experiment. Such a phenomenon was first described by Dyson [24] and was studied in further FMR experiments [25,26]. According to this theory, the symmetric Lorentz line is distorted due to the uneven distribution of the microwave field in the sample. The level of asymmetry, among many factors, depends on the relationship between the size of the sample and the depth of the skin current. The contribution of the asymmetric Dyson function to the overall FMR line is expected to be weaker for samples with a smaller diameter. Thermal treatment (nitriding, annealing) leads to an increase in the homogeneity of the magnetic material, which can be inferred from the increase in the symmetry of the observed FMR signal compared to the signal of the unmodified samples.

In the a–c inserts in Figure 5, the temperature dependencies of the FMR spectra of the AISI 1010 samples are shown for the initial sample (SS), the nitrided sample (1A), and the nitrided and annealed sample (11), respectively. As you can see, the shape and intensity of the FMR signal changed significantly when another process was used (compare Figure 5a–c). Moreover, it changed with increasing temperature.

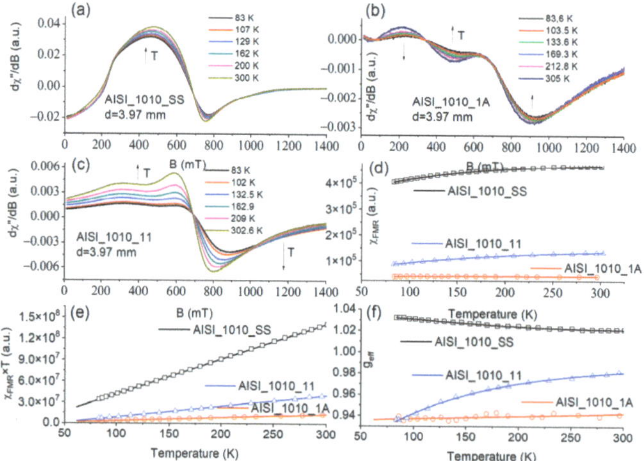

Figure 5. FMR spectra of the AISI_1010 sample; d = 3.97 mm. (**a–c**) Temperature dependence of FMR spectra of initial sample (SS), nitrided sample (1A), and nitrided and annealed sample (11), respectively; (**d–f**) temperature dependencies of the FMR magnetic susceptibility, χ_{FMR}; magnetic moment square, $\chi_{FMR} \times T$; and FMR signal resonant position, g_{eff}, respectively.

The d–f inserts in Figure 5 show the temperature dependencies of the FMR magnetic susceptibility, χ_{FMR}; the magnetic moment square, $\chi_{FMR} \times T$; and the FMR signal resonant position, respectively. The FMR magnetic susceptibility of all the samples (d) increased with temperature, although higher values were observed for the SS sample. This behavior was completely inconsistent with the Curie–Weiss relationship expected for simple paramagnetic species. The magnetic moments (e) decreased as the temperature decreased, suggesting antiferromagnetic interactions dominating the magnetic behavior of the samples. The resonant position of the FMR line (f) moved towards lower values for both of the nitrided and annealed samples.

The inserts in Figure 6a,c–e show the same relationships as those in Figure 5, but the magnetic susceptibility, magnetic moment, and resonance position are shown for the AISI 1085_SS_2.5 mm sample and not the AISI 1085_SS_5 mm sample. The FMR spectra of the second type of samples (SS, 5A, and 15) measured at T = 295 K are shown in Figure 6b.

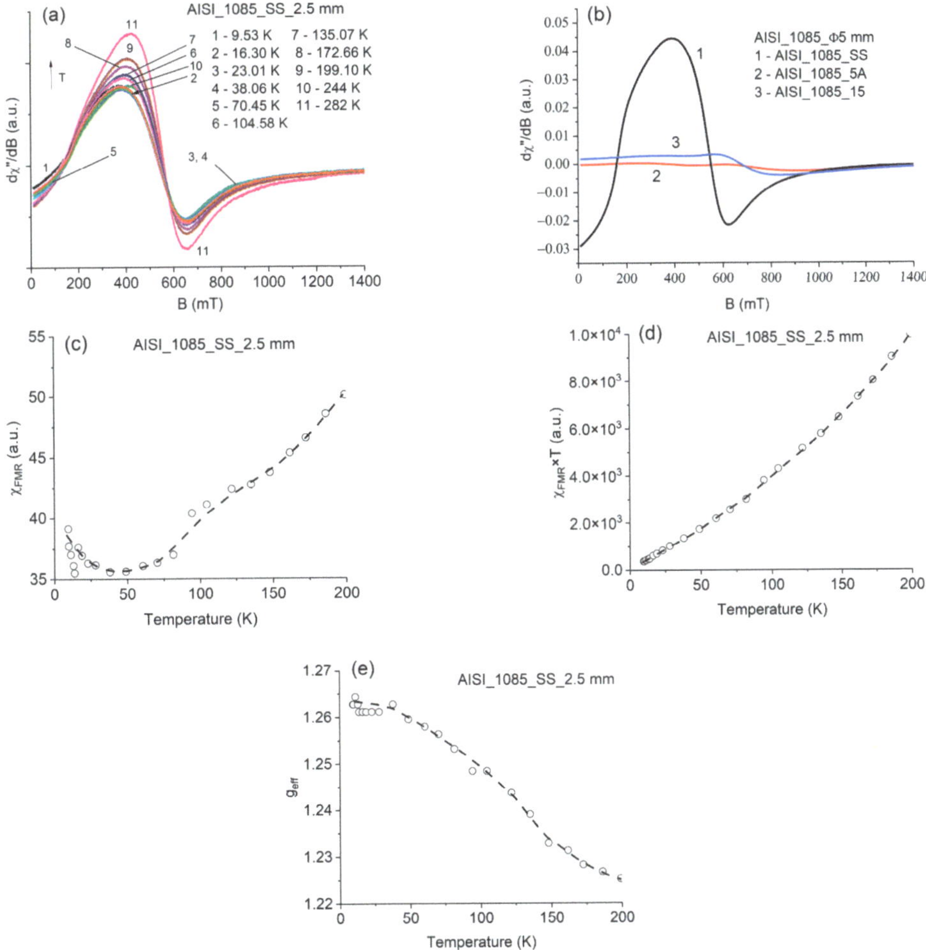

Figure 6. (**a**) The FMR spectra of the AISI 1085_SS_2.5 mm sample at several temperatures, (**b**) the FMR spectra of AISI 1085_SS_5 mm samples (SS, 5A, and 15) measured at room temperature, (**c**) χ_{FMR} of the AISI_11085_SS_2.5 mm sample, (**d**) $\chi_{FMR} \times T$ of the AISI_11085_SS_2.5 mm sample, and (**e**) g_{eff} of the AISI_11085_SS_2.5 mm sample.

Although the above temperature relationships (Figures 5 and 6) were obtained only in the range of 50–300 K, our previous tests in the range of 3–300 K [27] on the same samples tested for something different than the thermodynamical conditions confirmed the above conclusions. The question arises whether and how the nitriding process physically affected the magnetic properties of the tested steel ball samples. As can be seen in Figures 5 and 6, the nitriding process introduced a broad component into the FMR spectra. This component may be the result of the growth of a new phase, e.g., Fe_4N-γ' (see Table 3), which modified the surface of the ball at a different depth. In addition, the thermal nitriding processes led to a change in the shape of the FMR signal. This refers to various conditions, including the relaxation processes between the magnetic centers. On the one hand, the FMR signals recorded for the samples undergoing nitriding processes were transferred to higher magnetic fields, but, at the same time, they became more symmetrical and closer to the shape of the Lorentz line. This clearly suggests that both the temperature and the time of exposure to the nitrogen atmosphere had an impact on the magnetic properties of the tested steel materials. The observed shift of the resonant lines (g_{eff}) towards higher fields can be attributed to the decrease in the electron conductivity of the samples after the nitriding process. The nitriding process generally led to a decrease in the free-electron asymmetric Dysonian component in the FMR spectra. Thus, the general FMR signal consisted of a free-electron FMR signal (Dysonian-like) and a Lorentzian-like FMR signal assigned to the magnetic complexes.

3.3. Magnetic Susceptibility Measurements

Figure 7a, b shows the magnetization measurements of the initial samples, SS, of the AISI 1010 and AISI 1085 steels in a magnetic field of 100 Oe. As can be seen, the samples are characterized by high magnetic anisotropy at lower temperatures and an unusual dependence on temperature over the entire temperature range of 3–300 K. This confirms the previous FMR observations (see Figures 5d and 6c). Magnetic anisotropy tended to disappear at a temperature of about 300 K. Measurements were made in the FC and ZFC modes. For the tested samples, their magnetization was relatively weaker when the tests were carried out in a stronger magnetic field. Moreover, the magnetic susceptibility of this type of steel appears to be dependent on the carbon content.

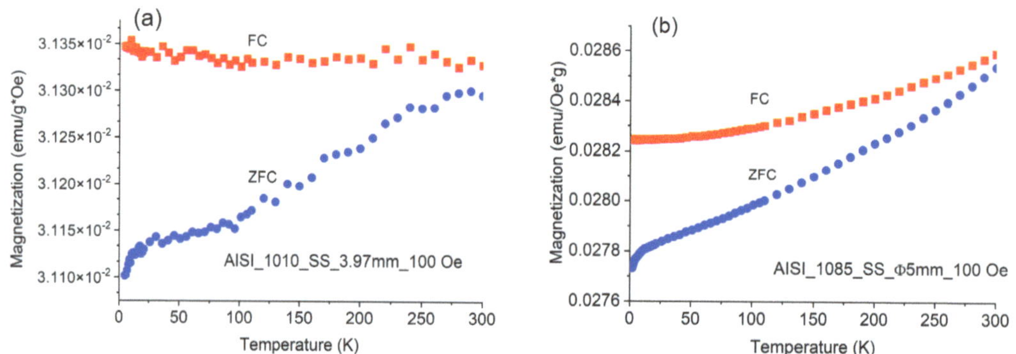

Figure 7. (**a**). Magnetization of the AISI 1010_SS sample over a temperature range of 3–300 K in a magnetic field of H = 100 Oe. (**b**) Magnetization of the AISI 1085_SS sample over a temperature range of 3–300 K in a magnetic field of H = 100 Oe.

Thus, the results obtained from the SQUID measurements, recording changes in magnetization, confirm the remarkable increase in magnetization with increasing temperature previously observed in the FMR measurements.

Figure 8 shows the hysteresis loops measured at 295 K. As you can see, they were very thin. They had a low coercive field, H_c, and low remnant parameters, B_r. They did not differ much from each other. Figure 8 indicates that all the tested samples revealed (anti)ferromagnetic interactions, regardless of the type of process, over the temperatures of 15–295 K. Only minor differences were observed for different processes. An increase in magnetization, as well as in FMR magnetic susceptibility, along with temperature can be observed in the antiferromagnetic structure. According to previous reports on a similar type of steel [28], we are dealing with a magnetic material, where a minority of ferromagnetic objects appear as an imperfection of the antiferromagnetic structure. Thus, the increase in temperature destroyed the antiferromagnetic order, and the overall magnetization increased. A similar behavior of magnetization with temperature dependence was observed in other materials containing Fe [29,30], where the increase in magnetization was explained by the formation of Fe^{2+} ions in a structure made of Fe^{3+}.

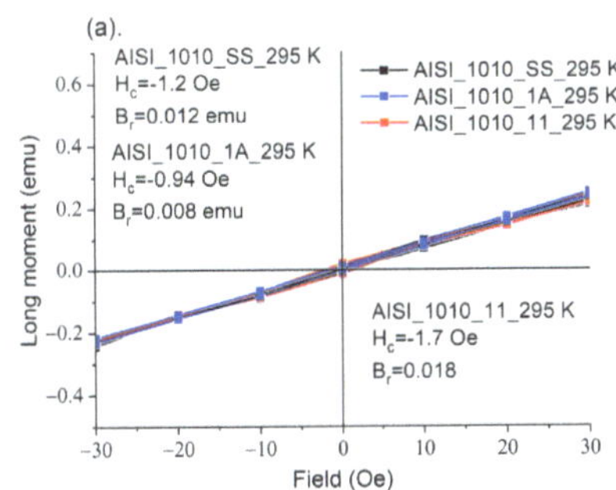

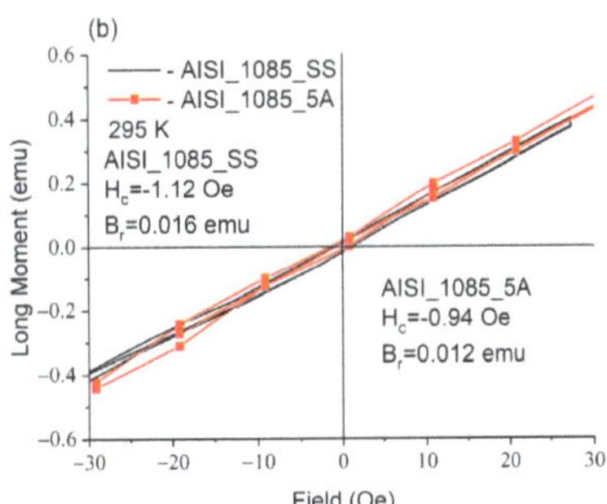

Figure 8. (**a**) Hysteresis loops of the AISI 1010 samples and (**b**) hysteresis loops of AISI 1085 samples; T = 295 K.

4. Conclusions

Our research shows that the iron nitrides formed during nitriding dissociated during annealing at 520 °C in an N_2/Ar atmosphere at 150 Pa. The ε phase was the first to dissociate according to the reaction $\varepsilon \rightarrow \gamma' + N_2\uparrow$. After complete ε phase decomposition, the γ' phase was also decomposed according to reaction $\gamma' \rightarrow \alpha Fe(N) + N_2\uparrow$. The nitrogen released during the dissociation of the iron nitrides caused a loss of mass in the annealed samples. The nitrogen released during the decomposition of the iron nitrides did not change the thickness of the iron nitride layer.

The FMR technique was sensitive enough to record the differences in the physical properties of the AISI steels subjected to different types of nitriding and annealing processes. The samples with a higher carbon content had more symmetrical FMR signals. This was assigned to the share of free electrons in the FMR line shape. Free electrons were observed as a Dyson line in the FMR spectra. In addition to the above line, the usual Lorentz line was observed and was attributed to iron complexes.

Our innovative methodology for investigating the changes in the magnetic properties of steel by observing the mechanical changes in the WL thickness combined with observations of the temperature dependencies of the FMR spectra of the spherical balls turned out to be a valid methodology independent of the type of steel tested; the geometric dimensions of the balls; and the heat treatment processes, including nitriding.

Author Contributions: Conceptualization, S.M.K., J.M. and T.F.; methodology, S.M.K. and J.M.; software, A.D. and G.L.; validation, J.M., S.M.K. and T.F.; formal analysis, S.M.K., J.M., T.F. and G.L.; investigation, S.M.K., A.D., T.F. and G.L.; resources, T.F.; data curation, S.M.K., A.D., H.F. and G.L.; writing—original draft preparation, S.M.K. and J.M.; writing—review and editing, S.M.K. and J.M.; visualization, S.M.K. and J.M.; supervision, S.M.K.; project administration, S.M.K., J.M. and T.F.; funding acquisition, S.M.K., A.D. and T.F. All authors have read and agreed to the published version of the manuscript.

Funding: This research received no external funding.

Data Availability Statement: The data can be found in the West Pomeranian University of Technology and Częstochowa University of Technology.

Conflicts of Interest: The authors declare no conflict of interest.

References

1. Borgioli, F. The Corrosion Behavior in Different Environments of Austenitic Stainless Steels Subjected to Thermochemical Surface Treatments at Low Temperatures: An Overview. *Metals* **2023**, *13*, 776. [CrossRef]
2. Godec, M.; Ruiz-Zepeda, F.; Podgoenik, B.; Donik, C.; Kocijan, A.; Balantic, D.A.S. The influence of the plasma-nitriding temperature on the microstructure evolution and surface properties of additive-manufactured 18Ni300 maraging steel. *Surf. Coat. Technol.* **2022**, *4333*, 12808. [CrossRef]
3. Sah, J.V.; Dwivedi, P.K.; Mukherjee, S.; Jhala, G.; Joseph, A. Influence of γ_N' and ε_N' phases on the properties of AISI 304L after low-temperature plasma nitrocarburizing. *J. Vac. Sci. Technol. A* **2023**, *41*, 33101. [CrossRef]
4. Hono, K.; Sepehri-Amin, H. Strategy for high-coercivity Nd–Fe–B magnet. *Scr. Mater.* **2012**, *67*, 530–535. [CrossRef]
5. Gupta, P.; Fiedler, H.; Rubanov, S.; Kennedy, J. Magnetisation and magnetic anisotropy of ion beam synthesized iron nitride. *J. Magn. Magn. Mat.* **2021**, *517*, 167388. [CrossRef]
6. Zhai, Y.; Zhu, D.; Duan, S.; Luo, F. Novel $Fe_{3-4}N@FCI$ particles with improved microwave absorption and antioxidation properties prepared by surface nitridation method. *Chem. Phys. Lett.* **2020**, *755*, 137803. [CrossRef]
7. Wojciechowski, P.; Lewandowski, M. Iron Nitride Thin Films: Growth, Structure, and Properties. *Cryst. Growth Des.* **2022**, *22*, 4618–4639. [CrossRef]
8. Yamaguchi, T.; Sakita, M.; Nakamura, M.; Kobira, T. Synthesis and Characteristics of Fe_4N Powders and Thin Films. *J. Magn. Magn. Mater.* **2000**, *215–216*, 529–531. [CrossRef]
9. Liu, J.; Guo, G.; Zhang, F.; Wu, Y.; Ma, B.; Wang, J.-P. Synthesis of α''-$Fe_{16}N_2$ Ribbons with a Porous Structure. *Nanoscale Adv.* **2019**, *1*, 1337–1342. [CrossRef]
10. Wang, X.; Zheng, W.T.; Tian, H.W.; Yu, S.S.; Xu, W.; Meng, S.H.; He, X.D.; Han, J.C.; Sun, C.Q.; Tay, B.K. Growth, Structural, and Magnetic Properties of Iron Nitride Thin Films Deposited by DC Magnetron Sputtering. *Appl. Surf. Sci.* **2003**, *220*, 30–39. [CrossRef]
11. Zheng, Q.; Yu, M.; Wen, W.; Liu, S.; Liang, X.; Wang, C.; Dai, Y.Y.; Xu, Y. Porous and flake-like γ' -$Fe_4N@iron$ oxides with enhanced microwave absorption performance. *Ceram. Internat.* **2021**, *47*, 8315–8321. [CrossRef]
12. Jiang, H.; Tao, K.; Li, H. The thermostability of the $Fe_{16}N_2$ phase deposited on a GaAs substrate by ion-beam-assisted deposition. *J. Phys. Condens. Matter* **1994**, *6*, L279. [CrossRef]
13. Yurovskikh, A.S.; Kardonina, N.I.; Kolpakov, A.S. Phase transformations in nitrided iron powders. *Met. Sci. Heat Treat.* **2015**, *57*, 507–514. [CrossRef]
14. Frączek, T.; Michalski, J.; Kucharska, B.; Opydo, M.; Ogorek, M. Phase transformations in the nitrided layer during annealing under reduced pressure. *Arch. Civ. Mech. Eng.* **2021**, *21*, 48. [CrossRef]
15. Liapina, T.; Leineweber, A.; Mittemeijer, E.J.; Kockelmann, W. The lattice parameters of ε-iron nitrides: Lattice strains due to a varying degree of nitrogen ordering. *Acta Mater.* **2004**, *52*, 173–180. [CrossRef]
16. Gupta, M.; Tayal, A.; Gupta, A.; Gupta, R.; Stahn, J.; Horisberger, M.; Wildes, A. Iron and Nitrogen Self-diffusion in Non-Magnetic Iron Nitrides. *J. Appl. Phys.* **2011**, *110*, 123518. [CrossRef]
17. Lei, X.; Wang, J.; Peng, R.; Wang, W. The controllable magnetic properties of Fe_3N nanoparticles synthesized by a simple urea route. *Mat. Res. Bull.* **2020**, *122*, 110662. [CrossRef]
18. Robbins, M.; White, J.G. Magnetic properties of epsilon-iron nitride. *J. Phys. Chem. Solids* **1964**, *25*, 717–720. [CrossRef]
19. Kucharska, B.; Michalski, J.; Wójcik, G. Mechanical and microstructural aspects of C20-steel blades subjected to gas nitriding. *Arch. Civ. Mech. Eng.* **2019**, *19*, 147–156. [CrossRef]
20. Michalski, J.; Tacikowski, J.; Wach, P.; Lunarska, E.; Baum, H. Formation of single-phase layer of γ'-nitride in controlled gas nitriding. *Met. Sci. Heat Treat.* **2005**, *47*, 516–519. [CrossRef]
21. Arabczyk, W.; Pelka, R.; Kocemba, I.; Brzoza-Kos, A.; Wyszkowski, A.; Lendzion-Bieluń, Z. Study of Phase Transformation Processes Occurring in the Nanocrystalline Iron/Ammonia/Hydrogen System by the Magnetic Permeability Measurement Method. *J. Phys. Chem. C* **2022**, *126*, 7704–7710. [CrossRef]
22. Małdziński, L. *Termodynamiczne i Technologiczne Aspekty Wytwarzania Warstwy Azotowanej na Żelazie i Stalach w Procesach Azotowania Gazowego (Thermodynamic and Technological Aspects of the Production of a Nitrided Layer on Iron and Steels in Gas Nitriding Processes)*; Wydawnictwo Politechniki Poznańskiej: Poznań, Poland, 2002. (In Polish)
23. Holst, A.; Kante, S.; Leineweber, A.; Buchwalder, A. Mechanism of Layer Formation during Gas Nitriding of Remelted Ledeburitic Surface Layers on Unalloyed Cast Irons. *Metals* **2023**, *13*, 156. [CrossRef]
24. Dyson, F.J. Electron spin resonance absorption in metals. II. Theory of electron diffusion and the skin effect. *Phys. Rev.* **1955**, *98*, 349. [CrossRef]
25. Bojanowski, B.; Kaczmarek, S.M. Electron spin resonance of $FeVO_4$. *Mater. Sci. Pol.* **2014**, *32*, 188–192. [CrossRef]

26. Kaczmarek, S.M.; Fuks, H.; Berkowski, M.; Głowacki, M.; Bojanowski, B. EPR properties of concentrated $NdVO_4$ single crystal system. *Appl. Magn. Res.* **2015**, *46*, 1023–1033. [CrossRef]
27. Kaczmarek, S.M.; Michalski, J.; Leniec, G.; Fuks, H.; Frączek, T.; Dudek, A. Applying the FMR technique to analyzing the influence of nitriding on the magnetic properties of steel. *Materials* **2022**, *15*, 4080. [CrossRef] [PubMed]
28. Crangle, J.; Fogarty, A.; Taylor, M.J. Weak ferromagnetism in 'non-magnetic' austenitic stainless steel. *J. Magn. Magn. Mater.* **1992**, *111*, 255–259. [CrossRef]
29. Jian, Z.; Kumar, N.P.; Zhong, M.; Yemin, H.; Reddy, P.V. Structural, Magnetic and Dielectric Properties of $Bi_{0.9}Re_{0.1}FeO_3$ (Re = La, Sm, Gd and Y). *J. Supercond. Nov. Magn.* **2015**, *28*, 2627–2635. [CrossRef]
30. Borgioli, F.; Galvanetto, E.; Bacci, T. Low temperature nitriding of AISI 300 and 200 series austenitic stainless steels. *Vacuum* **2016**, *127*, 51–60. [CrossRef]

Article

Electronic Structure and Hardness of Mn_3N_2 Synthesized under High Temperature and High Pressure

Shoufeng Zhang [1], Chao Zhou [1], Guiqian Sun [1], Xin Wang [1,*], Kuo Bao [1], Pinwen Zhu [1], Jinming Zhu [1], Zhaoqing Wang [1], Xingbin Zhao [2], Qiang Tao [1], Yufei Ge [1] and Tian Cui [1,2,*]

[1] State Key Laboratory of Superhard Materials, College of Physics, Jilin University, Changchun 130012, China
[2] Institute of High Pressure Physics, School of Physical Science and Technology, Ningbo University, Ningbo 315211, China
* Correspondence: wang-xin@jlu.edu.cn (X.W.); cuitian@jlu.edu.cn (T.C.)

Abstract: The hardness of materials is a complicated physical quantity, and the hardness models that are widely used do not function well for transition metal light element (TMLE) compounds. The overestimation of actual hardness is a common phenomenon in hardness models. In this work, high-quality Mn_3N_2 bulk samples were synthesized under high temperature and high pressure (HTHP) to investigate this issue. The hardness of Mn_3N_2 was found to be 9.9 GPa, which was higher than the hardness predicted using Guo's model of 7.01 GPa. Through the combination of the first-principle simulations and experimental analysis, it was found that the metal bonds, which are generally considered helpless to the hardness of crystals, are of importance when evaluating the hardness of TMLE compounds. Metal bonds were found to improve the hardness in TMLEs without strong covalent bonds. This work provides new considerations for the design and synthesis of high-hardness TMLE materials, which can be used to form wear-resistant coatings over the surfaces of typical alloy materials such as stainless steels. Moreover, our findings provide a basis for establishing a more comprehensive theoretical model of hardness in TMLEs, which will provide further insight to improve the hardness values of various alloys.

Keywords: hardness; Mn_3N_2; electronic structure; high temperature and high pressure synthesis

Citation: Zhang, S.; Zhou, C.; Sun, G.; Wang, X.; Bao, K.; Zhu, P.; Zhu, J.; Wang, Z.; Zhao, X.; Tao, Q.; et al. Electronic Structure and Hardness of Mn_3N_2 Synthesized under High Temperature and High Pressure. *Metals* **2022**, *12*, 2164. https://doi.org/10.3390/met12122164

Academic Editors: Shinichiro Adachi, Francesca Borgioli and Thomas Lindner

Received: 9 November 2022
Accepted: 13 December 2022
Published: 16 December 2022

Publisher's Note: MDPI stays neutral with regard to jurisdictional claims in published maps and institutional affiliations.

1. Introduction

Hardness is a well-known concept for which several models have been proposed [1–4], but the internal physics of this property are not well understood. This limited understanding of hardness considerably hinders the design and knowledge of superhard materials. Since the idea of combining high-bulk-elastic-modulus transition metals and high-shear-modulus light elements was proposed [5–7], many transition metal light element (TMLE) compounds have been synthesized and measured [2,8–11]. For TMLEs such as ReB [12–14], WC [15], and TiN [16], most of their measured hardness values deviate from the theoretically predicted values. Therefore, widely used hardness models might not fully describe all the important physical factors in these systems.

For the application of typical alloys limited by their hardness and chemical stability of the surface, novel wear-resistant and rust-resistant coating materials are required [17,18]. Transition metal nitrides have been extensively studied owing to their excellent physical and chemical properties [19]. In 2019, we found that the interactions between metal atoms might benefit the hardness of transition metal nitrides (TMNs) and TMLEs in our study on WN and CrN [20]. Compared with the outer valence electrons of Cr and W, both the semi-filled d orbital and fully filled s orbital electron clouds of Mn would better improve the interactions between metal atoms. Although the latest results show that Mn_3N_2 is mechanically unstable [21], the hardness of Mn_3N_2 could not be obtained from the elastic model of bulk materials [13], which are good candidates for investigating hardness. Based on the hardness results obtained from experiments, the existing hardness theories could be improved.

Many TMNs have been synthesized using methods such as film deposition [22], magnetron sputtering [23], and hermetic sintering [24], and acceptable results have been achieved via investigation of the obtained samples. However, the precise measurement of hardness has not yet been achieved because of the absence of high-quality synthesized bulk samples. The high temperature and high pressure (HTHP) technique is an effective synthesis approach to prepare high-quality bulk samples, especially TMNs. The triple covalent bond of N-N is one of the strongest chemical bonds, and it is difficult to synthesize TMNs until a mature high-pressure metathesis reaction has been achieved [20,25,26]. Meanwhile, the application of high pressure is effective in preventing nitrogen from escaping the experimental system. Further, HTHP synthesis is an effective approach to prepare fine bulk TMN samples [27,28]. After obtaining high-quality bulk samples, the precisely measured hardness of Mn_3N_2 could be easily obtained.

In this study, we investigate the preparation and simulation of Mn_3N_2 samples, discuss the experimental and simulated results, and then provide our conclusions concerning the hardness evaluation of this material.

2. Experiment and Simulation Details

We prepared our samples with Mn and NaN_3 powders (both 99.95% pure) at a molar ratio slightly lower than 3:2. NaN_3 was in excess in the raw material to ensure a sufficient N_2 atmosphere. During the grinding process, it should be noted that NaN_3 exhibits a certain explosiveness under severe vibrations, and intense impacts should be avoided. The mixture was pressed into cylinders before being sealed in *h*-BN capsules, and the cylinders were assembled and subjected to HTHP with a cubic China-type SPD6 × 600T apparatus. The sample was compressed under 5 GPa, heated to 1800 °C, and held under this thermal condition for 30 min. Then, the heating current was stopped, and the pressure was released to ambient conditions over approximately 9 min.

$$Mn + NaN_3 \rightarrow Mn_3N_2 + Na \tag{1}$$

The elemental substance Na very easily reacts with water and oxygen in air to generate sodium ions. To obtain samples with less impurities and better density, the primary samples were ground to powder and washed with water to eliminate ionic Na impurities. The centrifuged and dried powder was then subjected to second sintering with China-type SPD6 × 600T, where the powder was held at 5 GPa and 700 °C for 20 min.

The samples were characterized by X-ray diffraction (XRD) for phase identification on a Rigaku SmartLab SED/teX Ultra250 X-ray diffractometer ($CuK_{\alpha 1}$ radiation, $\lambda = 1.5404$ Å) in the 2θ range of 10° to 90° with a step size of 0.01°, and the X-ray photoelectron spectrum (XPS) measurements were conducted after it was cleaned by Ar+ sputtering within 180 s. The surface morphology of the sample was obtained by scanning electron microscopy (SEM) and energy-dispersive spectrometry (EDS). HV-1000 ZDT microhardness equipment was used to test the hardness of the sample after the second sintering.

Here, structural relaxation and property calculations were performed within a density functional theory (DFT) framework and the projector-augmented wave (PAW) [29], as implemented in the Vienna ab initio simulation package (VASP) [30]. Perdew–Burke–Ernzerhof (PBE) [31] generalized gradient approximation was used to process the exchange–correlation potential. We set an energy cutoff of 600 eV for plane-wave basis and gamma-centered k mesh spacing as $2\pi \times 0.03$ Å^{-1} in the calculations. The forces between the atoms were converged within 0.01 eV/Å in the relaxation. Mn $4s$, $3d$, and $3p$ electrons as well as N $2p$ and $2s$ electrons were explicitly calculated and not included in the pseudopotential. Crystal Orbital Hamilton Population (COHP) [32] and Crystal Orbital Bond Index (COBI) [33] were calculated using the LOBSTER package.

3. Results and Discussion

The powder XRD and Rietveld fitting [34] of our samples are shown in Figure 1. Comparing the sample with the standard crystal card, the XRD spectrum of the sample fits that of Mn_3N_2. The samples are crystalline and single-phase, and the confidence factors wRp and Rp are less than 3%. The structure of the sample was studied from two directions to investigate the connections between the internal parts of the lattice. As is evident from the structure shown in Figure 1, six Mn atoms are connected around each N atom, and N atoms are connected around each Mn atom. The chemical formula of the crystal is Mn_3N_2, and the space group of the crystal is *I4/mmm*, $a = 2.9769(1)$ Å, $b = 2.9769(1)$ Å, and $c = 12.2646(5)$ Å. More information concerning these crystals is given in Table 1.

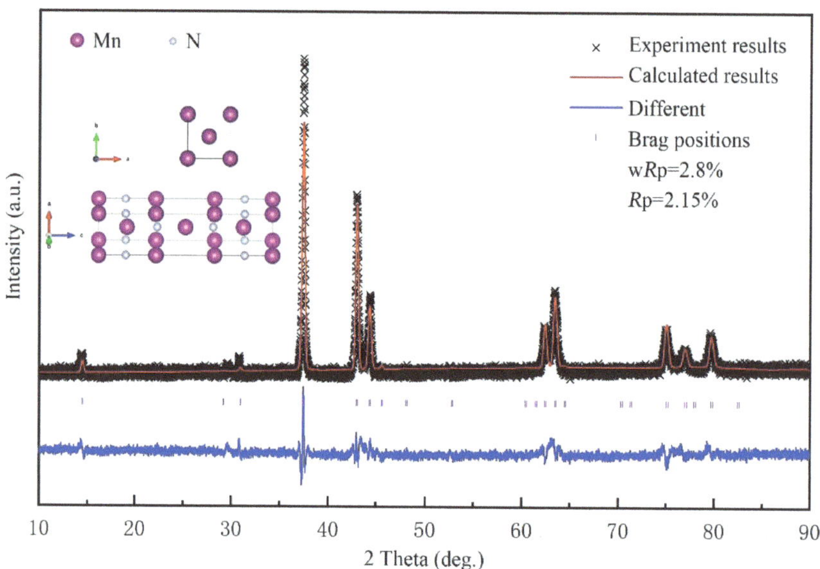

Figure 1. Powder (R3C4) XRD test pattern of crystal. × is the test value of the experiment, the red line is the calculated fitting value, and the blue line is the difference between the two.

Table 1. Experimental data and crystal details obtained from Rietveld refinement.

Identification Code	Mn_3N_2
Space group	*I4/mmm*
Unit cell dimensions (Å)	$a = 2.9769$ Å, $c = 12.2646$ Å, $a/b = 1$ $b/c = 0.2427$
Volume (Å^3)	108.69

The SEM evaluation of the morphology of the samples showed that our samples are compact and of sufficient quality for the hardness test, as shown in Figure 2. Here, (a,b) relate to the morphology of the samples synthesized during the first preparation, while (c,d) relate to the morphology of the samples sintered for a second time. Evidently, after the initial synthesis, the sample showed porosity and incompact characteristics, which may be due to the outgassing of NaN_3 during the initial reaction. In addition, after the initial synthesis of the sample, the crystallization degree of the sample was poor, and the crystallization was not obvious. After the secondary sintering treatment, the compactness of the sample was significantly improved, and no obvious pores were visible on the micro scale, which demonstrated that the samples met the standard for hardness measurements.

In addition, the density achieved after secondary sintering was approximately 20% higher than the primary synthesized samples. After secondary sintering, the density of the samples reached approximately 97%, which may improve the accuracy of the hardness test results.

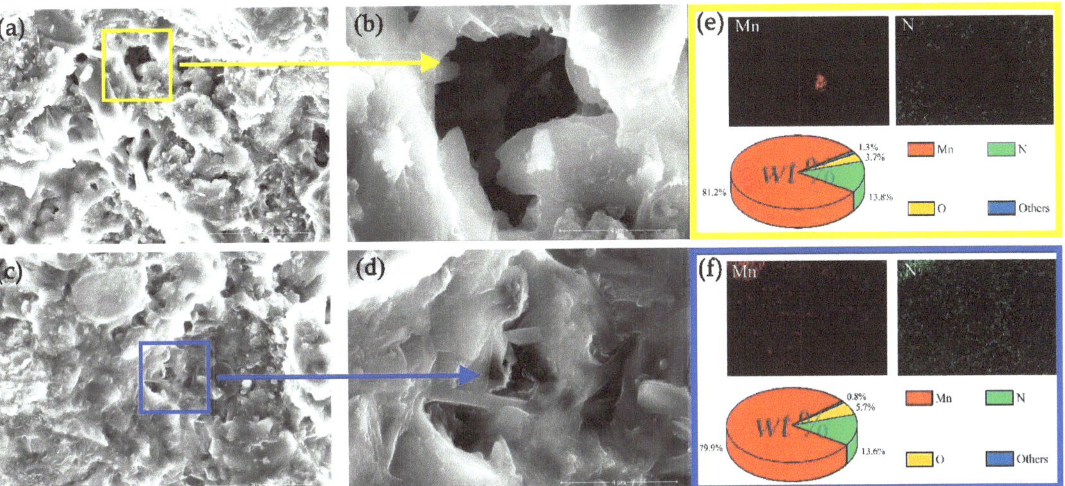

Figure 2. (**a–d**) show the low- and high-magnification SEM images of Mn_3N_2 bulk materials. (**a,b**) are the samples after primary synthesis, and (**c,d**) are the samples after secondary sintering. (**b**) is the magnification area of the yellow box in (**a**); (**d**) is the magnification area of the blue box in (**c**). (**e**) Element distribution diagram and element content diagram of yellow box of initial synthesis by EDS; (**f**) element distribution diagram and element content diagram of blue box of initial synthesis by EDS.

Meanwhile, the EDS analysis of the elemental distribution in the samples after secondary sintering indicated that the Mn and N atoms were more evenly distributed. Other elements were obviously reduced after secondary sintering. Considering the error range of the EDS analysis instrument, we believe that Na is no longer present in the samples after secondary sintering. However, a significant increase in O content was observed between the initial synthesis and secondary sintering samples. This increase in O content may be due to the grinding and washing of samples during secondary sintering.

To determine whether the sample was oxidized to form Mn-O, the XPS profiles were obtained, as shown in Figure 3. The N-1s spectrum and Mn-2p spectrum of Mn_3N_2 are plotted in Figure 3a,b, respectively. The fitted intensity demonstrates that the two peaks in the N-1s spectrum are located at binding energies of 397.18 and 397.28 eV. These two binding energies with a small difference reflect the binding state of N with Mn atoms at different positions in Mn_3N_2. The binding energies of 652.23 and 640.43 eV on the 2p orbital of metal Mn represent the split energy levels of Mn $2p^{1/2}$ and Mn $2p^{3/2}$, respectively. The binding energies of 638.73 and 652.43 eV show the change in the valence states of the Mn atoms, which clarifies the mechanism of electron transference. Therefore, the electrons of the Mn atoms correspond to different states when forming different bonds. The XPS results show that the Mn-N bond is the main bond and state in Mn_3N_2, and no O bonds are found, which indicates that this sample is antioxidative.

The Vickers hardness of the secondary sintered sample was evaluated, as shown in Figure 4. The applied load was increased from 0.49 N to 9.8 N. As the load increased, a downward trend was observed in the hardness. The trend slowed when the applied load reached 2.94 N and began to converge. As the external load was further increased, the hardness value slightly decreased but showed a generally moderate trend. Therefore, we believe that the intrinsic hardness of the sample was recorded under a 9.8 N load, and the hardness value is 9.9 GPa.

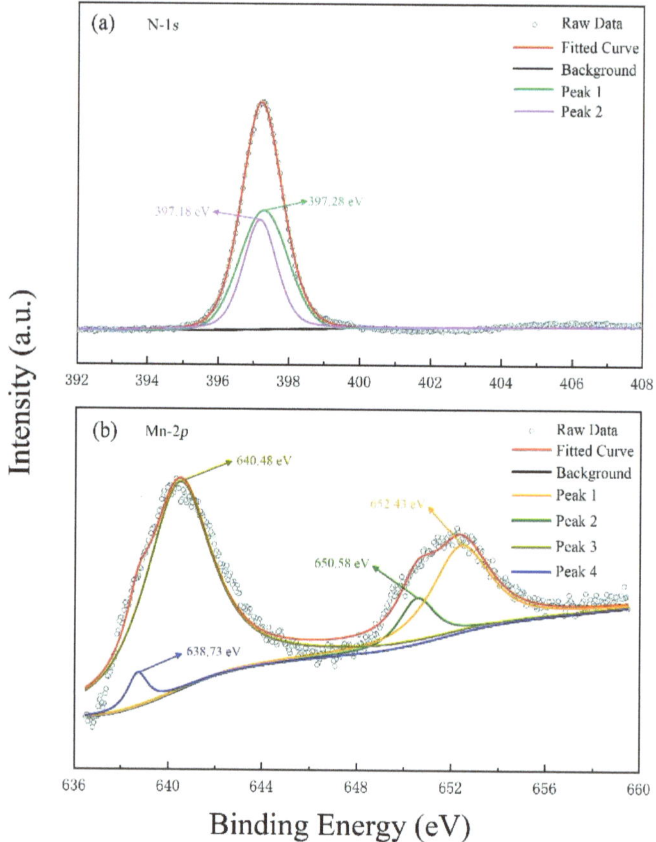

Figure 3. XPS spectra of the Mn_3N_2 sample. (**a**) N 1s spectrum with a binding energy of 392.0–408.0 eV; (**b**) Mn 2p spectrum with a binding energy of 636.48–659.48 eV.

The widely accepted hardness models by Gao et al. [2] and Guo et al. [3] are based on the assumption of the detailed analysis of chemical bonds. Guo further extended Gao's model by considering that the ionic bond in the crystal plays a major role in its hardness; therefore, we chose Guo's model to analyze the hardness of the sample. The following semi-empirical formula [3] was used:

$$Hv(\text{GPa}) = 1051 N_e^{2/3} d^{-2.5} e^{-1.191 f_i - 32.2 f_m^{0.55}}, \tag{2}$$

where N_e is the density of valence electrons in a single cell, $N_e = \frac{n_e}{v}$, n_e is the total number of valence electrons in a single cell, d is the distance between Mn and N in this system, f_i is the Phillips ionicity chemical bond, f_m is a defined factor of metal, $f_m = \frac{0.026 D_F}{n_e}$, and D_F is the number of electrons on the Fermi surface according to our total density of states (TDOS) calculation; that is, $D_F \approx 24.885$. Then, a hardness $Hv \approx 7.01$ GPa was obtained.

However, the measured Vickers hardness of Mn_3N_2 is 9.9 GPa, which is approximately 40% higher than the predicted value. There exists a noticeable underestimation guided by Guo's model. Thus, the electronic structures of Mn_3N_2 were investigated. Figure 5 provides the density of states (DOS) of the crystal. The TDOS at the Fermi level (E_f) indicated that Mn_3N_2 was a metal. Below the Fermi level, the TDOS for Mn_3N_2 peaks were in two energy regions. The first energy region, which was between -20 and -15 eV, was mainly induced by the *s* orbital bonding of N atoms with the *d* and *p* orbitals of Mn atoms. The second energy region, which was from -10 to 0 eV, was mainly due to the *p* orbital bonding of N atoms with the *d* orbitals of Mn. The contributions of the *p* orbitals of the Mn atoms were negligible. The *s* orbital of Mn was also ineffective for bonding. The major interaction in Mn_3N_2 was mainly the orbital hybridization of Mn and N atoms. At the Fermi level, the TDOS of Mn_3N_2 mainly comprised the *d* orbital electrons of Mn, and the electrons of N did not significantly contribute. Therefore, the *d* orbital electrons of the Mn atoms play a major role in this system, and the interaction of Mn–Mn contributes to the TDOS in Mn_3N_2.

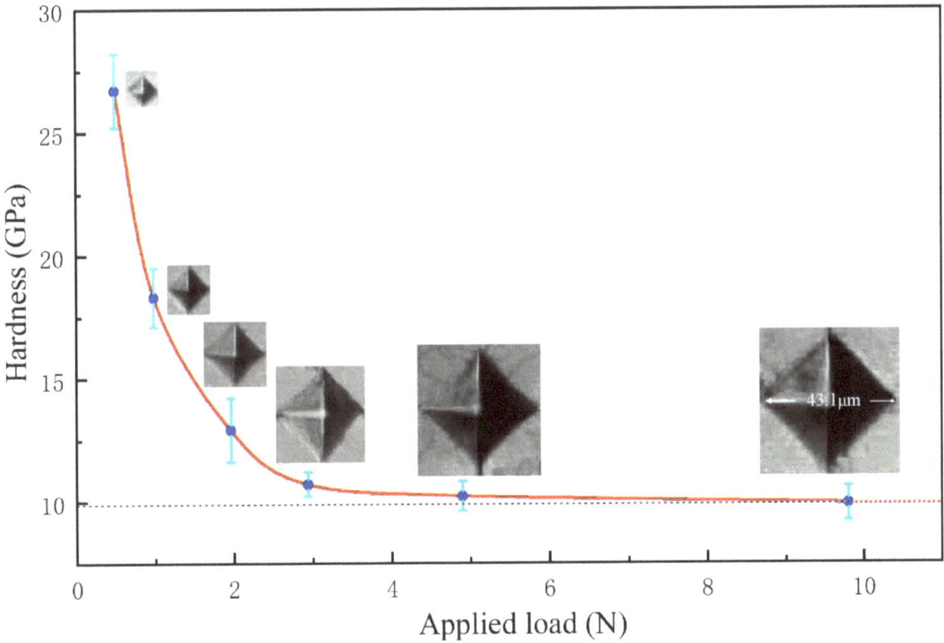

Figure 4. Vickers hardness for Mn_3N_2, with the applied load ranging from 0.49 N (low load) to 9.8 N (high load). Typical optical images of the indentation are shown in the inset.

Figure 6 shows a 2D plane of the crystal calculated via electronic localization function (ELF) and intercepted along different crystal plane directions. A strong interaction occurred between the N atoms and the nearby Mn atoms, and the electrons that induced this interaction were mainly concentrated around the N atoms, which indicates electron localization. Almost no electrons were localized between N–N in the crystal, which indicates the occurrence of no interactions between N–N and no N–N covalent bonding in the sample. Additionally, electron localization between Mn–Mn was not ignored. In Figure 6, there is a certain probability of occurrence for electrons between Mn atoms. An evident conductive path forms between the Mn atoms, indicating a strong interaction between these atoms. Additionally, there may be strong Coulombic interactions along this path. It is well known that Coulombic interactions are very strong and may strongly contribute to deformation resistance. According to these findings and the DOS analysis results, the Mn_3N_2 metal mainly originates from the *d* orbitals of the Mn atoms, while the conductive path in the sample was also the route of sample formation.

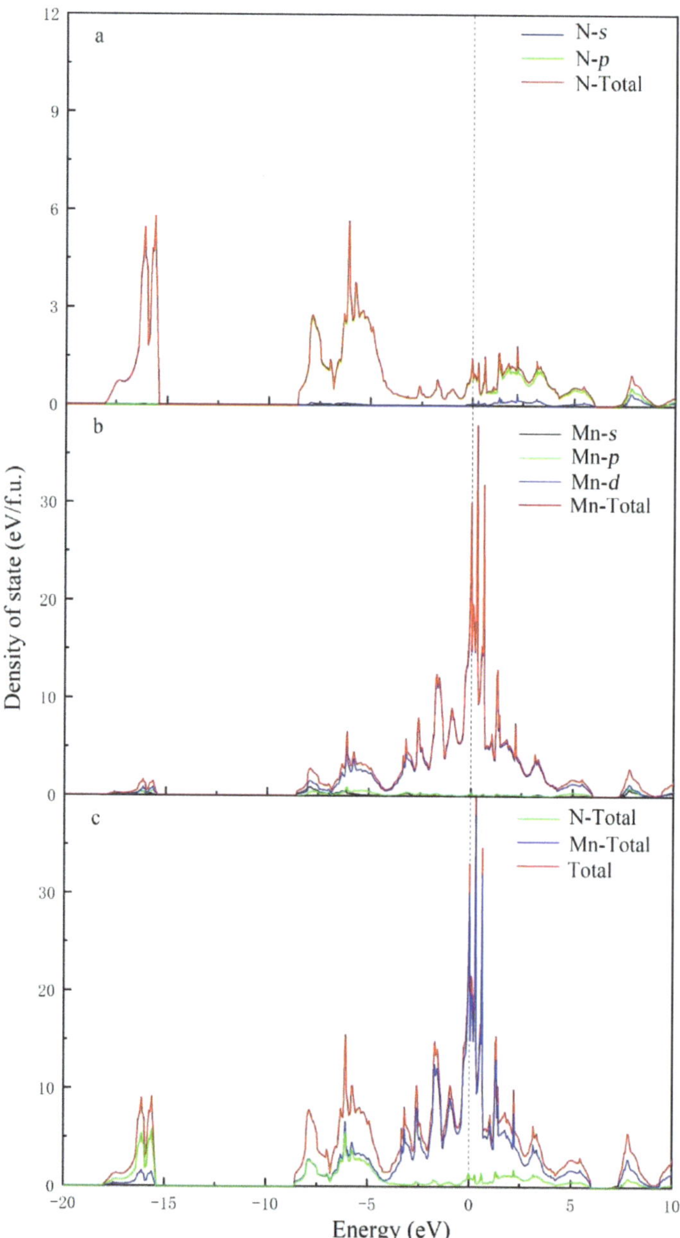

Figure 5. Total and partial density of states for Mn_3N_2. (**a**) The s and p orbitals, and TDOS of N atoms. (**b**) The s, p, and d orbitals, and TDOS of Mn atoms. (**c**) The TDOS of Mn atoms, the total TDOS, and the TDOS of N atoms.

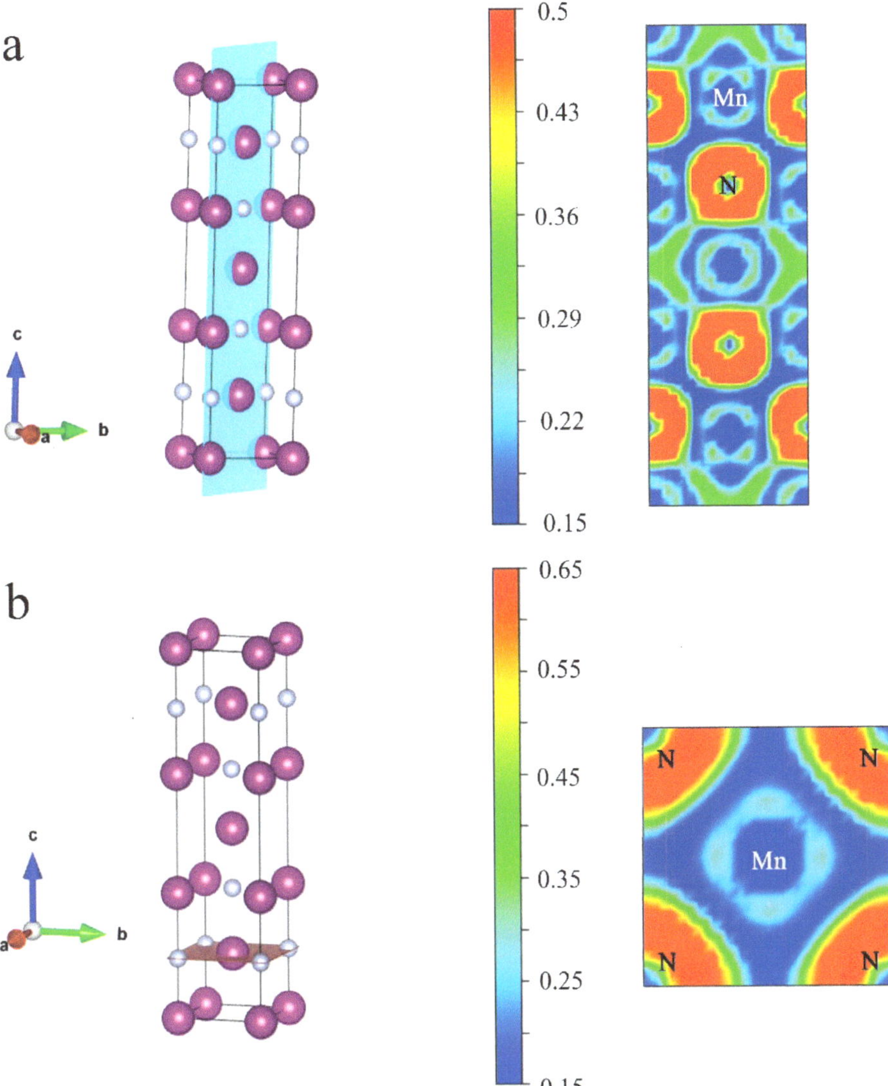

Figure 6. Electronic localization function for Mn_3N_2. (**a**) Cutting along the (110) crystal plane with cyan and (**b**) cutting along the (001) crystal plane with caramel. Purple balls are Mn atoms, white ball are N atoms.

The COHP results used to investigate the bond strength of each specific bond in the crystal are shown in Figure 7, and the -ICOHP results are shown in Table 2. Owing to the different distances of Mn atoms among the different positions in the crystal, the interaction between the first neighboring atoms was stronger than that between the other atoms. Therefore, we only considered the nearest and next-nearest Mn atoms, and the nearest neighbor atoms between N-N were also considered.

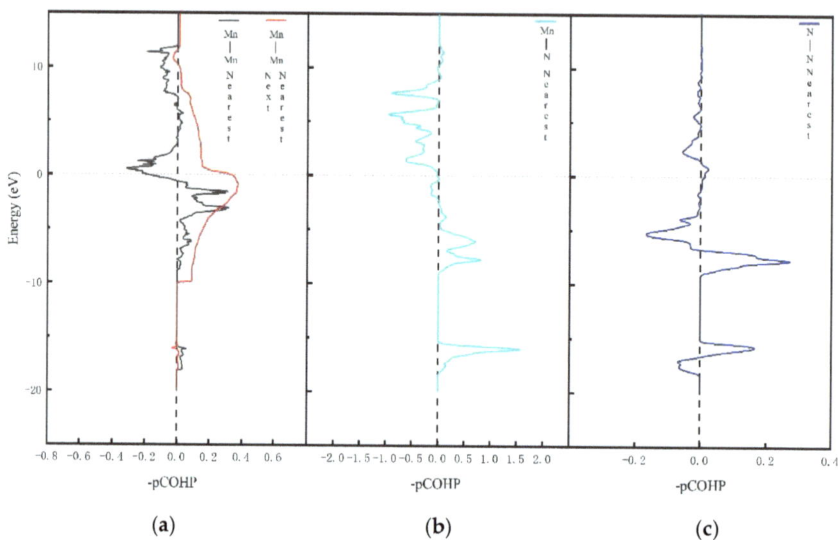

Figure 7. COHPs for Mn_3N_2. The negative and positive COHP values denote bonding and antibonding interactions, respectively. (**a**) shows the COHPs for Mn−Mn nearest and next nearest; (**b**) is the COHP for Mn−N nearest; (**c**) is the COHP for N−N nearest.

Table 2. -ICOHP of bonds in Mn_3N_2.

	Mn-N	Mn-Mn Nearest Mn-Mn Next-Nearest	N-N
−ICOHP (eV/pair)	3.0303	0.7319 0.5650	0.0975

The strength of the N-N bond in the crystal was near zero because the value of -ICOHP was only 0.097 eV/pair. The weak bonding of the N-N bond was mainly explained by the extremely long distance between these atoms such that the corresponding effect was difficult to achieve. Meanwhile, the DOS and ELF results showed that there was nearly no N–N covalent bond in the crystal, and the N-N covalent bonds could therefore be ignored in the subsequent analysis. The strongest bonds in the crystal were the ionic bonds between the Mn–N atoms, and the value of -ICOHP was 3.03 eV/pair. The bonding between Mn–N was extensive even at deep energy levels less than -15.0 eV, and almost no anti-bonding state was found below the Fermi surface, thereby indicating that the bonding between Mn–N was strong and the ionic bond of Mn–N played a crucial role in the crystal structure. The bonding between Mn–Mn in the crystal could also not be ignored. The -ICOHP values of Mn atoms at different positions and distances differed, but the approximate range was between 0.5 and 0.8 eV/pair. Moreover, most of the atoms were bonded below the Fermi level, which was conducive to the interactions between the Mn atoms. In the region of -10.0 to 0.0 eV, the bonding state between the Mn–Mn atoms was significant. Considering these results, we can provide a clear hypothesis regarding the bonding in the crystal. The Mn–N ionic bond plays a major role in the crystal hardness. Similar to metal crystals, electrons exist between Mn atoms, and the bonding state is strong. Moreover, no covalent N-N bonds are present in the crystal.

Different values between the theoretical hardness obtained using Guo's semi-empirical formula and the actual hardness obtained by Hv testing will likely be observed for numerous TMLEs. For Mn_3N_2, attention should be paid to the metal bonds originally considered helpless to the hardness of TMLEs when evaluating the hardness of these materials, as indicated by the results of the above investigations [2,3,35]. First-principle calculations showed that the interactions between the Mn atoms in the Mn_3N_2 sample play an important role in the large DOS. Therefore, when an external shear force is applied, the common electrons in this crystal may slip. At this time, the electrons between the Mn–Mn metal atoms exhibit covalent bond-like properties, which strengthen the bonding between the atoms, and consequently improve the hardness and resistance to the external shear stress. Therefore, the interaction between the metal atoms is similar to that between atoms in pure metal crystals. The evaluated material would undergo permanent deformation because of the interactions between the metal atoms. The hardness of pure metals is often low at approximately 1–2 GPa because the lack of strong ionic or covalent bonds in a metal limits the slipping region of electrons, which also leads to limited hardness improvement resulting from metal bonds in metal crystals.

4. Conclusions

Mn_3N_2 gives us a perspective to understand the difference between the predictions by the hardness models and the tested results. Here, we synthesized high-quality, single-phase bulk Mn_3N_2 samples using HTHP synthesis, and the measured Vickers hardness was 9.9 GPa. By comparing the predicted value of hardness model with the experimental value, we found that the actual hardness of the sample was higher than the theoretical prediction. We investigated the electronic structures of Mn_3N_2 using first-principle simulations and found that the d orbital electrons in Mn–Mn form a path for the current and can effectively improve the hardness of this crystal. Therefore, the commonly observed electron slippage between TM–TM enhances the hardness of the crystals owing to its covalent bonding-like properties, on the premise that ionic bonds act as a "skeleton" in the crystals of TMLEs. The underestimated impact of the TM–TM bonds on the hardness and methods for optimizing these bonds by introducing LE atoms may assist in the design of modern alloy materials with enhanced hardness values.

Author Contributions: Data curation, S.Z., G.S. and Z.W.; Formal analysis, X.W., K.B., P.Z., X.Z., Q.T., Y.G. and T.C.; Funding acquisition, T.C.; Resources, P.Z.; Software, C.Z. and J.Z.; Writing—original draft, S.Z.; Writing—review and editing, S.Z. All authors have read and agreed to the published version of the manuscript.

Funding: This research was funded by the National Key R&D Program of China (2018YFA0703404 and 2017YFA0403704), the National Natural Science Foundation of China (11774121 and 91745203), the National Key Research and Development Program of China (2016YFB0201204), and the Program for Changjiang Scholars and Innovative Research Team in University (IRT_15R23).

Data Availability Statement: The data that support the findings of this study are available from the corresponding author upon reasonable request.

Acknowledgments: Parts of calculations were performed in the High Performance Computing Center (HPCC) of Jilin University.

Conflicts of Interest: The authors declare no conflict of interest.

References

1. Chen, X.Q.; Niu, H.; Li, D.; Li, Y. Modeling hardness of polycrystalline materials and bulk metallic glasses. *Intermetallics* **2011**, *19*, 1275–1281. [CrossRef]
2. Gao, F.M.; He, J.; Wu, E.; Liu, S.; Yu, D.; Li, D.; Zhang, S.; Tian, Y. Hardness of covalent crystals. *Phys. Rev. Lett.* **2003**, *91*, 015502. [CrossRef] [PubMed]
3. Guo, X.J.; Li, L.; Liu, Z.; Yu, D.; He, J.; Liu, R.; Xu, B.; Tian, Y.; Wang, H.-T. Hardness of covalent compounds: Roles of metallic component and d valence electrons. *J. Appl. Phys.* **2008**, *104*, 023503. [CrossRef]
4. Li, Q.; Zhou, D.; Zheng, W.; Ma, Y.; Chen, C. Anomalous Stress Response of Ultrahard WBn Compounds. *Phys. Rev. Lett.* **2015**, *115*, 185502. [CrossRef]
5. Chung, H.Y.; Weinberger, M.B.; Yang, J.M.; Tolbert, S.H.; Kaner, R.B. Correlation between hardness and elastic moduli of the ultraincompressible transition metal diborides RuB_2, OsB_2, and ReB_2. *Appl. Phys. Lett.* **2008**, *92*, 261904. [CrossRef]
6. Weinberger, M.B.; Levine, J.B.; Chung, H.; Cumberland, R.W.; Rasool, H.I.; Yang, J.-M.; Kaner, R.B.; Tolbert, S.H. Incompressibility and Hardness of Solid Solution Transition Metal Diborides: $Os_{1-x}Ru_xB_2$. *Chem. Mater.* **2009**, *21*, 1915–1921. [CrossRef]

7.   Liang, Y.; Zhang, B. Mechanical and electronic properties of superhard ReB$_2$. *Phys. Rev. B* **2007**, *76*, 132101. [CrossRef]

8.   Zhang, Y.; Zhao, D.Q.; Pan, M.X.; Wang, W.H. Glass forming properties of Zr-based bulk metallic alloys. *J. Non-Cryst. Solids* **2003**, *315*, 206–210. [CrossRef]

9.   Gou, H.Y.; Hou, L.; Zhang, J.; Gao, F. Pressure-induced incompressibility of ReC and effect of metallic bonding on its hardness. *Appl. Phys. Lett.* **2008**, *92*, 241901. [CrossRef]

10. Fu, H.Z.; Peng, W.M.; Gao, T. Structural and elastic properties of ZrC under high pressure. *Mater. Chem. Phys.* **2009**, *115*, 789–794. [CrossRef]

11. Simunek, A. Vickers hardness of Ru-, Ta-, W-, Re-, Os-, and Ir carbides and nitrides calculated by the bond strength method. *MRS Commun.* **2022**, *12*, 759–761. [CrossRef]

12. Zhang, R.F.; Legut, D.; Niewa, R.; Argon, A.S.; Veprek, S. Shear-induced structural transformation and plasticity in ultraincompressible ReB$_2$ limit its hardness. *Phys. Rev. B* **2010**, *82*, 104104. [CrossRef]

13. Chen, X.Q.; Fu, C.L.; Krčmar, M.; Painter, G.S. Electronic and structural origin of ultraincompressibility of 5d transition-metal diborides MB$_2$ (M = W, re, os). *Phys. Rev. Lett.* **2008**, *100*, 196403. [CrossRef] [PubMed]

14. Chung, H.Y.; Weinberger, M.B.; Levine, J.B.; Kavner, A.; Yang, J.M.; Tolbert, S.H.; Kaner, R.B. Synthesis of ultra-incompressible superhard rhenium diboride at ambient pressure. *Science* **2007**, *316*, 436–439. [CrossRef] [PubMed]

15. Haines, J.; Leger, J.M.; Bocquillon, G. Synthesis and design of superhard materials. *Annu. Rev. Mater. Res.* **2001**, *31*, 1–23. [CrossRef]

16. Yao, H.Z.; Ouyang, L.Z.; Ching, W.Y. Ab initio calculation of elastic constants of ceramic crystals. *J. Am. Ceram. Soc.* **2007**, *90*, 3194–3204. [CrossRef]

17. Akrami, S.; Edalati, P.; Fuji, M.; Edalati, K. High-entropy ceramics: Review of principles, production and applications. *Mater. Sci. Eng. R-Rep.* **2021**, *146*, 100644. [CrossRef]

18. Miracle, D.B.; Senkov, O.N. A critical review of high entropy alloys and related concepts. *Acta Mater.* **2017**, *122*, 448–511. [CrossRef]

19. Ningthoujam, R.S.; Gajbhiye, N.S. Synthesis, electron transport properties of transition metal nitrides and applications. *Prog. Mater. Sci.* **2015**, *70*, 50–154. [CrossRef]

20. Feng, X.; Bao, K.; Tao, Q.; Li, L.; Shao, Z.; Yu, H.; Xu, C.; Ma, S.; Lian, M.; Zhao, X.; et al. Role of TM-TM Connection Induced by Opposite d-Electron States on the Hardness of Transition-Metal (TM = Cr, W) Mononitrides. *Inorg. Chem.* **2019**, *58*, 15573–15579. [CrossRef]

21. Yu, R.; Chong, X.; Jiang, Y.; Zhou, R.; Yuan, W.; Feng, J. The stability, electronic structure, elastic and metallic properties of manganese nitrides. *RSC Adv.* **2015**, *5*, 1620–1627. [CrossRef]

22. Zhang, Z.; Mi, W. Progress in ferrimagnetic Mn$_4$N films and its heterostructures for spintronics applications. *J. Phys. D Appl. Phys.* **2021**, *55*, 013001. [CrossRef]

23. Lei, L.; Hong, X.; Fangfang, L.; Yongqiang, Z.; Bin, B.; Kezhao, L. Preparation and Characteristics of Uranium Nitride Thin Films. *Rare Met. Mater. Eng.* **2015**, *44*, 1975–1978.

24. Rognerud, E.G.; Rom, C.L.; Todd, P.K.; Singstock, N.R.; Bartel, C.J.; Holder, A.M.; Neilson, J.R. Kinetically Controlled Low-Temperature Solid-State Metathesis of Manganese Nitride Mn$_3$N$_2$. *Chem. Mater.* **2019**, *31*, 7248–7254. [CrossRef]

25. Yin, W.W.; Lei, L.; Jiang, X.; Liu, P.; Liu, F.; Li, Y.; Peng, F.; He, D. High pressure synthesis and properties studies on spherical bulk epsilon-Fe3N. *High Press. Res.* **2014**, *34*, 317–326. [CrossRef]

26. Lei, L.; Zhang, L.; Gao, S.; Hu, Q.; Fang, L.; Chen, X.; Xia, Y.; Wang, X.; Ohfuji, H.; Kojima, Y.; et al. Neutron diffraction study of the structural and magnetic properties of epsilon-Fe$_3$N$_{1.098}$ and epsilon-Fe$_{2.322}$CO$_{0.678}$N$_{0.888}$. *J. Alloy. Compd.* **2018**, *752*, 99–105. [CrossRef]

27. Wang, S.M.; Yu, X.; Zhang, J.; Wang, L.; Leinenweber, K.; He, D.; Zhao, Y. Synthesis, Hardness, and Electronic Properties of Stoichiometric VN and CrN. *Cryst. Growth Des.* **2016**, *16*, 351–358. [CrossRef]

28. Chen, M.; Wang, S.; Zhang, J.; He, D.; Zhao, Y. Synthesis of Stoichiometric and Bulk CrN through a Solid-State Ion-Exchange Reaction. *Chem. A Eur. J.* **2012**, *18*, 15459–15463. [CrossRef]

29. Blochl, P.E. Projector augmented-wave method. *Phys. Rev. B Condens. Matter* **1994**, *50*, 17953–17979. [CrossRef]

30. Kresse, G.; Furthmuller, J. Efficient iterative schemes for ab initio total-energy calculations using a plane-wave basis set. *Phys. Rev. B Condens. Matter* **1996**, *54*, 11169–11186. [CrossRef]

31. Perdew, J.P.; Burke, K.; Ernzerhof, M. Generalized gradient approximation made simple. *Phys. Rev. Lett.* **1996**, *77*, 3865–3868. [CrossRef] [PubMed]

32. Steinberg, S.; Dronskowski, R. The Crystal Orbital Hamilton Population (COHP) Method as a Tool to Visualize and Analyze Chemical Bonding in Intermetallic Compounds. *Crystals* **2018**, *8*, 225. [CrossRef]

33. Mueller, P.C.; Ertural, C.; Hempelmann, J.; Dronskowski, R. Crystal Orbital Bond Index: Covalent Bond Orders in Solids. *J. Phys. Chem. C* **2021**, *125*, 7959–7970. [CrossRef]

34. Toby, B.H. EXPGUI, a graphical user interface for GSAS. *J. Appl. Crystallogr.* **2001**, *34*, 210–213. [CrossRef]

35. Zhang, L.J.; Wang, Y.; Lv, J.; Ma, Y. Materials discovery at high pressures. *Nat. Rev. Mater.* **2017**, *2*, 17005. [CrossRef]

MDPI AG
Grosspeteranlage 5
4052 Basel
Switzerland
Tel.: +41 61 683 77 34

Metals Editorial Office
E-mail: metals@mdpi.com
www.mdpi.com/journal/metals